Methods in Molecular Biology

Series Editor
John M. Walker
School of Life and Medical Sciences
University of Hertfordshire
Hatfield, Hertfordshire, UK

For further volumes:
http://www.springer.com/series/7651

For over 35 years, biological scientists have come to rely on the research protocols and methodologies in the critically acclaimed *Methods in Molecular Biology* series. The series was the first to introduce the step-by-step protocols approach that has become the standard in all biomedical protocol publishing. Each protocol is provided in readily-reproducible step-by-step fashion, opening with an introductory overview, a list of the materials and reagents needed to complete the experiment, and followed by a detailed procedure that is supported with a helpful notes section offering tips and tricks of the trade as well as troubleshooting advice. These hallmark features were introduced by series editor Dr. John Walker and constitute the key ingredient in each and every volume of the *Methods in Molecular Biology* series. Tested and trusted, comprehensive and reliable, all protocols from the series are indexed in PubMed.

Tissue Proteomics

Methods and Protocols

Edited by

Taufika Islam Williams

Department of Chemistry, North Carolina State University, Raleigh, NC, USA

Humana Press

Editor
Taufika Islam Williams
Department of Chemistry
North Carolina State University
Raleigh, NC, USA

ISSN 1064-3745 ISSN 1940-6029 (electronic)
Methods in Molecular Biology
ISBN 978-1-0716-4300-6 ISBN 978-1-0716-4298-6 (eBook)
https://doi.org/10.1007/978-1-0716-4298-6

This Humana imprint is published by the registered company Springer Science+Business Media, LLC, part of Springer Nature.
The registered company address is: 1 New York Plaza, New York, NY 10004, U.S.A.

Preface

The interrogation of biology through an OMICS or "systems biology" lens has become the cornerstone of understanding the living world. A key component of this strategy is the interrogation of proteins, the true vehicles of change in living organisms. Among the diverse applications of proteomics, tissue proteomics stands at the forefront, offering unparalleled insights into the molecular underpinnings of health and disease. Gone are the days when, even recently, obtaining a differential list of proteins from the target tissue compared to a matched control was the *state of the art* of what proteomics could deliver. We now dig ever deeper, interrogating sub-proteomes, interactomes, pathways, glycosylation patterns, phosphorylation sites, and the like at vanishingly low limits of detection and quantification. Furthermore, the information we glean from proteomics investigations should now be tied into the global OMICS of the target organism to better understand the biology we wish to study. We have come a long way since the exciting days of deciphering the human genome, but every door opened into biological function leads to many, many more doors to knock on. What a great time in the scientific inquiry of life.

For the scientist new to the field of proteomics, the relevant scientific literature presents a daunting landscape. This book serves as a roadmap to that landscape, offering starting points and directions to channel one's learning. Proteomics methods in the literature can get quite complicated, and many are tailored to specific inquiries. In this book, we have compiled germane and well-established workflows for biological interrogations. These methodologies encompass strategies for general proteomic sample processing, protein depletion, working in the low-abundance proteome, analysis of proteomes in diverse tissue types and organisms, post-translational modification analysis, protein-protein interactions, tissue imaging, proteomics of cellular compartments, as well as computational tools for data treatment. The chapters in this book are designed to be not just informative but accessible as well, elaborating step-by-step protocols, troubleshooting guidelines, and insights into cutting-edge technology. Whether the reader is just starting out on a proteomics journey or seeking an expansion of knowledge and techniques, this book will serve as a valuable resource to inspire new discoveries in the dynamic field of tissue proteomics.

The book opens with chapters on key strategies for sample preparation and bottom-up tissue proteomics. Here we boil down to the bare basics that must be considered when designing a comprehensive, bottom-up proteomics workflow. Chapter 1 takes the reader into the world of immunoaffinity depletion to dig deeper into the low abundance sub-proteome in biological fluids. Owing to the tremendous dynamic range often seen in the proteome of these sample types, low abundance proteins are consistently underrepresented in such proteomics data sets. Chapter 2 discusses an optimized sample preparation workflow for in-depth analysis of samples with low cell numbers by applying tandem mass tagging (TMT) for relative quantification of proteins. In Chapter 3, the reader is asked to consider peptide and protein fractionation as a strategy to further shine a light on that low-abundance proteome, significantly enriching the data that can be obtained in comparison to a single-shot analysis.

The next few chapters of the book are home to methods that provide a general overview of the proteomics landscape through the lens of important tissue types and organisms. Chapter 4 discusses the interrogation of adipose tissue and challenges associated with lipid-heavy samples. Chapter 5 focuses on the interrogation of any type of microbial proteome. The hair proteome is explored in Chapter 6, with implications in forensics, cosmetics, toxicology, developmental biology, among other spaces of inquiry. Insect proteomics is covered in Chapter 7 (whole insect body) and 8 (brain tissue only). In Chapter 9, the reader explores complex samples with proteomes from multiple organisms.

In the following group of chapters, the reader is asked to consider diverse matrices. Chapter 10 explores biological fluid proteomics, both liquid and dried formats. Chapter 11 also focuses on the analysis of various diverse biological fluids, such as urine, plasma, serum, cerebrospinal fluid (CSF), and even milk. These types of fluids are indeed key targets for biomarker discovery. Other matrices explored include extracellular matrix or ECM (Chapter 12) with its own set of unique challenges, and even commercial white wines (Chapter 13) for the purposes of quality control, varietal differentiation, allergen analysis, and authentication. The multi-enzyme protocol discussed in Chapter 12 highlights the fact that all proteins are not created equal and different protein types may require different treatments for proteolytic digestion. While proteomics remains squarely in the realm of sample interrogation in the living world, Chapter 13 alerts us to the fact that proteomics methodologies can be applied to other protein-containing matrices as well.

Chapters 14 and 15 focus on compartmental proteomics. In Chapter 14, the reader explores the analysis of the sub-proteome of synaptic vesicles originating from neurospheres derived from induced pluripotent stem cells (iPSCs). Chapter 15 engages the reader to consider comprehensive translational proteomics in studying extracellular vesicles (EVs) from brain tissue.

Chapter 16 takes a look at a crosslinking mass spectrometry methodology to study the plasma membranome, which can allow for the examination of such processes as protein networking, transport, signal transduction, secretion, interaction dynamics, and endocytosis. Additional aspects of protein dynamics are explored in post-translational modifications (PTMs) such as phosphoproteomics (Chapter 17) and glycoproteomics (Chapter 18).

The reader finds a more nuanced world in Chapter 19 where multi-OMIC workflows and mass spectrometry imaging (MSI) of the N-glycome and ECM-targeted proteome in single cells are investigated. Spatial proteomics with laser-capture microdissected (LCM) tissues is detailed in Chapter 20. The book draws to a close with a decision tree for computationally investigating protein biomarker candidates for clinical applications (Chapter 21) and an exploration of proteomics and computational tools to identify and monitor prognostic, diagnostic, and therapeutic markers for studying such diseases as systemic amyloidosis (Chapter 22).

There are a myriad of different tissues and protein containing matrices one might encounter in a proteomics experiment. This book offers carefully curated chapters with proteomics methods that are tried and true and applicable to diverse tissue formats. The reader is invited to carefully identify as many factors of a proteomics experiment as possible that might impact the quality of results obtained to answer the research question at hand. This book presents a survey of broad and more nuanced approaches in tissue proteomics to assist the reader with critical decision-making for experimental design. We hope this material assists exploration of the true scope and breadth of this very exciting field.

Sincerest thanks to all the contributors who generously shared their expertise and experiences, as well as to the researchers and students whose curiosity and dedication continue to drive the advancement of proteomics research. It is our wish that this book will not only facilitate scientific progress in the field of OMICS but also foster collaboration and dialogue among proteomics scientists across the globe.

Raleigh, NC, USA *Taufika Islam Williams*

Contents

Contributors

MICHAEL AFFOLTER • *Proteomics, Bioanalytics Department, Nestlé Institute of Food Safety & Analytical Sciences, Nestlé Research, Lausanne, Switzerland*

ZESHAN ALI • *School of Natural Sciences, Macquarie University, North Ryde, NSW, Australia*

MARCOS ROGÉRIO ANDRÉ • *Laboratory of Immunoparasitology, Department of Pathology, Reproduction and One Health, Faculdade de Ciências Agrárias e Veterinárias, Universidade Estadual Paulista (FCAV/UNESP), Jaboticabal, SP, Brazil; Intracellular Pathogens Research Laboratory, Department of Clinical Sciences, The Comparative Medicine Institute, College of Veterinary Medicine, North Carolina State University, Raleigh, NC, USA*

PEGGI M. ANGEL • *Department of Pharmacology & Immunology, Medical University of South Carolina, Charleston, SC, USA*

JEAN ARMENGAUD • *Département Médicaments et Technologies pour la Santé (DMTS), SPI, Université Paris-Saclay, CEA, INRAE, Bagnols-sur-Cèze, France*

HONGXIA BAI • *Department of Chemistry, North Carolina State University, Raleigh, NC, USA*

IKRAM BELGHIT • *Institute of Marine Research, Bergen, Norway*

ANDREEA BIEHL • *Joint Department of Biomedical Engineering, North Carolina State University & University of North Carolina at Chapel Hill, Raleigh, NC, USA; Comparative Medicine Institute, North Carolina State University, Raleigh, NC, USA*

FRANCESCA BRAMBILLA • *Institute for Biomedical Technologies – National Research Council (ITB-CNR), Segrate, Milan, Italy*

CAROLINE BRANDÃO-TELES • *Laboratory of Neuroproteomics, Department of Biochemistry and Tissue Biology, Institute of Biology, University of Campinas, Campinas, Brazil*

EDWARD B. BREITSCHWERDT • *Intracellular Pathogens Research Laboratory, Department of Clinical Sciences, The Comparative Medicine Institute, College of Veterinary Medicine, North Carolina State University, Raleigh, NC, USA*

ROQUE BRU-MARTÍNEZ • *Plant Proteomics and Functional Genomics Group, Department of Biochemistry and Molecular Biology and Soil and Agricultural Chemistry, Faculty of Science, University of Alicante, Alicante, Spain; Allergy Group, Alicante Institute for Health, and Biomedical Research (ISABIAL), Alicante, Spain*

KYLE BURK • *Department of Environmental Toxicology, The University of California, Davis, Davis, CA, USA*

MARIANNA CATERINO • *Department of Molecular Medicine and Medical Biotechnology, School of Medicine, University of Naples Federico II, Naples, Italy; CEINGE—Biotecnologie Avanzate Franco Salvatore Scarl, Naples, Italy*

SAYANTANI CHATTERJEE • *Department of Biochemistry, Center for Biomedical Mass Spectrometry, Boston University School of Medicine, Boston, MA, USA*

XIULAN CHEN • *Laboratory of Proteomics, Institute of Biophysics, Chinese Academy of Sciences, Beijing, China; University of Chinese Academy of Sciences, Beijing, China*

JAYAPRADA R. CHUNDURI • *Freelance Researcher and Reviewer, Mumbai, India; Prarambh Science Academy, Sangli, Maharashtra, India*

LEONARD B. COLLINS • *Molecular Education, Technology and Research Innovation Center (METRIC), North Carolina State University, Raleigh, NC, USA*
VICTOR CORASOLLA CARREGARI • *Laboratory of Neuroproteomics, Department of Biochemistry and Tissue Biology, Institute of Biology, University of Campinas, Campinas, Brazil*
MICHELE COSTANZO • *Department of Molecular Medicine and Medical Biotechnology, School of Medicine, University of Naples Federico II, Naples, Italy; CEINGE—Biotecnologie Avanzate Franco Salvatore Scarl, Naples, Italy*
FERNANDA CRUNFLI • *Laboratory of Neuroproteomics, Department of Biochemistry and Tissue Biology, Institute of Biology, University of Campinas, Campinas, Brazil; Department of Clinical Medicine, Translational Neuropsychiatry Unit, Aarhus University, Aarhus, Denmark*
PHILLIP DANIELSON • *The University of Denver, Department of Biological Sciences, Denver, CO, USA*
LOÏC DAYON • *Proteomics, Bioanalytics Department, Nestlé Institute of Food Safety & Analytical Sciences, Nestlé Research, Lausanne, Switzerland; Institut des Sciences et Ingé nierie Chimiques, Ecole Polytechnique Fédérale de Lausanne (EPFL), Lausanne, Switzerland*
ÉRICA DE ALMEIDA DUQUE • *Department of Pharmacology, Institute of Biomedical Science, University of São Paulo, São Paulo, Brazil*
SABRINA GRIZZI DE OLIVEIRA • *School of Natural Sciences, Macquarie University, North Ryde, NSW, Australia*
DARIO DI SILVESTRE • *Institute for Biomedical Technologies – National Research Council (ITB-CNR), Segrate, Milan, Italy*
AMÉLIE DENDOOVEN • *Department of Pathology, Ghent University Hospital, Ghent, Belgium; Department of Diagnostic Sciences, Ghent University, Ghent, Belgium*
SIMON DEVOS • *VIB-UGent Center for Medical Biotechnology, Ghent, Belgium; Department of Biomolecular Medicine, Ghent University, Ghent, Belgium; VIB Proteomics Core, Ghent, Belgium*
PAULA DÍEZ • *Department of Immunology, Leiden University Medical Center (LUMC), Leiden, The Netherlands; Sarcomas and Experimental Therapeutics Laboratory, Health Research Institute of Asturias (ISPA) and Asturias Central University Hospital (HUCA), Oviedo, Spain; Department of Physical and Analytical Chemistry, Faculty of Chemistry, University of Oviedo, Oviedo, Spain*
RICHARD R. DRAKE • *Department of Pharmacology & Immunology, Medical University of South Carolina, Charleston, SC, USA*
SARA DUFOUR • *VIB-UGent Center for Medical Biotechnology, Ghent, Belgium; Department of Biomolecular Medicine, Ghent University, Ghent, Belgium; VIB Proteomics Core, Ghent, Belgium*
JACLYN B. DUNNE • *Department of Pharmacology & Immunology, Medical University of South Carolina, Charleston, SC, USA*
MARION ERNEST • *Proteomics, Bioanalytics Department, Nestlé Institute of Food Safety & Analytical Sciences, Nestlé Research, Lausanne, Switzerland*
DONALD FREYTES • *Joint Department of Biomedical Engineering, North Carolina State University & University of North Carolina at Chapel Hill, Raleigh, NC, USA; Comparative Medicine Institute, North Carolina State University, Raleigh, NC, USA*
MARCELO GANZELLA • *Laboratory of Neurobiology, Max-Planck-Institute for Biophysical Chemistry, Göttingen, Germany*

Zachary C. Goecker • *University of California – Davis, Department of Environmental Toxicology, Davis, CA, USA*
Paolo Grumati • *Telethon Institute of Genetics and Medicine (TIGEM), Pozzuoli, Italy; Department of Clinical Medicine and Surgery, University of Naples, Naples, Italy*
Joel P. A. Gummer • *ChemCentre, Forensic Science Laboratory, Bentley, WA, Australia*
Fatemeh Habibpourmehraban • *School of Natural Sciences, Macquarie University, North Ryde, NSW, Australia*
Paul A. Haynes • *School of Natural Sciences, Macquarie University, North Ryde, NSW, Australia*
Francis Impens • *VIB-UGent Center for Medical Biotechnology, Ghent, Belgium; Department of Biomolecular Medicine, Ghent University, Ghent, Belgium; VIB Proteomics Core, Ghent, Belgium*
Reinhard Jahn • *Laboratory of Neurobiology, Max-Planck-Institute for Biophysical Chemistry, Göttingen, Germany*
Sara Kassem • *Department of Immunology, Leiden University Medical Center (LUMC), Leiden, The Netherlands; Department of Immunology, ErasmusMC, Rotterdam, The Netherlands*
Romy E. Keane • *ChemCentre, Forensic Science Laboratory, Bentley, WA, Australia*
Jane I. Khudyakov • *Department of Biological Sciences, University of the Pacific, Stockton, CA, USA*
Pey Yee Lee • *UKM Medical Molecular Biology Institute (UMBI), Universiti Kebangsaan Malaysia, Kuala Lumpur, Malaysia*
Kevin Legg • *Protein Metrics, Cupertino, CA, USA*
Michelle V. Lihon • *Department of Biochemistry, Purdue University, West Lafayette, IN, USA*
Mark J. Lim • *AmberGen, Inc., Billerica, MA, USA*
Teck Yew Low • *UKM Medical Molecular Biology Institute (UMBI), Universiti Kebangsaan Malaysia, Kuala Lumpur, Malaysia*
Clément Lozano • *Département Médicaments et Technologies pour la Santé (DMTS), SPI, Université Paris-Saclay, CEA, INRAE, Bagnols-sur-Cèze, France*
Teresa Mendes Maia • *VIB-UGent Center for Medical Biotechnology, Ghent, Belgium; Department of Biomolecular Medicine, Ghent University, Ghent, Belgium; VIB Proteomics Core, Ghent, Belgium*
Daniel Martins-de-Souza • *Laboratory of Neuroproteomics, Department of Biochemistry and Tissue Biology, Institute of Biology, University of Campinas, Campinas, Brazil; Experimental Medicine Research Cluster (EMRC), University of Campinas, Campinas, SP, Brazil; D'Or Institute for Research and Education (IDOR), São Paulo, Brazil; INCT in Modelling Human Complex Diseases with 3D Platforms (Model3D), Conselho Nacional de Desenvolvimento Científico e Tecnológico, São Paulo, Brazil*
Farhad Masoomi-Aladizgeh • *School of Natural Sciences, Macquarie University, North Ryde, NSW, Australia*
Pierluigi Mauri • *Institute for Biomedical Technologies – National Research Council (ITB-CNR), Segrate, Milan, Italy; Institute of Life Sciences, Sant'Anna School of Advanced Studies, Pisa, Italy*
Nicole M. McLoughlin • *California Department of Fish and Wildlife, Wildlife Forensic Laboratory, Sacramento, CA, USA*
Anand Mehta • *Department of Pharmacology & Immunology, Medical University of South Carolina, Charleston, SC, USA*

GIAMPAOLO MERLINI • *Amyloidosis Research and Treatment Centre, Fondazione IRCCS Policlinico San Matteo and University of Pavia, Pavia, Italy*
CAMILO MORA-NAVARRO • *Department of Chemical Engineering, University of Puerto Rico-Mayaguez, Mayaguez, PR, USA*
DYLAN H. MULTARI • *School of Natural Sciences, Macquarie University, North Ryde, NSW, Australia*
CAROLINA DEMARCHI MUNHOZ • *Department of Pharmacology, Institute of Biomedical Science, University of São Paulo, São Paulo, Brazil*
LINDA OLSTHOORN • *Laboratory of Neurobiology, Max-Planck-Institute for Biophysical Chemistry, Göttingen, Germany*
JUNAIDA OSMAN • *UKM Medical Molecular Biology Institute (UMBI), Universiti Kebangsaan Malaysia, Kuala Lumpur, Malaysia*
EYSTEIN OVELAND • *Institute of Marine Research, Bergen, Norway*
MAGNUS PALMBLAD • *Center for Proteomics and Metabolomics, Leiden University Medical Center, Leiden, The Netherlands*
GLENDON J. PARKER • *University of California – Davis, Department of Environmental Toxicology, Davis, CA, USA*
ANE LAURA PEDERSEN • *Proteomics, Bioanalytics Department, Nestlé Institute of Food Safety & Analytical Sciences, Nestlé Research, Lausanne, Switzerland*
SARA PETROSINO • *Telethon Institute of Genetics and Medicine (TIGEM), Pozzuoli, Italy*
JOSEF D. RASINGER • *Institute of Marine Research, Bergen, Norway*
ROBERT H. RICE • *University of California – Davis, Department of Environmental Toxicology, Davis, CA, USA*
KENNETH J. ROTHSCHILD • *AmberGen, Inc., Billerica, MA, USA; Department of Physics and Photonics Center, Boston University, Boston, MA, USA*
MARGHERITA RUOPPOLO • *Department of Molecular Medicine and Medical Biotechnology, School of Medicine, University of Naples Federico II, Naples, Italy; CEINGE—Biotecnologie Avanzate Franco Salvatore Scarl, Naples, Italy*
SIDDHARTH P. SAGAR • *Freelance Researcher and Reviewer, Mumbai, India; Prarambh Science Academy, Sangli, Maharashtra, India*
ANTONIO SAMPER-HERRERO • *Plant Proteomics and Functional Genomics Group, Department of Biochemistry and Molecular Biology and Soil and Agricultural Chemistry, Faculty of Science, University of Alicante, Alicante, Spain; Allergy Group, Alicante Institute for Health, and Biomedical Research (ISABIAL), Alicante, Spain*
LUCIA SANTORELLI • *Telethon Institute of Genetics and Medicine (TIGEM), Pozzuoli, Italy*
MICHELE SANTORO • *Telethon Institute of Genetics and Medicine (TIGEM), Pozzuoli, Italy*
SUSANA SELLÉS-MARCHART • *Plant Proteomics and Functional Genomics Group, Department of Biochemistry and Molecular Biology and Soil and Agricultural Chemistry, Faculty of Science, University of Alicante, Alicante, Spain; Research Technical Facility, Proteomics and Genomics Division, University of Alicante, Alicante, Spain*
MANVEEN K. SETHI • *Department of Biochemistry, Center for Biomedical Mass Spectrometry, Boston University School of Medicine, Boston, MA, USA*
MOHADESEH MONTAZERI SHATOURI • *School of Natural Sciences, Macquarie University, North Ryde, NSW, Australia*
ASHLEY M. SPICER • *California Department of Fish and Wildlife, Wildlife Forensic Laboratory, Sacramento, CA, USA*

W. Andy Tao • *Department of Biochemistry, Purdue University, West Lafayette, IN, USA; Department of Chemistry, Purdue University, West Lafayette, IN, USA; Department of Medicinal Chemistry and Molecular Pharmacology, Purdue University, West Lafayette, IN, USA; Purdue Institute for Cancer Research, Purdue University, West Lafayette, IN, USA; Tymora Analytical Operations, West Lafayette, IN, USA*

Harrison B. Taylor • *Department of Pharmacology & Immunology, Medical University of South Carolina, Charleston, SC, USA*

Rebecca J. Tidy • *ChemCentre, Forensic Science Laboratory, Bentley, WA, Australia*

Nathan A. Tuchscherer • *Department of Biochemistry, Purdue University, West Lafayette, IN, USA*

Delphi Van Haver • *VIB-UGent Center for Medical Biotechnology, Ghent, Belgium; Department of Biomolecular Medicine, Ghent University, Ghent, Belgium; VIB Proteomics Core, Ghent, Belgium*

Malaïka Van der Linden • *Department of Pathology, Ghent University Hospital, Ghent, Belgium; Department of Diagnostic Sciences, Ghent University, Ghent, Belgium*

Madhushri S. Varunjikar • *Institute of Marine Research, Bergen, Norway*

Katharina Waury • *Department of Computer Science, Vrije Universiteit Amsterdam, Amsterdam, The Netherlands*

Shasha Wei • *National Local Joint Engineering Research Center for Precision Surgery & Regenerative Medicine, Shaanxi Provincial Center for Regenerative Medicine and Surgical Engineering, First Affiliated Hospital of Xi'an Jiaotong University, Xi'an, Shaanxi Province, China*

Taufika Islam Williams • *Department of Chemistry, North Carolina State University, Raleigh, NC, USA; Molecular Education, Technology and Research Innovation Center (METRIC), North Carolina State University, Raleigh, NC, USA*

Ying Xi • *Department of Chemistry, North Carolina State University, Raleigh, NC, USA*

Gargey Yagnik • *AmberGen, Inc., Billerica, MA, USA*

Fuquan Yang • *Laboratory of Proteomics, Institute of Biophysics, Chinese Academy of Sciences, Beijing, China; University of Chinese Academy of Sciences, Beijing, China*

Joseph Zaia • *Department of Biochemistry, Center for Biomedical Mass Spectrometry, Boston University School of Medicine, Boston, MA, USA; Boston University Bioinformatics Program, Boston University, Boston, MA, USA*

Giuliana S. Zuccoli • *Laboratory of Neuroproteomics, Department of Biochemistry and Tissue Biology, Institute of Biology, University of Campinas, Campinas, Brazil*

Chapter 1

Immunoaffinity Depletion of High-Abundance Proteins from Serum/Plasma for Proteomic Analysis

Pey Yee Lee, Junaida Osman, and Teck Yew Low

Abstract

Mass spectrometry-based proteomics is widely applied to human blood serum or plasma in the search of biomarkers for various diseases. However, the enormous complexity and dynamic range of protein concentrations in these samples render a significant analytical challenge, particularly for detecting low-abundance candidate biomarkers. As a result, strategies for enriching low-abundance proteins and improving their identification in serum or plasma proteomics are commonly used. Here, we describe an immunodepletion technique that is routinely used in our lab for removing high-abundance proteins from human serum/plasma.

Key words Immunoaffinity depletion, High-abundance proteins, Serum, Plasma, Proteomics

1 Introduction

Blood is the most common biofluid sample routinely used in clinical tests [1]. This is because blood accumulates a diverse array of biochemical molecules while passing through multiple organs and tissues. These biomolecules can serve as markers that reflect the physiological or pathological status of the body [2]. Hence, analysis of plasma or serum is of great interest for the discovery of clinical biomarkers [3]. Most biomarker studies have focused on detecting the presence of proteins in plasma or serum whose variations in concentration or posttranslational modifications (PTMs) are associated with the underlying conditions [4].

Proteomics refers to the qualitative and quantitative analysis of the complete set of proteins that are expressed at a particular time in a living system, which encompasses the study of protein expression, function, structure, PTMs, protein-protein interactions, or even protein-drug interactions [4–10]. Solution-based digestion strategies such as filter-aided sample preparation (FASP) and single-pot, solid-phase-enhanced sample preparation (SP3) streamline

Taufika Islam Williams (ed.), *Tissue Proteomics: Methods and Protocols*, Methods in Molecular Biology, vol. 2884,
https://doi.org/10.1007/978-1-0716-4298-6_1,

proteomic sample preparation by efficiently removing contaminants and detergents while retaining proteins for enzymatic digestion. The traditional technologies used for separating complex protein samples include two-dimensional gel electrophoresis (2-DE) and two-dimensional differential gel electrophoresis (2D-DIGE) techniques [12], whereas liquid chromatography coupled with tandem mass spectrometry (LC-MS/MS) is the workhorse for modern proteomics [11]. Other cutting-edge technologies for proteomics include protein microarray, aptamers, and proximity extension assay. The SOMAscan proteomic platform employs the modified DNA aptamers known as slow off rate modified aptamer (SOMAmer) as the protein-capture reagents for highly multiplexed measurement of a broad range of proteins targets for biomarker discovery [13]. Proximity extension assay (PEA) is a unique technology developed by Olink Proteomics based on a dual recognition of a targeted biomarker through a matched pair of antibodies labelled with unique DNA oligonucleotides using multiplex immunoassay PCR, which allows high-throughput protein biomarker analysis [14]. Developments in mass spectrometry technologies have led to advances in the field of single-cell proteomics and spatial proteomics, which enables the characterization of proteomes of individual cells and spatial distribution of proteins at high resolution, respectively [15–17].

Nonetheless, proteomic studies of plasma/serum face several challenges. First, there is significant individual variation in the content of the plasma/serum proteome attributable to both genetic and environmental factors [18], which hinders robust measurement. Moreover, the enormous dynamic concentration range (over ten orders of magnitude) of the constituent protein/peptide species also results in biased measurement toward high-abundance proteins as typically, a few extremely high-abundance serum/plasma proteins tend to dominate the blood proteome, thus masking lower abundant protein species [19]. For instance, serum albumin accounts for approximately 55% of the total protein mass in plasma, whereas the top 22 most abundant proteins make up 99% of the total protein mass in the human plasma proteome [19]. Therefore, there is a significant hurdle for current proteomics technologies (i.e., mass spectrometry) to detect potential biomarkers that account for less than 1% of all proteins—since their concentrations are below the detection threshold of the instruments [20].

Consequently, strategies to separate or eliminate high-abundance proteins from serum/plasma are needed to improve the detection of low-abundance protein species, as well as increase proteome coverage. This includes pre-fractionation techniques such as multidimensional liquid chromatography and electrophoresis, not to mention selective enrichment of the low-abundance proteins or removal of the major proteins [21, 22]. One of the most commonly used methods for serum/plasma proteomics is high-abundance protein depletion using affinity resins [23]. The immunoaffinity depletion technique is based on polyclonal antibodies that are cross-linked to either chromatographic matrices or microbeads to immobilize high-abundance target proteins. Many immunoaffinity columns are now commercially available for the removal of up to 20 high-abundance proteins. However, a typical issue with the immunoaffinity approach is the possible loss of nontarget proteins due to nonspecific binding to the affinity matrices or by interactions between targeted and nontargeted proteins [24]. Nevertheless, these antibody-based depletion techniques are effective in eliminating the desired high-abundance proteins with good reproducibility and result in higher numbers of identified proteins and hence remain widely used for proteomics analysis [25, 26]. Here we present a protocol for immunoaffinity depletion of high-abundance proteins from serum/plasma for proteomics analysis.

2 Materials

2.1 Serum/Plasma Sample Collection

1. Butterfly needles.
2. Syringes.
3. Tourniquets.
4. Alcohol swab.
5. Sterile gauze pads.
6. BD Vacutainer Serum Separation Tubes (SST) (Becton Dickinson).
7. BD Vacutainer Plasma Preparation Tubes (PPT) (Becton Dickinson).
8. Refrigerated centrifuge.

2.2 Immunoaffinity Depletion Using Spin Column

1. High-Select™ Top14 Abundant Protein Depletion Mini Spin Columns (Thermo Scientific) (Fig. 1) (Table 1).
2. 2-mL collection tubes.
3. 15-mL collection tubes.
4. Microcentrifuge.
5. Mixer.

2.3 Protein Measurement Using Nanodrop

1. Nanodrop (Thermo Fisher Scientific) (*see* **Note 1**).
2. Dry laboratory wipes.
3. Bovine serum albumin (BSA) standard solutions.

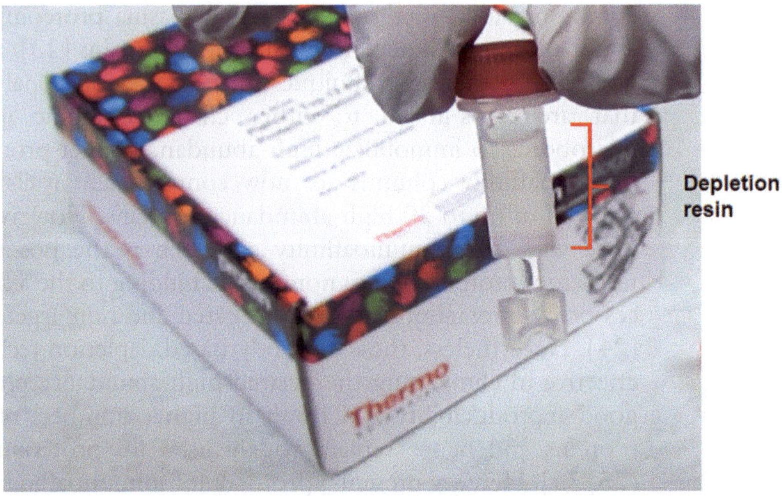

Fig. 1 High-Select™ Top14 Abundant Protein Depletion Mini Spin Columns. The column contains resin with highly specific, immobilized antibodies for depletion of high-abundance proteins

Table 1
The top 14 proteins removed by the depletion column

Depleted proteins	
1. Albumin	8. Alpha-1-acid glycoprotein
2. IgA	9. Alpha-1-antitrypsin
3. IgD	10. Alpha-2-macroglobulin
4. IgE	11. Apolipoprotein A1
5. IgG	12. Fibrinogen
6. IgG (light chain)	13. Haptoglobin
7. IgM	14. Transferrin

2.4 Sodium Dodecyl Sulfate Polyacrylamide Gel Electrophoresis (SDS-PAGE)

1. SDS-PAGE gel casting and electrophoresis apparatus.
2. Staining tray.
3. Ultrapure water.
4. Acrylamide/bis, 30% 37.5:1.
5. Resolving gel buffer: 1.5 M Tris–HCl, pH 8.8.
6. Stacking gel buffer: 0.5 M Tris–HCl, pH 6.8.
7. 10% SDS.
8. N,N,N′,N′-tetramethylethylene-diamine (TEMED).
9. Ammonium persulfate: 10% solution in water.
10. Laemmli sample buffer (2×): 125 mM Tris–HCl, pH 6.8, 4% SDS, 20% glycerol, 0.02% bromophenol blue, 100 mM dithiothreitol (DTT), or 2-mercaptoethanol.

11. Isopropanol.
12. Molecular mass protein markers.
13. SDS-PAGE running buffer: 0.025 M Tris–HCl, pH 8.3, 0.192 M glycine, 0.1% SDS.
14. Coomassie blue or silver stain.

3 Methods

3.1 Serum/Plasma Sample Collection (See Note 2)

1. For serum samples, collect blood by venipuncture into BD Vacutainer Serum Separation Tubes (SST) and allow to clot for 30 min (*see* **Note 3**). Centrifuge at 1300× *g* for 10 min at 4 ° C. Aliquot the supernatant and store serum at −80 °C.
2. For plasma samples, collect blood into BD Vacutainer Plasma Preparation Tubes (PPT) coated with EDTA anticoagulant (*see* **Note 4**). Centrifuge at 1300× *g* for 10 min at 4 °C within 30 min of collection. Aliquot the supernatant and store plasma at −80 °C (*see* **Note 5**).

3.2 Immunoaffinity Depletion Using Spin Column (See Note 6)

1. Equilibrate the depletion spin column to room temperature.
2. Remove the column screw cap and add up to 10 μL of the sample directly to the resin slurry in the column.
3. Cap the column and invert the column several times until the resin is completely homogenous in the solution.
4. Incubate the mixture in the column with gentle end-over-end mixing for 10 min at room temperature. Make sure the sample mixes with the resin during the incubation period. Alternatively, gently vortex every few minutes.
5. After incubation, snap off the bottom closure and loosen the top cap. Place the mini column into a 2-mL collection tube and centrifuge at 1000× *g* for 2 min.
6. Discard the column containing the resin.
7. Filtrate contains sample with the high-abundance proteins removed. Use for further processing or store at −20 °C for later use. The depleted sample will be in 10 mM PBS and 0.02% sodium azide, pH 7.4 (*see* **Note 7**).

3.3 Protein Measurement Using Nanodrop (See Note 8)

1. Select the Protein A280 application from the main menu.
2. Select the type of sample (i.e., protein) to be measured from the drop-down list. Choose the concentration units from the drop-down list.
3. Establish a blank using an appropriate buffer. Pipette 2 μL of the appropriate blanking solution onto the bottom pedestal, lower the arm, and click Blank.

4. Wipe the blank solution from the lower and upper pedestals using a dry laboratory wipe.
5. Enter the values for each standard concentration in the right pane table. Under the Standards tab, highlight the desired standard (e.g., BSA) and pipette 2 μL of the desired standard onto the lower pedestal. Lower the arm and click Measure. Repeat the process for all standards. Be sure to measure all standards prior to measuring samples. To view standard curve, click View Standard Curve (*see* **Note 9**).
6. Enter a sample ID in the appropriate field, load 2 μL of the first sample, and click Measure. A fresh 2 μL aliquot of sample should be used for each measurement.
7. Remove the sample from the lower and upper pedestals using a dry laboratory wipe.

3.4 SDS-PAGE (See Note 10)

1. Prepare the separation gel (10%). Mix in the following order as indicated in Table 2.
2. Pour gel, leaving ~2 cm below the bottom of the comb for the stacking gel. Make sure to remove bubbles.
3. Layer the top of the gel with isopropanol. This will help to remove bubbles at the top of the gel and will also keep the polymerized gel from drying out.
4. Leave for ~30 min to allow the gel to polymerize.
5. Remove the isopropanol and wash out the remaining traces of isopropanol with distilled water.
6. Prepare the stacking gel (4%). Mix in the following order as indicated in Table 3.
7. Pour stacking gel on top of the separating gel.
8. Add combs to make wells. In ~30 min, the stacking gel should become completely polymerized.

Table 2
Preparation of separation gel

Reagents	Volume
Ultrapure water	4.1 mL
Acrylamide/bis (30% 37.5:1)	3.3 mL
Tris–HCl (1.5 M, pH 8.8)	2.5 mL
SDS, 10%	100 μL
N,N,N′,N′-tetramethylethylene-diamine (TEMED)	10 μL
Ammonium persulfate (APS), 10%	32 μL

Table 3
Preparation of stacking gel

Reagents	Volume
Ultrapure water	6.1 mL
Acrylamide/bis (30% 37.5:1)	1.3 mL
Tris–HCl (0.5 M, pH 6.8)	2.5 mL
SDS, 10%	100 μL
N,N,N′,N′-tetramethylethylene-diamine (TEMED)	10 μL
Ammonium persulfate (APS), 10%	100 μL

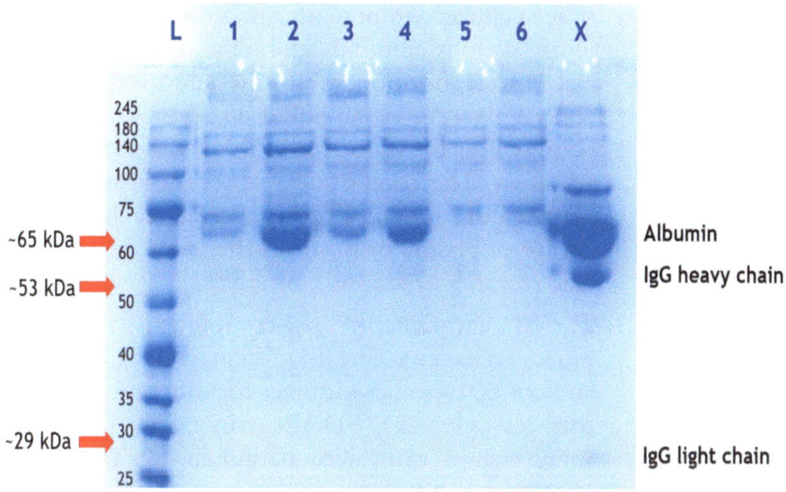

Fig. 2 Representative image of evaluation of the high-abundance protein depletion efficiency by SDS-PAGE and Coomassie blue staining. Lanes L, Pre-stained Protein Ladder (SMOBIO); 1–6, depleted serum; X, crude serum. Arrows indicate albumin, IgG heavy chain and light chain

9. Clamp gel into the apparatus and fill both buffer chambers with gel running buffer according to the instructions for the specific apparatus.
10. Prepare protein samples by boiling equal volumes in 2× Laemmli sample buffer for 5 min.
11. Load samples and molecular mass protein markers into wells for separation by electrophoresis.
12. Run the gel at 100 V until the dye front migrates into the running gel (~10 min) and increase to 200 V until the dye front reaches the bottom of the gel.
13. Remove the gel from the apparatus and remove the spacers and glass plates. Place the gel onto a small tray.
14. Perform staining using Coomassie blue or silver stain according to the manufacturer's protocol (Fig. 2).

4 Notes

1. We use the Nanodrop spectrophotometer from Thermo Fisher Scientific for total protein measurement. However, it is important to note that similar workflows can also be applied using equipment and reagents from various manufacturers.
2. Plasma and serum are derived from the blood following different biochemical processes. Plasma is the liquid component of blood that remains when an anticoagulant is used to prevent clotting. The EDTA chelates the calcium required for coagulation and is the most frequently used anticoagulant in clinical diagnostic labs. Other anticoagulants in use include heparin and citrate. Plasma is devoid of the blood cells, platelets, and other cellular components and consists of mostly water, proteins (such as albumin, globulins, and fibrinogen), as well as electrolytes, hormones, enzymes, and waste products. On the contrary, serum is the liquid component of blood that is left behind after the blood has clotted without the presence of any additives. The composition of serum is similar to that of the plasma but without clotting factors, such as fibrinogen. The choice of plasma or serum is therefore dependent on the specific research needs and objectives.
3. The BD Vacutainer offers a comprehensive line of products for blood collection. The BD Vacutainer Tubes are color coded and contain specific additives for different analytes and laboratory tests (Fig. 3). BD Vacutainer SST Tubes (gold cap) are spray coated with silica to aid in the clotting process and contain a polymer gel for serum separation. The gel helps to form a stable barrier between the serum and clotted blood, hence improving sample stability and quality. Alternatively, you may use BD Vacutainer Serum Tubes (red cap) that contain the additive spray-coated silica without the polymer gel.
4. BD Vacutainer PPT Tubes (pearlescent white cap) are spray dried with K_2EDTA anticoagulant and contain an inert gel for plasma separation. The gel creates a physical barrier between the undiluted plasma and the cellular components upon centrifugation. Alternatively, you may use BD Vacutainer EDTA tubes (lavender cap) that are coated with K_2EDTA anticoagulant or BD Vacutainer Heparin Tubes (green cap) that contain either lithium heparin or sodium heparin for plasma sample.
5. It is generally recommended to keep plasma/serum samples at −80 °C for long-term storage whereby certain proteins have been demonstrated to be relatively stable for up to 7 years [27]. Recently, there is also suggestion for sample storage at −70 °C rather than −80 °C as supported by the literature demonstrating little difference in the sample preservation

BD Vacutainer® Venous Blood Collection

Tube Guide

For the full line of BD Vacutainer® Specimen Collection Products, visit www.bd.com/vacutainer.

BD Vacutainer® Tubes with BD Hemogard™ Closure	BD Vacutainer® Tubes with Conventional Stopper	Additive	Inversions at Blood Collection*	Laboratory Use
Gold	Red/ Black	• Clot activator and gel for serum separation	5	For serum determinations in chemistry. May be used for routine blood donor screening and diagnostic testing of serum for infectious disease.** Tube inversions ensure mixing of clot activator with blood. Blood clotting time: 30 minutes.
Light Green	Green/ Gray	• Lithium heparin and gel for plasma separation	8	BD Vacutainer® PST Tube For plasma determinations in chemistry. Tube inversions prevent clotting.
Red	Red	• None (glass) • Clot activator (plastic)	0 5	For serum determinations in chemistry. May be used for routine blood donor screening and diagnostic testing of serum for infectious disease.** Tube inversions ensure mixing of clot activator with blood. Blood clotting time: 60 minutes.
Orange		• Thrombin	8	For stat serum determinations in chemistry. Tube inversions ensure complete clotting, which usually occurs in less than 5 minutes.
Royal blue		• Clot activator (plastic serum) • K_2EDTA (plastic)	8 8 0 5 8	For trace-element, toxicology, and nutritional-chemistry determinations. Special stopper formulation provides low levels of trace elements (see package insert).
Green		• Sodium heparin • Lithium heparin	8 8	For plasma determinations in chemistry. Tube inversions prevent clotting.
Gray		• Potassium oxalate/sodium fluoride • Sodium fluoride/Na_2 EDTA • Sodium fluoride (serum tube)	8 8 8	For glucose determinations. Oxalate and EDTA anticoagulants will give plasma samples. Sodium fluoride is the antiglycolytic agent. Tube inversions ensure proper mixing of additive and blood.
Tan		• K_2EDTA (plastic)	8	For lead determinations. This tube is certified to contain less than .01 µg/mL(ppm) lead. Tube inversions prevent clotting.
	Yellow	• Sodium polyanethol sulfonate (SPS) • Acid citrate dextrose additives (ACD): **Solution A -** 22.0 g/L trisodium citrate, 8.0 g/L citric acid, 24.5 g/L dextrose **Solution B -** 13.2 g/L trisodium citrate, 4.8 g/L citric acid, 14.7 g/L dextrose	8 8 8	SPS for blood culture specimen collections in microbiology. Tube inversions prevent clotting. ACD for use in blood bank studies, HLA phenotyping, and DNA and paternity testing.
Lavender		• Liquid K_3EDTA (glass) • Spray-coated K_2EDTA (plastic)	8 8	K_2EDTA and K_3EDTA for whole blood hematology determinations. K_2EDTA may be used for routine immunohematology testing, and blood donor screening.*** Tube inversions prevent clotting.
White		• K_2EDTA and gel for plasma separation	8	For use in molecular diagnostic test methods (such as, but not limited to, polymerase chain reaction [PCR] and/or branched DNA [bDNA] amplification techniques.)
Pink	New	• Spray-coated K_2EDTA (plastic)	8	For whole blood hematology determinations. May be used for routine immunohematology testing and blood donor screening.*** Designed with special cross-match for patient information required by the AABB. Tube inversions prevent clotting.
Light Blue	Clear	• Buffered sodium citrate • Citrate, theophylline, adenosine, dipyridamole (CTAD)	3-4 3-4	For coagulation determinations. CTAD for selected platelet function assays and routine coagulation determination. Tube inversions prevent clotting.
Clear	New Red/ Gray	• None (plastic)	0	For use as a discard tube or secondary specimen tube.

Fig. 3 Types of collection tubes for different laboratory tests. The type of additive and whether separator gel is present or not are indicated by the color of the cap that covers each type of tube

between the two temperatures, and setting the freezer to −70 °C is also more energy-saving and environment friendly [28].

6. The spin column is available in a mini format and midi format for the depletion of up to 10 or 100 μL of human serum or plasma, respectively. The immunoaffinity column is designed to specifically remove the 14 high-abundance proteins listed in Table 1. In terms of performance, existing literature show that these immunodepletion kits in general exhibited high specificity (>90% efficiency) [22]. Several problems that can occur during immunoaffinity depletion of serum/plasma samples and potential solutions include:

   (i) Nonspecific Binding: Nonspecific binding of proteins to the affinity resin can result in the loss of target proteins and contamination of the depleted sample with unwanted proteins. This can be identified by analyzing the depleted sample for the presence of nontarget proteins using mass spectrometry or other analytical techniques. To overcome this issue, optimize the wash steps to minimize nonspecific binding, use blocking agents to reduce background binding, and choose affinity resins with low nonspecific binding properties.

   (ii) Inefficient Depletion: Incomplete depletion of high-abundance proteins can limit the detection of low-abundance proteins in the sample. This can be identified by comparing the protein profiles of depleted and non-depleted samples using gel electrophoresis or other protein separation techniques. To improve depletion efficiency, optimize the affinity resin conditions, increase the sample-to-resin ratio, and ensure proper sample mixing and incubation times.

   (iii) Protein Degradation: Prolonged exposure of the sample to harsh conditions during depletion, such as high temperatures or extreme pH, can result in protein degradation and loss of sample integrity. This can be identified by monitoring changes in protein size or abundance using gel electrophoresis or mass spectrometry. To prevent protein degradation, optimize the depletion conditions to minimize exposure to denaturing conditions and perform the depletion process at lower temperatures if possible.

   (iv) Sample Loss: Loss of sample during handling or transfer steps can lead to reduced sample recovery and inaccurate quantification of proteins. This can be identified by comparing the input and output sample volumes before and after depletion. To minimize sample loss, handle the samples carefully, avoid unnecessary transfers between tubes or containers, and use low-retention surfaces to maximize sample recovery.

(v) Antibody Performance: Poor antibody quality or specificity can result in inefficient depletion of target proteins and increased background binding. This can be identified by testing the antibody performance using positive and negative control samples or by comparing the depletion efficiency of different antibody batches. To ensure optimal antibody performance, validate the antibody specificity and affinity using appropriate controls and quality control measures.

7. It is optional to collect a small aliquot of each fraction and analyze it by SDS-PAGE to evaluate the efficiency and reproducibility of the bound and unbound proteins.
8. This is a modification to the conventional protein concentration measurement technique for microvolume protein quantification by absorbance at 280 nm using a Nanodrop instrument. The Nanodrop features innovative sample retention technology that operates based on the surface tension of the liquid between two measurement surfaces, which enables quick quantification with lower amounts of sample.
9. The standards should be measured in replicates. It is recommended that appropriate standards be included as necessary to cover the expected assay concentration range.
10. After running a gel using SDS-PAGE, you may perform additional downstream analyses to further characterize the separated proteins. For example, you may excise the protein bands of interest from the gel and subject them to mass spectrometry analysis for protein identification or perform Western blotting analysis for the detection of specific proteins using antibodies.

Acknowledgments

The authors would like to thank the Ministry of Higher Education Malaysia (MOHE) which supported us via the Higher Institution Centre of Excellence (HiCOE) grant (JJ-2021-004).

References

1. Végvári A, Welinder C, Lindberg H, Fehniger TE, Marko-Varga G (2011) Biobank resources for future patient care: developments, principles and concepts. J Clin Bioinforma 1:24
2. Liotta LA, Ferrari M, Petricoin E (2003) Clinical proteomics: written in blood. Nature 425: 905
3. Veenstra TD, Conrads TP, Hood BL, Avellino AM, Ellenbogen RG, Morrison RS (2005) Biomarkers: mining the biofluid proteome. Mol Cell Proteomics 4:409–418
4. Geyer PE, Kulak NA, Pichler G, Holdt LM, Teupser D, Mann M (2016) Plasma proteome profiling to assess human health and disease. Cell Syst 2:185–195
5. Low TY, Syafruddin SE, Mohtar MA, Vellaichamy A, A Rahman NS, Pung YF, CSH T (2021) Recent progress in mass

spectrometry-based strategies for elucidating protein-protein interactions. Cell Mol Life Sci 78:5325–5533

6. Low TY, Mohtar MA, Lee PY, Omar N, Zhou H, Ye M (2021) Widening the bottleneck of phosphoproteomics: evolving strategies for phosphopeptide enrichment. Mass Spectrom Rev 40:309–333
7. Low TY, Magliozzi R, Guardavaccaro D, Heck AJ (2013) Unraveling the ubiquitin-regulated signaling networks by mass spectrometry-based proteomics. Proteomics 13:526–537
8. de Souza N, Picotti P (2020) Mass spectrometry analysis of the structural proteome. Curr Opin Struct Biol 60:57–65
9. Le Sueur C, Hammarén HM, Sridharan S, Savitski MM (2022) Thermal proteome profiling: insights into protein modifications, associations, and functions. Curr Opin Chem Biol 71:102225
10. Lee PY, Yeoh Y, Low TY (2023) A recent update on small-molecule kinase inhibitors for targeted cancer therapy and their therapeutic insights from mass spectrometry-based proteomic analysis. FEBS J 290:2845–2286
11. Aebersold R, Mann M (2003) Mass spectrometry-based proteomics. Nature 422: 198–207
12. Lee PY, Saraygord-Afshari N, Low TY (2020) The evolution of two-dimensional gel electrophoresis – from proteomics to emerging alternative applications. J Chromatogr A 1615: 460763
13. Gold L, Walker JJ, Wilcox SK, Williams S (2012) Advances in human proteomics at high scale with the SOMAscan proteomics platform. New Biotechnol 29:543–549
14. Smith JG, Gerszten RE (2017) Emerging affinity-based proteomic technologies for large-scale plasma profiling in cardiovascular disease. Circulation 135:1651–1664
15. Vistain LF, Tay S (2021) Single-cell proteomics. Trends Biochem Sci 46:661–672
16. Lundberg E, Borner GHH (2019) Spatial proteomics: a powerful discovery tool for cell biology. Nat Rev Mol Cell Biol 20:285–302
17. Lee PY, Yeoh Y, Omar N, Pung YF, Lim LC, Low TY (2021) Molecular tissue profiling by MALDI imaging: recent progress and applications in cancer research. Crit Rev Clin Lab Sci 58:513–529
18. Liu Y, Buil A, Collins BC, Gillet LCC, Blum L, Cheng LY et al (2015) Quantitative variability of 342 plasma proteins in a human twin population. Mol Syst Biol 11:786
19. Anderson NL, Anderson NG (2002) The human plasma proteome: history, character, and diagnostic prospects. Mol Cell Proteomics 1:845–867
20. Crutchfield CA, Thomas SN, Sokoll LJ, Chan DW (2016) Advances in mass spectrometry-based clinical biomarker discovery. Clin Proteomics 13:1
21. Cao Z, Tang HY, Wang H, Liu Q, Speicher DW (2012) Systematic comparison of fractionation methods for in-depth analysis of plasma proteomes. J Proteome Res 11:3090–3100
22. Lee PY, Osman J, Low TY, Jamal R (2019) Plasma/serum proteomics: depletion strategies for reducing high-abundance proteins for biomarker discovery. Bioanalysis 11:1799–1812
23. Gong Y, Li X, Yang B, Ying W, Li D, Zhang Y et al (2006) Different immunoaffinity fractionation strategies to characterize the human plasma proteome. J Proteome Res 5:1379–1387
24. Bellei E, Bergamini S, Monari E, Fantoni LI, Cuoghi A, Ozben T et al (2011) High-abundance proteins depletion for serum proteomic analysis: concomitant removal of non-targeted proteins. Amino Acids 49:145–156
25. Smith MP, Wood SL, Zougman A, Ho JT, Peng J, Jackson D et al (2011) A systematic analysis of the effects of increasing degrees of serum immunodepletion in terms of depth of coverage and other key aspects in top-down and bottom-up proteomic analyses. Proteomics 11:2222–2235
26. Echan LA, Tang HY, Ali-Khan N, Lee K, Speicher DW (2005) Depletion of multiple high-abundance proteins improves protein profiling capacities of human serum and plasma. Proteomics 5:3292–3303
27. Valo E, Colombo M, Sandholm N, McGurnaghan SJ, Blackbourn LAK, Dunger DB et al (2022) Effect of serum sample storage temperature on metabolomic and proteomic biomarkers. Sci Rep 12:4571
28. Drahl C (2018) A matter of degree. ACS Cent Sci 4:1294–1297

Chapter 2

Optimal Sample Preparation Workflow for Quantitative Mass Spectrometry-Based Studies in the Low Range

Sara Kassem and Paula Díez

Abstract

Mass spectrometry (MS) enables the high-throughput characterization of thousands of proteins in one single run. Nowadays, advances in the field have allowed for deep analysis of low cell numbers, even reaching the single-cell level. However, standardized protocols are still needed in this regard, especially oriented for non-proteomics specialists. Here we detail a reproducible, easy-to-implement workflow for cell-limited sample preparation, from cell storage to peptide labelling, for further TMT-based relative quantitative MS-based analysis.

Key words Quantity-limited sample, Proteomics sample preparation, Cell storage, Cell lysis, Protein quantification, SP3 technology, Tandem mass tag, Mass spectrometry

1 Introduction

High-throughput analysis of cell proteomes can be achieved by using mass spectrometry (MS)-based techniques [1, 2]. Advances in the field have allowed the development of new equipment for the simultaneous identification and quantification of more than 10,000 proteins in one single sample [3, 4]. However, the performance of the sample preparation method is also essential to produce optimal outcomes, especially in a quantity-limited setting. Any sample loss (from the starting cell material to the proteins and peptides generated for the final analysis) might lead to poor proteomic characterizations [5]. Thus, multiple approaches have been developed over the years to improve protein and peptide recoveries when working with low cell numbers. However, such a plethora of sample preparation methods might result in greater confusion, especially for nonexperts in the field [5]. Here, we propose a reproducible, easy-to-implement, and complete workflow for sample processing for MS-based studies in the low range, from sample collection to peptide labelling for quantitative analysis. We evaluated the

Taufika Islam Williams (ed.), *Tissue Proteomics: Methods and Protocols*, Methods in Molecular Biology, vol. 2884, https://doi.org/10.1007/978-1-0716-4298-6_2,

feasibility of this approach in 38 different cell types (for a total of 194 samples), even using only 300 cells or 2 μg protein as starting material amounts, quantifying >8300 proteins [6].

2 Materials

Unless otherwise indicated, all solutions requiring water can be prepared with distilled water grade or above, but we highly recommend using at least Milli-Q water grade. To prevent keratin contamination, always wear gloves, avoid wool-based clothing while sample processing, and dedicate glassware for proteomics use only. Inform yourself beforehand about the waste disposal regulations in your institution to safely dispose of the leftover solutions.

2.1 Cell Sample Storage

1. Washing buffer: PBS Na^+/K^+, pH 7.4 (137 mM NaCl, 2.7 mM KCl, 10 mM Na_2HPO_4, 1.8 mM KH_2PO_4) supplemented with 1% (v/v) protease and 1% (v/v) phosphatase inhibitor cocktails (*see* **Note 1**). Resuspend 15 μL 100% (v/v) protease and 15 μL 100% (v/v) phosphatase inhibitor cocktails in 1470 μL of PBS. Prepare fresh and store at 4 °C.

2.2 Cell Lysis

1. 200 mM 4-(2-hydroxyethyl)piperazine-1-ethanesulfonic acid (HEPES), pH 8.0: Weigh 24 g of HEPES and transfer to a glass beaker. Add 450-mL water, mix, and adjust the pH with HCl. Fill the buffer up with water to a volume of 500 mL and transfer it to a bottle, filtering the solution using a 0.22 μm filter. Store at 4 °C for up to 6 months.
2. 125 mM sodium pyrophosphate solution ($Na_4P_2O_7$): Weigh 1.2 g of $Na_4P_2O_7$ and add 20 mL of water. Store at 4 °C for up to 1 month.
3. 1 M β-glycerophosphate: Weigh 1.1 g of β-glycerophosphate and transfer it to a conical tube containing 5-mL water. Leave one aliquot at 4 °C for current use and store the remaining aliquots (100 μL each) at −20 °C for up to 6 months (*see* **Note 2**).
4. 100 mM sodium orthovanadate (Na_3VO_4), pH 10.0: Resuspend 1.84 g of Na_3VO_4 in 90 mL of water. Adjust the pH to 10 with NaCl (the solution will turn yellow). Then boil the solution at around 45 °C on a heat block until it turns colorless and allow it to cool down to room temperature (RT) on ice. After cooling, readjust the pH to 10 if needed. If adjusted, repeat the heating and cooling steps until the pH is stable at 10. Add water up to a volume of 100 mL, aliquot (1 mL per aliquot, leave one aliquot at 4 °C for current use), and store at −20 °C for up to 6 months (*see* **Note 2**).

5. 10 mg/mL DNase I: Dissolve 50 mg of DNase I in 5 mL of water, make aliquots (100 μL each, leave one aliquot at 4 °C for current use), and store at −20 °C (*see* **Note 2**).
6. Cell lysis buffer: 9 M urea, 20 mM HEPES, 2.5 mM sodium pyrophosphate, 1 mM β-glycerophosphate, 1 mM sodium orthovanadate, 1% (v/v) protease inhibitor cocktail, 1% (v/v) phosphatase inhibitor cocktail, and 10 μg/mL DNase I. For preparing 1 mL, weigh 0.54 g of urea, transfer to a tube, and add 100 μL of 200 mM HEPES, 20 μL of 125 mM sodium pyrophosphate, 1 μL of 1 M β-glycerophosphate, 10 μL of 100 mM sodium orthovanadate, 10 μL of protease inhibitor cocktail, 10 μL of phosphatase inhibitor cocktail, and 1 μL of 10 mg/mL DNase I. Add water up to 1 mL. Prepare fresh and keep at RT.

2.3 Protein Quantification

1. Qubit™ working solution (1:200 (v/v) dilution, Thermo Fisher): Mix 6 μL of Qubit™ protein reagent and 1194 μL of Qubit™ protein buffer in a light-protected tube (*see* **Note 3**).

2.4 Sera-Mag Carboxylate-Modified Magnetic Beads Preparation

Prepare the bead working solution by combining carboxylate hydrophilic and hydrophobic bead stocks at a final bead/protein ratio of 10:1 (wt/wt), as described by Hughes et al. [7].

1. Bead working solution: Mix 20 μL of carboxylate hydrophilic beads and 20 μL of carboxylate hydrophobic beads (both stocks at 50 mg/mL) in a 0.2-mL tube (*see* **Notes 4** and **5**), fill it up to 150 μL with ultrapure water, and mix. Place the tube in a magnetic rack for 2 min and discard the supernatant (*see* **Note 6**). Wash the beads again (off the magnet) by adding 150 μL of ultrapure water and place the tube back on the magnetic rack for 2 min. Repeat this washing step once more. Finally, resuspend the pelleted washed beads in 100 μL of ultrapure water (10 mg/mL bead concentration) and store them at 4 °C for up to 1 month.

2.5 Protein Reduction and Alkylation

1. 0.5 M dithiothreitol (DTT): Weigh 0.15 g of DTT and resuspend in 2 mL of water. Filter this solution through a 0.22 μm syringe filter, aliquot (50 μL each, leave one aliquot at 4 °C for current use), and store at −20 °C for up to 1 year (*see* **Note 2**).
2. 400 mM iodoacetamide (IAA): Dissolve 0.15 g of IAA in 2 mL of ultrapure water. Prepare fresh and keep in the dark at RT.

2.6 SP3 Protein Cleanup and Digestion

1. Bead rinse solution (70% ethanol): Add 70 mL of 100% ethanol in a glass bottle and fill it up until 100 mL with ultrapure water. Store at 4 °C.
2. 50 mM ammonium bicarbonate (AB), pH 8: Weigh 0.16 g of AB and add 30 mL of ultrapure water. Adjust the pH to 8 and

fill it up to 40 mL with water. This solution can be stored for up to a week at 4 °C.

3. Protein digestion solution: Resuspend the trypsin/LysC mix (1 μg total enzyme) in 50 μL of 50 mM AB to achieve a 0.02 μg/μL concentration.
4. Elution solution (2% dimethyl sulfoxide, DMSO): Mix 40 μL of 100% DMSO in 1960 μL of ultrapure water. Prepare fresh and keep in the dark at RT.

2.7 Tandem Mass Tag (TMT)-Based Peptide Labelling

1. Isobaric stable isotope labeling using TMT tags is a recent approach for relative quantitative mass spectrometry analysis [8, 9]. 40 mM HEPES, pH 8.5: Mix 10 mL of 200 mM HEPES (*see* Subheading 2.2) and 30 mL of water, adjust the pH to 8.5, and fill up to 50 mL with water. Prepare fresh and store at RT.
2. 5% hydroxylamine: Mix 100 μL of 50% hydroxylamine with 900 μL of water. Prepare fresh and store at RT.

3 Methods

Unless otherwise indicated, all steps are performed at RT. Figure 1 summarizes the main steps of the sample preparation protocol for quantitative mass spectrometry-based studies described in this chapter.

3.1 Cell Sample Storage

1. Once the cells of interest are collected in a tube, wash them three times by adding 400 μL of ice-cold washing buffer and centrifuging at 500 *g* for 5 min at 4 °C (*see* **Notes 7** and **8**).
2. After the last washing step, snap freeze the cell pellet by placing the tube in dry ice for at least 3 min. Immediately proceed with Subheading 3.2 or store the pellet at −80 °C until further processing.

3.2 Cell Lysis

1. Thaw the snap-frozen cell pellet rapidly with gentle agitation of the tube in lukewarm water. Then place the sample tube on ice.
2. Resuspend the cell pellet in cell lysis buffer (buffer volume will be determined by sample cell number; *see* **Note 9**) and mix gently by finger flicking.
3. Sonicate the sample by placing the tube in an ice-water bath (15 W, 3 pulses of 5 s each). Then centrifuge the sample at 21,000 *g* for 15 min.
4. Collect the supernatant containing the proteins in a new tube.
5. Prepare two aliquots of 4 μL each (*see* **Notes 10–12**) for the protein quantification step (see Subheading 3.3). The aliquots

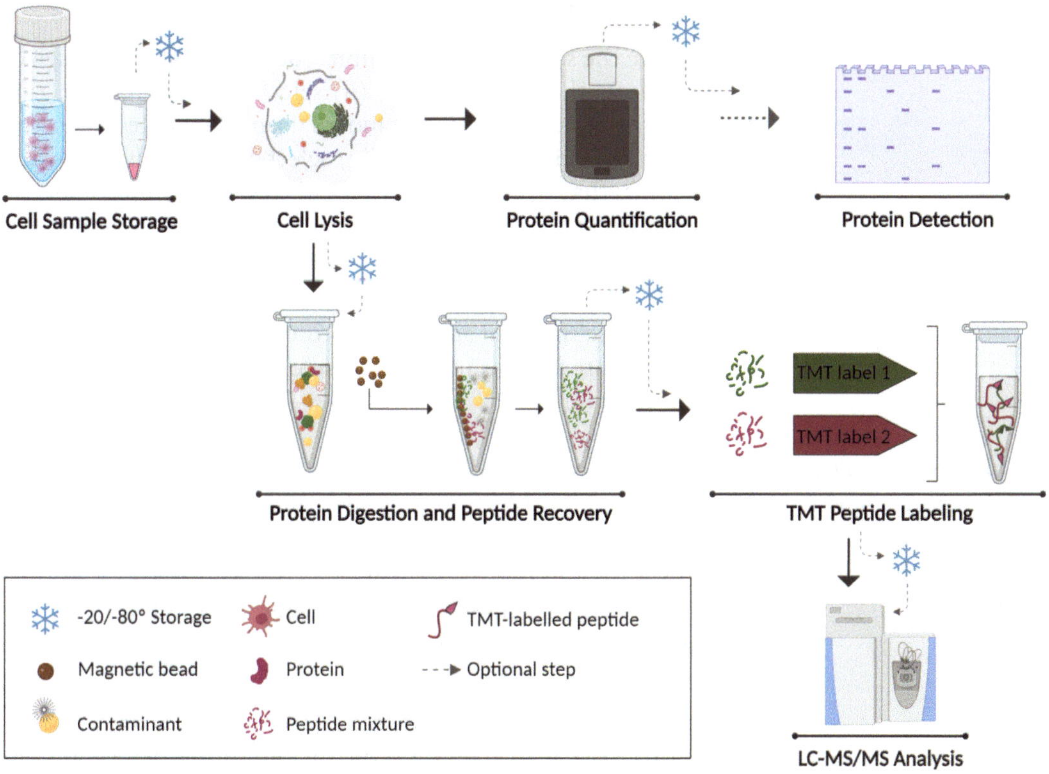

Fig. 1 Quantitative mass spectrometry (MS)-oriented sample preparation procedure. The main steps for sample processing of quantity-limited starting material are described, including optional steps. *TMT*, tandem mass tag; nano*LC-MS/MS*, liquid chromatography with tandem mass spectrometry. Created with BioRender.com

can be used immediately or stored at −20 °C for later processing.

6. Snap freeze the remaining protein sample and store it at −80 °C until further usage.

3.3 Protein Quantification

This step might be performed using any fluorescent-based methodology as they offer superior sensitivity [5].

1. Label three Qubit™ assay tubes (one tube per standard sample: 0, 200, and 400 ng/μL) on their lids (*see* **Note 11**). Add 10 μL of standard reagent to its corresponding labelled tube. Add 4 μL of cell lysis buffer to correct for the background signal (if less sample volume was aliquoted for quantification in **step 5** of Subheading 3.2, adjust the volume here accordingly).
2. Collect the protein sample tubes from **step 5** of Subheading 3.2.
3. Add the Qubit™ working solution to the tubes containing the samples and standards up to a total of 200 μL.

4. Incubate in the dark for 15 min.
5. Measure the samples on the Qubit™ device as follows: (i) select *Protein* on the home screen, (ii) select *Read standards*, (iii) insert the tube containing standard 1 and press *Read* (do the same for standards 2 and 3), (iv) select *Sample volume* (4 μL) and *units* (μg/mL or mg/mL) for the output sample concentration on the screen, and (v) insert the tube containing the sample and press *Read*. Once finished, the instrument displays the results of the assay on screen.
6. To calculate the total amount of protein in each sample, multiply the protein concentration value obtained in **step 5** by the sample volume.

3.4 Protein Digestion and Peptide Recovery

Protein digestion and peptide recovery steps are performed in a vertical flow cabinet to avoid potential sample contamination.

3.4.1 Protein Reduction and Alkylation

1. Preheat the thermo-block mixer to 45 °C.
2. Get the protein sample tubes from **step 6** of Subheading 3.2 and place them on dry ice. Thaw the samples rapidly, gently mix them in lukewarm water, and keep them on ice.
3. Transfer the volume corresponding to the protein amount you want to work with (based on the quantification determination performed in Subheading 3.3) (*see* **Note 13**) to a 0.2-mL low-absorption tube (*see* **Note 14**). [As an example, for this protocol, we will process 5 μg of protein contained in 10 μL of sample volume].
4. Add 0.2 μL of 0.5 M DTT to reduce the disulfide bonds of the proteins (final concentration: 10 mM). Mix tube by finger flicking (*see* **Note 15**) and incubate for 30 min at 45 °C without agitation. When incubation is finished, briefly spin down the tube to collect the condensation that may have formed.
5. Add 1.02 μL of 400 mM IAA to alkylate the reduced disulfide bonds (final concentration: 40 mM) and mix by finger flicking. Incubate in the dark for 30 min (*see* **Note 16**).
6. Quench the alkylation reaction by adding 0.45 μL of 0.5 M DTT (final concentration: 20 mM) by gently mixing by finger flicking and incubating in the dark for 5 min.

3.4.2 Protein Digestion and Peptide Recovery Using Carboxylate Beads

1. Allow the acetonitrile (ACN), bead rinse, and bead working solutions to reach RT for at least 10 min and vortex the beads before use (*see* **Note 4**).
2. Add 5 μL of bead working solution (final bead/protein ratio: 10:1 (wt/wt)) to the sample tube from **step 6** Subheading 3.4.1. Do not mix.

3. Add 38.9 μL of ACN (final concentration: 70% (v/v)) and mix gently by pipetting (*see* **Note 17**).
4. Incubate for 20 min off the magnetic rack (e.g., DynaMag™-PCR Magnet, from Thermo).
5. Place the tube in the magnetic rack and incubate for 2 min or until the solution looks clear (*see* **Note 18**).
6. Carefully remove the supernatant while keeping the beads on the magnet (*see* **Notes 6** and **19**).
7. Add 200 μL of bead rinse solution to the tube containing the protein-bound beads and rinse by pipetting up and down gently (*see* **Note 17**).
8. Place the tube on the magnetic rack for 30 s.
9. Carefully remove and discard the supernatant.
10. Repeat once **steps 7–9**.
11. Add 180 μL of ACN and incubate for 15 s on the magnetic rack.
12. Carefully remove and discard the supernatant.
13. Take the tube off the magnetic rack and air-dry the beads for 30 s.
14. Reconstitute the beads in 10 μL of protein digestion solution (1:25 enzyme/protein ratio, *see* **Notes 20** and **21**) by adding it just above the beads without touching them. Do not pipette.
15. Incubate for 16 h at 37 °C.
16. Slowly resuspend the beads by vortexing.
17. Add 190 μL of ACN (placing the pipette at a 15° angle, in one fluent movement; final concentration: ≥95% (v/v)) and incubate for 8 min off the magnetic rack.
18. Place the tube on the magnetic rack and incubate for 2 min.
19. Carefully remove the supernatant while keeping the beads on the magnetic rack (*see* **Notes 6** and **22**).
20. Add 180 μL of ACN and incubate for 15 s on the magnetic rack.
21. Carefully remove the supernatant while keeping the beads on the magnetic rack (*see* **Notes 6** and **22**). Take the tube off the magnetic rack.
22. Elute the peptides bound to the beads by adding 100 μL of elution solution, mix gently by pipetting, and incubate for 5 min.
23. Place the tube on the magnetic rack for 2 min and collect the supernatant containing the purified peptides in a new 0.5-mL low-absorption tube (*see* **Notes 8** and **23**).

24. Place the purified peptide-containing sample tube in liquid nitrogen and lyophilize it in a vacuum freeze dryer.
25. Once dried out, proceed immediately with Subheading 3.4.3 or store the peptide sample at −20 °C.

3.4.3 Tandem Mass Tag (TMT) Labeling of Peptides

1. Equilibrate the TMT labels (0.8 mg) to RT. If there is condensation on the outside of the tubes, remove it using a tissue (*see* **Note 24**). Resuspend in 160 μL of ACN and aliquot (20 μL/vial, 100 μg each). Lyophilize and store at −80 °C.
2. Dissolve each TMT label aliquot in 25 μL of ACN (final concentration: 4 μg/μL) by pipetting and incubating for 5 min.
3. Resuspend each lyophilized peptide sample (from **step 25** of Subheading 3.4.2) in 25 μL of 40 mM HEPES pH 8.5 and mix carefully by finger flicking.
4. Briefly centrifuge the tubes to gather the solutions.
5. Add 5 μL of TMT label (1:4 sample/tag (wt/wt) ratio) to the corresponding peptide sample and mix by finger flicking.
6. Incubate for 1 h.
7. Quench the reaction by adding 6 μL of 5% hydroxylamine (final concentration: 1%) to each tagged sample and incubate for 15 min.
8. Combine TMT-tagged samples in a new low-absorption tube and lyophilize the pooled sample.
9. Store the lyophilized sample at −20 °C until the nanoLC-MS/MS analysis.

After **step 9**, TMT-based discovery proteomics by nanoLC separation and data-dependent acquisition (DDA) can be performed, followed by TMT-based discovery proteomics data interrogation.

4 Notes

1. Protease and phosphatase inhibitors are added to prevent protein degradation and loss of phosphorylation sites, respectively.
2. Avoid freeze-thaw cycles since the performance/activity of the reagent might be negatively affected.
3. Prepare the Qubit™ working solution just before starting the protein quantification protocol since its fluorescence signal is only stable for 3 h at RT.
4. The Sera-Mag carboxylate beads have to be warmed up to RT for at least 10 min and homogenized very well before usage. Improper homogenization can reduce protein recovery.

5. If you want to prepare more beads (>200 μL), make sure to use multiple 0.2-mL tubes (scaling up in one single tube is not advised). Only in the final step, all 0.2-mL bead vials can be pooled.
6. Be careful not to touch the beads with your pipet tip. The best way to remove the supernatant is to hold your pipet tip on the other side to where the beads migrated to in the wall of the tube. If by accident beads are taken off the magnet while removing the supernatant, pipet the supernatant back in the tube (while on the magnetic rack) and incubate for 2 min so that beads migrate back.
7. Centrifugation speed must be adapted to the cell type. Overall, the widely accepted recommendation for mammalian cells is to centrifuge at 300–500 *g* for 5 min to avoid cell disruption. However, it is important to mention that this very low speed might result in loose cell pellets and subsequent sample loss, so extreme caution must be taken while washing.
8. We highly recommend washing and storing the cells, proteins, and peptide samples in tubes presenting low hydrophobicity (minimizing protein binding) known as low-absorption tubes (e.g., protein LoBind tubes from Eppendorf®). Moreover, we recommend choosing tubes at low surface area-to-solution volume ratios.
9. We have experimentally determined the optimal cell lysis buffer volume for each cell number range, to obtain the maximum protein extraction and minimum sample dilution [1–2500 cells: 40 μL; 2501–10,000 cells: 50 μL; 10,001–50,000 cells: 70 μL; 50,001–100,000 cells: 80 μL; 100,001–500,000 cells: 90 μL; 500,001–10 × 10^6 cells: 100 μL]. When lysing more than one million cells, extra DNase I solution volume should be added accordingly to the cell pellet size (e.g., 20 μL extra for 1–10 million cells).
10. For protein quantification using the Qubit system, 4 μL/sample is the maximum volume that can be employed for measurement since the cell lysis buffer described in this protocol includes components at concentrations not compatible with the system. For this reason, keeping the minimum sample dilution (as indicated in **Note 9**) is also critical.
11. Do not label the Qubit™ assay tubes on the side as it could interfere with the sample measurement.
12. Based on our experience, 4 μL/sample is enough to obtain a protein concentration value for low cell numbers when following the protocol described here. However, lower sample volumes can also be used when using greater cell numbers (i.e., > 100,000 cells). Method development might be also required for certain cell lines to define the optimal volumes.

13. The sample volume can vary from sample to sample due to different concentration values. Therefore, when processing multiple samples in the same experiment, adjust all volumes to the highest one by adding cell lysis buffer. Thus, all samples will have the same starting volume, facilitating the processing.
14. If the starting volume is >47 μL, transfer your protein sample to a 1.5- or 2-mL tube. Later in the protocol, the addition of reagents will result in a volume above 200 μL that would not fit in a 0.2-mL tube.
15. We highly recommend mixing by finger flicking to avoid sample losses that might occur if pipetting.
16. IAA is light sensitive. Therefore, work at all times in the dark until the reaction is quenched. Do not forget to turn off the light in the flow cabinet during the incubation step or cover the tube with aluminum foil.
17. Avoid using the vortex as mixing at a high speed may cause protein-bound beads to stick to the upper parts of the walls or lid of the tube, reducing the sample recovery.
18. After 2-min incubation, check the sample. A correct magnet-bead binding will be observed as a distributed spread of beads on the wall and a clear and colorless supernatant.
19. For optimal bead-protein binding, we highly recommend re-incubating this supernatant again with 2.5 μL of beads (final bead/protein ratio: 5:1 (wt/wt)) for 20 min off the magnet to ensure all proteins are bound to beads. After this second incubation, transfer the supernatant to the tube containing the beads from the first incubation and place it for 2 min on the magnet. Carefully remove the supernatant and collect it in a low-absorption tube. We also recommend saving this tube at −20 °C until obtaining the MS/MS results in case the proteins were not properly attached to the beads.
20. The enzyme-to-protein ratio can be optimized for each sample type if needed. We recommend starting with a 1:25 enzyme/protein ratio.
21. Trypsin and LysC enzymes can be used separately for protein digestion. In this case, incubate first with LysC enzyme for 4 h and then overnight with trypsin, both at 37 °C.
22. Carefully remove the supernatant and collect it in a low-absorption tube. We recommend saving this tube at −20 °C until obtaining the MS/MS results in case the peptides were not properly attached to the beads.
23. To eliminate any potential bead carryover, we recommend re-incubating (2 min on the magnetic rack) the purified peptide solution collected in **step 23** of Subheading 3.4.2.

24. The TMT reagents are highly moisture sensitive and therefore must be equilibrated to RT before opening.

Acknowledgments

This work was supported by the European Research Council under the European Union's Horizon 2020 Research and Innovation Programme with an ERC Advanced Grant (ERC-2015-AdG 695655, TiMaScan). PD is supported by a Sara Borrell postdoctoral fellowship (Institute of Health Carlos III, Spanish Government, CD22/00009).

References

1. Ahrens CH, Brunner E, Qeli E et al (2010) Generating and navigating proteome maps using mass spectrometry. Nat Rev Mol Cell Biol 11:789–801. https://doi.org/10.1038/nrm2973
2. Aebersold R, Mann M (2016) Mass-spectrometric exploration of proteome structure and function. Nature 537:347–355. https://doi.org/10.1038/nature19949
3. Kawashima Y, Nagai H, Konno R et al (2022) Single-shot 10K proteome approach: over 10,000 protein identifications by data-independent acquisition-based single-shot proteomics with ion mobility spectrometry. J Proteome Res 21:1418–1427. https://doi.org/10.1021/acs.jproteome.2c00023
4. Zhong C-Q, Wu J, Qiu X et al (2020) Generation of a murine SWATH-MS spectral library to quantify more than 11,000 proteins. Sci Data 7:104. https://doi.org/10.1038/s41597-020-0449-z
5. Kassem S, van der Pan K, de Jager AL et al (2021) Proteomics for low cell numbers: how to optimize the sample preparation workflow for mass spectrometry analysis. J Proteome Res 20:4217–4230. https://doi.org/10.1021/acs.jproteome.1c00321
6. van der Pan K, Kassem S, Khatri I et al (2022) Quantitative proteomics of small numbers of closely-related cells: selection of the optimal method for a clinical setting. Front Med (Lausanne) 9. https://doi.org/10.3389/fmed.2022.997305
7. Hughes CS, Moggridge S, Müller T et al (2019) Single-pot, solid-phase-enhanced sample preparation for proteomics experiments. Nat Protoc 14:68–85. https://doi.org/10.1038/s41596-018-0082-x
8. Zecha J, Satpathy S, Kanashova T et al (2019) TMT labeling for the masses: a robust and cost-efficient, in-solution labeling approach. Mol Cell Proteomics 18:1468–1478. https://doi.org/10.1074/mcp.TIR119.001385
9. Li J, Vranken JGV, Vaites LP et al (2020) TMTpro reagents: a set of isobaric labeling mass tags enables simultaneous proteome-wide measurements across 16 samples. Nat Methods 17:399–404. https://doi.org/10.1038/s41592-020-0781-4

Chapter 3

In-Gel and In-Solution Approaches to Protein and Peptide Fractionation

Dylan H. Multari, Mohadeseh Montazeri Shatouri, Sabrina Grizzi de Oliveira, Zeshan Ali, Fatemeh Habibpourmehraban, Farhad Masoomi-Aladizgeh, and Paul A. Haynes

Abstract

Identifying proteins from living organisms helps us understand the biological functions of cells, discover new molecular mechanisms, and interrogate known mechanisms for improving our understanding. For a comprehensive understanding of cellular functions, identifying the whole protein content, or proteome, of a cell is desirable but challenging. Here, we describe in detail two methods of proteome fractionation at either the protein (SDS-PAGE) or peptide (high-pH reversed-phase fractionation) level, which can be used to maximize the identification of proteins from complex biological samples. We apply these methods to two different sample types commonly processed in our laboratory to demonstrate the versatility of these protocols. These methods produce many more peptide identifications when compared to conventional single-shot analysis and can also be used in combination to generate larger complementary datasets with greater depth of proteome coverage.

Key words Sodium dodecyl-sulfate polyacrylamide gel electrophoresis (SDS-PAGE), Filter-aided sample preparation (FASP), High-pH reversed-phase fractionation, Proteomics, In-gel digestion

1 Introduction

One of the inherent limitations of proteomic analysis is sample complexity. When dealing with complex mixtures such as a cell or tissue lysate, there is a chance that peptides from lower abundance proteins will be masked during mass spectrometric analysis. This can result in data that is not fully representative of the whole protein expression profile of the sample. One widely used approach to improve data acquisition of complex samples is protein or peptide fractionation [1–3].

Fractionation is advantageous for complex samples since a more discrete signal acquisition reduces the chances of interference between peptides and improves signal-to-noise ratios in peptide

Taufika Islam Williams (ed.), *Tissue Proteomics: Methods and Protocols*, Methods in Molecular Biology, vol. 2884, https://doi.org/10.1007/978-1-0716-4298-6_3,

detection [4]. There are a multitude of fractionation techniques available, but some commonly employed methods include SDS-PAGE for separation of proteins according to molecular weight and high-pH reversed-phase fractionation for separation of peptides according to hydrophobicity [5, 6]. These approaches are orthogonal to each other; either can be used as the first dimension of a shotgun proteomics experiment, but both can also be used in combination to produce complementary datasets that further enhance depth of proteome coverage [3]. Fractionation involves more sample handling, with more opportunities for potential losses, so these types of approaches are best suited to samples which are not strictly limited in the amount available.

In this chapter, the methods we use for both SDS-PAGE and high-pH reversed-phase fractionation are described in detail and applied to different sample types analyzed in our laboratory to illustrate how fractionation can be employed in a range of proteomic studies.

2 Materials

Prepare all solutions with ultrapure water and analytical-grade reagents. All reagents can be prepared and stored at room temperature unless otherwise noted, and it is best to use the highest-grade analytical quality available.

2.1 SDS-Page

1. SDS-PAGE running buffer (10×): 250 mM Tris–HCl, 1.92 M glycine, 1% (w/v) SDS (pH 8.3) (*see* **Note 1**).
2. SDS-PAGE sample loading buffer (5×): 250 mM Tris–HCl (pH 6.8), 10% (w/v) SDS, 50% (v/v) glycerol, 5% (v/v) β-mercaptoethanol, 62.5 mM EDTA, 0.1% bromophenol blue (*see* **Note 2**).
3. Coomassie fixative solution: 50% (v/v) ethanol, 10% (v/v) acetic acid.
4. Colloidal Coomassie staining solution: 0.05% (w/v) Coomassie Brilliant Blue G-250, 40% (v/v) ethanol, 10% (v/v) acetic acid.
5. Coomassie destaining solution: 25% methanol in ultrapure water.
6. Precast SDS-PAGE gel (10% Tris–HCl 10 × 50 μL well gel).
7. Protein molecular weight markers.
8. 100 mM NH_4HCO_3 (*see* **Note 3**).
9. 50% acetonitrile (ACN)/50 mM NH_4HCO_3.
10. 100% ACN, analytical grade.

11. Reducing solution: 10 mM dithiothreitol (DTT) in 50 mM NH_4HCO_3.
12. Alkylating solution: 55 mM iodoacetamide (IAA) in 50 mM NH_4HCO_3.
13. Trypsin solution: 12.5 ng/μL sequencing-grade modified porcine trypsin in 50 mM NH_4HCO_3 (*see* **Note 4**).
14. 50% ACN/2% formic acid (FA).
15. 0.1% trifluoroacetic acid (TFA).

2.2 Peptide Cleanup and Purification

1. SDB-RPS (3 M Empore) Stage tips.
2. 80% ACN/5% ammonium hydroxide.
3. 0.1% TFA.
4. 0.1% formic acid (FA).

2.3 High-pH Fractionation

1. High-pH reversed-phase peptide fractionation kit (Thermo Fisher, San Jose, CA).
2. 0.1% triethylamine (TEA).
3. 100% acetonitrile.

2.4 Samples Analyzed

1. Tuna meal waste.
2. *Eucalyptus grandis* (flooded gum) leaves collected from mature trees on the campus of Macquarie University campus (North Ryde, NSW, Australia).

2.5 In-Solution Digestion

1. Tris–HCl 100 mM buffer (pH 8.8).
2. Trypsin enzyme: 0.2 μg/μL sequencing-grade modified porcine trypsin.

2.6 Eucalyptus grandis Leaf Protein Extraction

1. 10% TCA, 2% β-mercaptoethanol, in acetone (*see* **Note 5**).
2. 1% SDS, 0.15 M Tris–HCl (pH 8.8), 0.1 M DTT, 1 mM EDTA solution (*see* **Note 6**).
3. 0.5 M Tris base.
4. Tris-buffered phenol solution, pH 7.8–8.0 (stored at 4 °C).
5. Wash buffer solution: 10 mM Tris–HCl (pH 8.0), 1 mM EDTA, 0.7 M sucrose.
6. 0.1 M ammonium acetate in methanol (stored at −20 °C).
7. 10 mM DTT in 80% acetone (*see* **Note 6**).
8. 8 M urea.

2.7 Protein and Peptide Assay

BCA Protein and Micro BCA Protein Assay Kits (bicinchoninic acid assay).

3 Methods

3.1 Sodium Dodecyl Sulphate-Polyacrylamide Gel Electrophoresis (SDS-PAGE) Protein Fractionation and In-Gel Digestion

1. To obtain approximately 100 μg of protein, add 50 μL of 1× SDS-PAGE sample loading buffer and vortex before heating at 95 °C for 5 min (*see* **Note 7**).
2. After heating, centrifuge the samples at 14,000 rpm for 5 min and transfer the clarified supernatant to fresh microcentrifuge tubes for loading onto the SDS-PAGE gel.
3. Into a precast SDS-PAGE gel (*see* **Note 8**), load 10 μL of protein marker into the first well. Then proceed to load 30 μL of sample into separate lanes until all samples are loaded (*see* **Note 9**).
4. Once samples are loaded into the gel, fill the gel tank with 1× SDS-PAGE running buffer until the wells of the gel are completely submerged in liquid. Connect lid and power supply to gel tank and then run the gel at 60 V constant voltage for 10 min to ensure proteins are entering into the top of the gel.
5. After 10 min, adjust the voltage to 170 V constant voltage and allow the gel to run until the dye front reaches the bottom of the gel (*see* **Note 10**).
6. Once electrophoretic separation is complete, disassemble the gel tank and carefully pry the gel cassette open with a spatula (*see* **Note 11**).
7. Transfer gel to a container and rinse briefly with ultrapure water.
8. Discard water and replace with Coomassie fixative solution. Allow gel to fix for 1 h on a shaker.
9. Discard fixative solution and replace with colloidal Coomassie staining solution. Allow gel to stain overnight on a shaker.
10. Discard the staining solution and rinse the gel and container thoroughly with ultrapure water to remove excess stain (*see* **Note 12**).
11. To the gel, add enough Coomassie destaining solution to cover and return to shaker (*see* **Note 13**).
12. Once the gel is clear enough to see the marker lane and protein samples, discard the destaining solution and rinse gel with ultrapure water.
13. Transfer the gel carefully to a clean glass plate and use a scalpel to excise your sample lanes. Divide each excised lane into eight equal-sized fractions (*see* **Note 14**).
14. Finely chop each fraction into approximately 1 mm × 1 mm-sized pieces and transfer each fraction to a new 1.5-mL microcentrifuge tube.

15. Equilibrate gel pieces briefly with 200 μL of 100 mM NH_4HCO_3 and discard the liquid.
16. Add 200 μL of 50% ACN/100 mM NH_4HCO_3, vortex, and then incubate at room temperature for 10 min before removing the liquid. Repeat until gel pieces are clear.
17. Wash gel pieces with 50 μL of 100% ACN, vortex, and allow to incubate for 5 min (*see* **Note 15**).
18. Decant the ACN and allow the gel pieces to air-dry for 10 min.
19. Rehydrate gel pieces with 50 μL of 10 mM DTT in 50 mM NH_4HCO_3. Incubate gel pieces at room temperature for 1 h to reduce disulfide bonds.
20. Remove excess DTT solution and add 50 μL of 55 mM IAA in 50 mM NH_4HCO_3. Incubate gel pieces in the dark at room temperature for 1 h to alkylate cysteine residues.
21. Remove IAA solution and wash as described in **steps 15–17**.
22. Once gel pieces are dehydrated, rehydrate the gel pieces with 30 μL of trypsin solution (or enough to cover) for 30 min at 4 °C (*see* **Note 16**).
23. Allow the gel pieces to digest overnight at 37 °C.
24. Following digestion, transfer any supernatant into new 1.5-mL microcentrifuge tubes (*see* **Note 17**).
25. Add 50 μL of 50% ACN/2% FA to the gel pieces and incubate for 15 min. Vortex and spin briefly and then transfer supernatants to the corresponding initial digest supernatants. Repeat two more times for a total volume of ~150 μL of digest.
26. Vacuum centrifuge digest extracts to dryness and reconstitute each fraction in 10 μL of 0.1% TFA. Fractions are now ready for cleanup and purification prior to mass spectrometric analysis.

3.2 High-pH Reversed-Phase Peptide Fractionation

1. Resuspend 100 μg of peptides in 300 μL of 0.1% TFA (*see* **Note 18**).
2. Prepare a step gradient consisting of eight solutions of differing concentrations of ACN for column elution as per Table 1 (*see* **Note 19**).
3. Place the high-pH column into a 2-mL microcentrifuge tube and centrifuge at 2500 rpm for 2 min (*see* **Note 20**). Discard any flowthrough and transfer the spin column into a fresh 2-mL microcentrifuge tube.
4. Remove the screw cap and load 300 μL of 100% ACN into the column. Replace cap and centrifuge at 2500 rpm for 2 min. Discard flowthrough and repeat once more.
5. Wash the column twice with 0.1% TFA as described in **step 3**.

Table 1
High-pH reversed-phase elution solutions for eight fractions of peptides

Fraction no.	Acetonitrile (%)	Acetonitrile (μL required to make up 1 mL)	Triethylamine 0.1% solution (μl required to make up 1 mL)
1	5	50	950
2	7.5	75	925
3	10	100	900
4	12.5	125	875
5	15	150	850
6	17.5	175	825
7	20	200	800
8	50	500	500

6. Transfer the spin column into a fresh 2-mL microcentrifuge tube. Load 300 μL of sample into the column and centrifuge at 2500 rpm for 2 min (*see* **Note 21**).
7. Transfer the spin column into a fresh 2-mL microcentrifuge tube. Load 300 μL of ultrapure water into the column and centrifuge at 2500 rpm for 2 min (*see* **Note 22**).
8. Transfer the spin column into a fresh 2-mL microcentrifuge tube. Load 300 μL of 5% ACN in 0.1% TEA (fraction 1 elution solution as outlined in Table 1) and centrifuge at 2500 rpm for 2 min.
9. Continue peptide elution using a new 2-mL microcentrifuge tube for each fraction while increasing the acetonitrile concentration from 5% to a final concentration of 50% (as described in Table 1). Each step is performed as described in **step 8**.
10. Once all fractions are collected, evaporate fractions to dryness in a vacuum centrifuge and resuspend each in 20 μL of 0.1% FA for mass spectrometric analysis.

3.3 Peptide Cleanup and Purification

This cleanup procedure is broadly applicable across a wide range of sample types in preparation for single-shot proteomic analysis. Tryptic peptides from in-solution digests of tuna meal and *Eucalyptus grandis* leaves were desalted as follows.

1. Prepare Stage tips by packing 200 μL pipette tips with SDB-RPS membrane (Fig. 1) (*see* **Note 23**).
2. Equilibrate the tip with 100 μL of 0.1% TFA and centrifuge at 2500 rpm for 30 min (*see* **Note 24**).
3. Discard flowthrough and add 100 μL of sample to the tip. Centrifuge at 2500 rpm for 30 min (*see* **Note 25**).

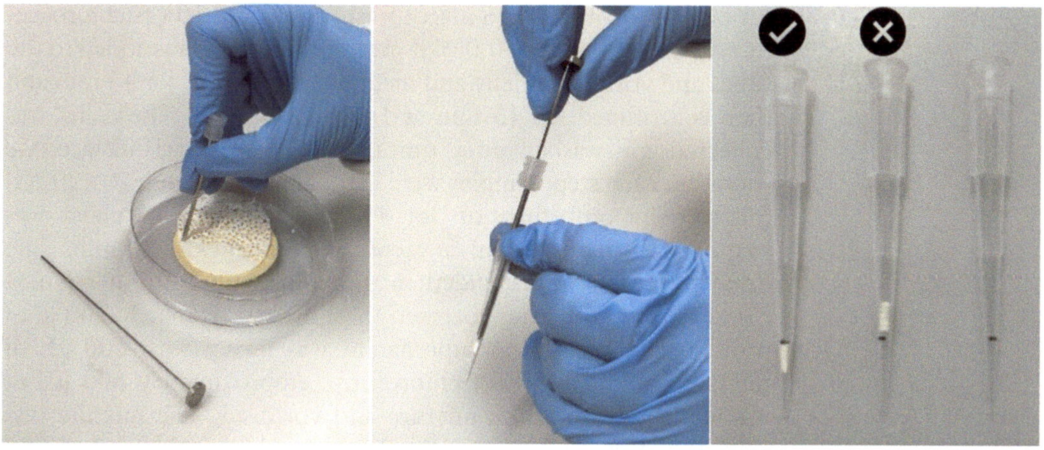

Fig. 1 Packing of SDB-RPS Stage tip. Left: Manual extraction of SDB membrane using a blunt-ended 18-gauge syringe. Middle: Ejection of SDB membrane from syringe using syringe plunger. Right: Correct packing depth of SDB membrane in standard 200 μL pipette tip

4. Repeat **step 3** until all initial sample has been passed through the tip.
5. Desalt peptides with 100 μL of 0.1% TFA and centrifuge at 2500 rpm for 30 min. Discard flowthrough and repeat.
6. Discard wash flowthrough and then transfer Stage tips to fresh 1.5-mL microcentrifuge tubes. Add 150 μL of freshly prepared 80% ACN/5% ammonium hydroxide and centrifuge at 2500 rpm for 30 min (*see* **Note 26**).
7. Vacuum centrifuge desalted peptides to dryness and reconstitute in 20 μL of 0.1% FA for mass spectrometric analysis.

3.4 Example Applications: Characterization of Tuna Meal Waste and Eucalyptus grandis Leaves

In order to illustrate the protein and peptide fraction procedures outlined above, proteomic analysis on fractionated and unfractionated material was performed for two different samples: tuna meal waste and *Eucalyptus grandis* leaves.

3.5 Example Application 1: Tuna Meal Waste Proteome Characterization

3.5.1 Tuna Meal Waste Protein Extraction and Tryptic Digestion

Tuna meal waste is a by-product of tuna processing and consists of a complex dry mixture of bone, scale, skin, and muscle containing approximately 5% moisture. Proteins were first extracted and quantified and then digested and fractionated using high-pH reversed-phase fractionation. Proteomic analysis was performed on the fractionated and unfractionated material to allow for comparison between approaches.

Approximately 1 g of tuna meal waste was homogenized with liquid nitrogen in a mortar and pestle, and then six aliquots of 50 mg were weighed and transferred to 2-mL microcentrifuge tubes. Approximately 1.5 mL of freshly prepared 10% trichloroacetic acid in acetone, plus 0.07% β-mercaptoethanol, was added to the tubes and vortexed briefly and incubated at −20°C for 45 min and then centrifuged for 15 min at 16,000 g at 4°C. The pellet was collected and washed three times with 1.5 mL of 100% acetone (between each step, samples were incubated for 10 min at −20 °C) followed by centrifugation for 15 min at 4°C. The pellet was evaporated to dryness in a vacuum centrifuge for 5 min. The dried pellet was resuspended in 400 μL of 3% SDS in 50 mM Tris–HCl (pH 8.8) and vortexed for 2 h. Approximately 200 μl of protein extract solution supernatant was mixed with 800 μL of methanol and 200 μL of chloroform. Approximately 600 μL of water was added to the mixture and vortexed. The mixture was centrifuged at 12,000 rpm for 2 min, and the upper phase was removed. Approximately 600 μL of methanol was added to the mixture and centrifuged at 12,000 rpm for 2 min. The supernatant was removed, and the pellet was air dried and solubilized in 80 μL of 8 M urea in 100 mM Tris–HCl (pH 8.8).

Approximately 100 μg of protein was diluted with 100 mM Tris–HCl (pH 8.0) to a final urea concentration of 1.6 M before reduction using DTT and alkylation using IAA, followed by the addition of trypsin (1:100 enzyme/substrate) and incubation at 37 °C overnight. The pH was adjusted to 3.0 by addition of formic acid before peptides were purified using SDB-RPS (3 M Empore) Stage tips (*see* Subheading 3.3). Desalted samples were dried and reconstituted in 20 μL of 0.1% FA. To illustrate the difference between fractionated and non-fractionated samples, three replicates of non-fractionated sample were analyzed on by nano-LC-MS/MS, followed by three replicates of fractionated sample (Subheading 3.2).

3.5.2 Mass Spectrometric Analysis of Tuna Meal Waste by nanoLC-MS/MS

Aliquots of 100 μg of peptides were analyzed using an Easy-nanoLC 1000 high-performance liquid chromatography system coupled to a Thermo Q Exactive Orbitrap mass spectrometer with a 60-min gradient, starting with 100% buffer A (0.1% FA), using a linear solvent ramp with steps from 2% to 30% of buffer B (99.9% (v/v) ACN, 0.1% (v/v) FA) for 50 min and 30% to 85% of buffer B for 10 min. Reversed-phased chromatographic separation was performed on an in-house packed reversed-phase column (75 μm ID ×10 cm, Halo 2.7 μm 160 Å ES-C18 beads). Peptides were separated for 80 min using a gradient starting at 99% buffer A (0.1% formic acid (FA)) and then 1–35% of buffer B (99.9% (v/v) acetonitrile, 0.1% (v/v) formic acid) over 60 min, 35–95% of buffer B over 10 min, and 95% buffer B for 10 min. The mass spectrometer was operated in data-dependent acquisition (DDA) mode to acquire MS/MS spectra of the top ten most abundant ions. Xcalibur software (version 2.06) was used to perform spectral acquisition over the mass range of 350–1500 m/z, by detecting of ions in the orbitrap with a resolution of 70,000 after accumulation to an automated gain control (AGC) target value of 1×10^6 ions and 1.4 *m/z* isolation window width. The maximum injection time for target ions selected for MS/MS was 120 ms. The lock mass option was enabled using the polysiloxane ion (445.12003) as an internal calibrant for accurate mass measurement.

3.5.3 Bioinformatic Analysis of Tuna Meal Waste Mass Spectrometric Data

Peptide to spectrum matching of the raw files obtained from nanoLC-MS/MS was performed using MaxQuant 1.6.14.0, using a *Thunnus* proteome fasta file downloaded from NCBI (102,910 proteins, July 2022). Carbamidomethylation of cysteine was set as fixed modification, and oxidation of methionine and acetylation of protein N-terminal were defined as variable modification. Trypsin/P was chosen as the enzyme specificity. Data processing of MaxQuant outputs was performed in Perseus version 2.0.7.0, and contaminants, reversed proteins, and proteins "identified only by side" were removed. Data were transformed to a logarithmic scale (base 2), and the protein groups identified in less than three replicates were removed.

3.5.4 Tuna Meal Proteome Characterization Results

Table 2 provides information on the analysis of tuna meal samples, comparing results between single-shot and high-PH fractionation method. The data includes the number of identified peptides and proteins, as well as the relative standard deviation (RSD) for peptides and proteins. The "reproducibly identified proteins" column shows the count of proteins consistently identified across all three replicates (R1, R2, R3) for each method [7, 8]. The results indicate that high-pH reversed-phase fractionation yields higher numbers of peptides and proteins, with lower RSD values for both peptides and proteins compared to the single-shot method.

These results are also presented below in graphical format. Figure 2 shows that fractionation resulted in identification of approximately twice as many proteins and peptides when compared to the analysis of unfractionated material. Figure 3 shows the

Table 2
Summary of protein identification data for single-shot and high-pH fractionation for tuna meal waste

		Peptides			Proteins			Peptide RSD[b]	Protein RSD	
Row	Tuna meal	R1[a]	R2	R3	R1	R2	R3	(%)	(%)	Reproducibly identified proteins
1	Single shot	3364	3621	3525	257	262	268	3.7	2.09	262
2	High-pH eight fractionations	7056	7259	6951	586	612	591	2.2	2.31	596

[a]R1,2,3 refer to biological replicates
[b]*RSD* relative standard deviation

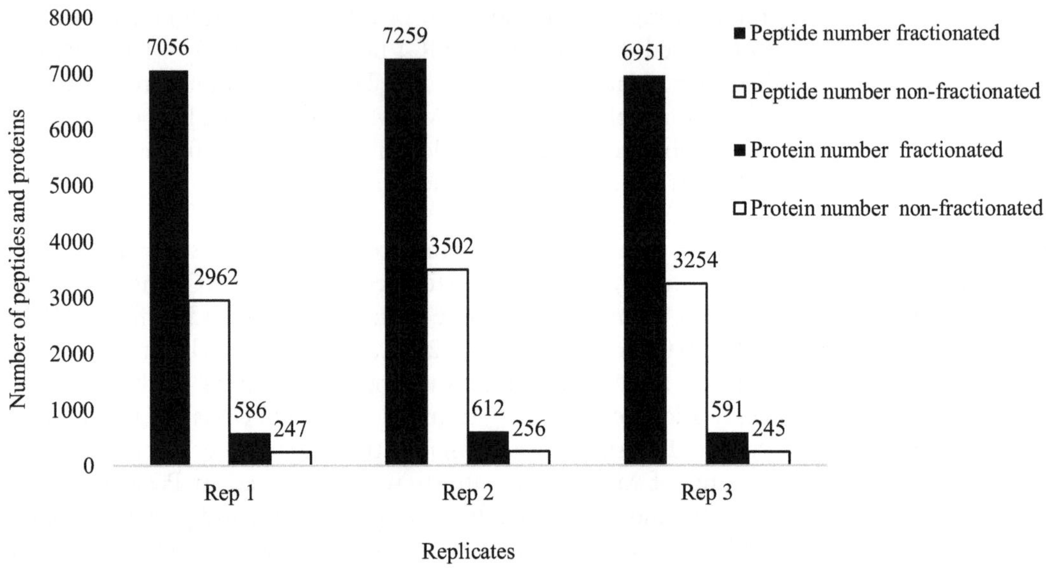

Fig. 2 Comparison between the number of peptides and proteins identified from non-fractionated and fractionated tuna meal waste

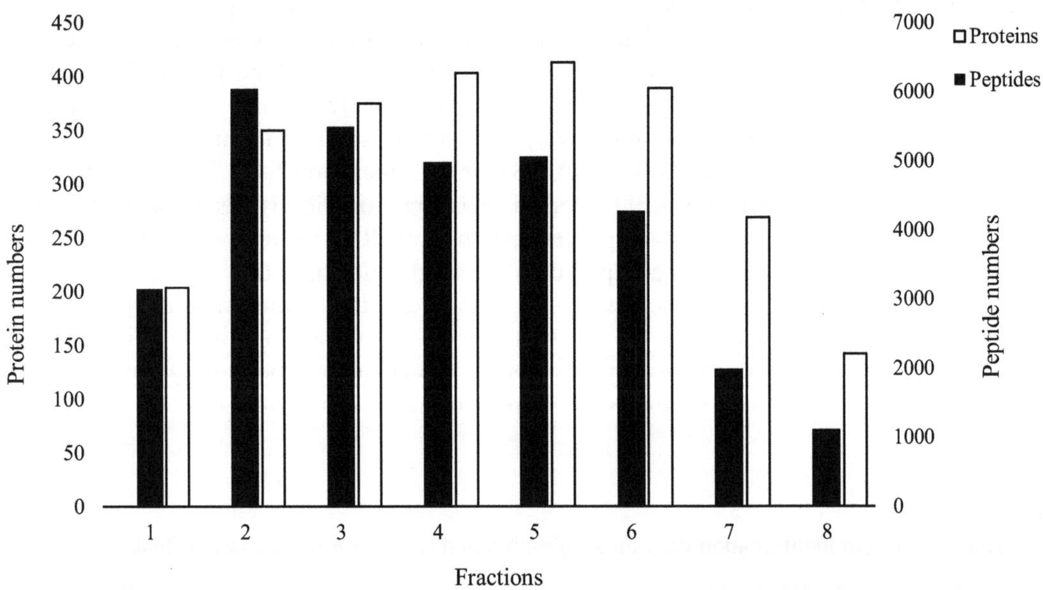

Fig. 3 Number of peptides and proteins identified in each fraction for tuna meal waste

distribution of peptides identified in each fraction and illustrates that, in this case, there is a relatively even distribution of peptides across each fraction.

3.6 Example Application 2: *Eucalyptus grandis* Leaf Proteome Characterization

Proteomic analysis was performed on both the fractionated and unfractionated *E. grandis* leaf proteins to allow for comparisons between both sample types and approaches. Aliquots of approximately 5 g of fresh *E. grandis* leaf were ground to a fine powder using a mortar and pestle and liquid nitrogen. Aliquots of 200 mg of ground leaf were transferred to 2-mL microcentrifuge tubes, and 1.5 mL of 10% of TCA in acetone, plus 2% β-mercaptoethanol, was added. The tubes were vortexed and centrifuged for 5 min at 15,000× *g* at 4 °C, and the supernatant was discarded. This step was repeated twice. To the pellet, 1.5 mL of 10% of TCA in acetone was added, vortexed, and centrifuged as described above. The supernatant was discarded, and this step was repeated, and the pellet was evaporated to dryness in a vacuum centrifuge.

3.6.1 *Eucalyptus grandis* Leaf Protein Extraction and Tryptic Digestion

To the dried pellet, 1.5 mL of 1% SDS, 0.15 M Tris–HCl (pH 8.8), 0.1 M DTT, and 1 mM EDTA solution were added and stirred at room temperature at 1 h prior to centrifugation at 15,000× *g* for 10 min at RT. The supernatant was transferred to a new 2-mL microcentrifuge tube, and Tris-buffered phenol solution was added in a ratio 1:1 to the supernatant. The mixture was vortexed for 5 min and centrifuged at 15,000× *g* for 5 min at RT. The lower phase was collected and transferred a new 2-mL microcentrifuge tube. An aliquot of 10 mM Tris–HCl (pH 8.0), 1 mM EDTA, and 0.7 M sucrose solution was added in a ratio of 1: 1, and the mixture was vortexed for 5 min and centrifuged at 15,000× *g* for 5 min at RT. The upper phase was collected and transferred to a new 15-mL Falcon tube, and five volumes of 0.1 M ammonium acetate in methanol were added. This mixture was vortexed and incubated overnight at −20 °C and then centrifuged at 15,000× *g* for 10 min at 4 °C, with the supernatant discarded. This protein pellet was then further washed to remove any remaining phenol. To the precipitated pellet, 1 mL of 0.1 M of ammonium acetate in methanol was added, vortexed, and centrifuged at 15,000× *g* for 5 min at 4 °C, with the supernatant discarded. Approximately 1 mL of 80% acetone and 10 mM DTT were was added and vortexed and then centrifuged at 15,000× *g* for 5 min at 4 °C, with the supernatant discarded. This last step was repeated. The pellet was air-dried in a fume work and then solubilized in 80 μL of a solution containing 8 M urea in 100 mM Tris–HCl (pH 8.8). The protein content was estimated with bicinchoninic acid (BCA) assay kit.

For digestion, 100 μg of protein was diluted in Tris–HCl 100 mM buffer (pH 8.0), to a final urea concentration of 1.6 M before reduction with DTT and alkylation with IAA, followed by the addition of trypsin (1:100 enzyme/substrate) and digestion at 37 °C overnight. The pH was adjusted to 3.0 with addition of formic acid, and peptides were purified using SDB-RPS (3 M Empore) Stage tips (*see* Subheading 3.3). Desalted samples were dried and reconstituted in 20 μL of 0.1% FA. Three replicates of

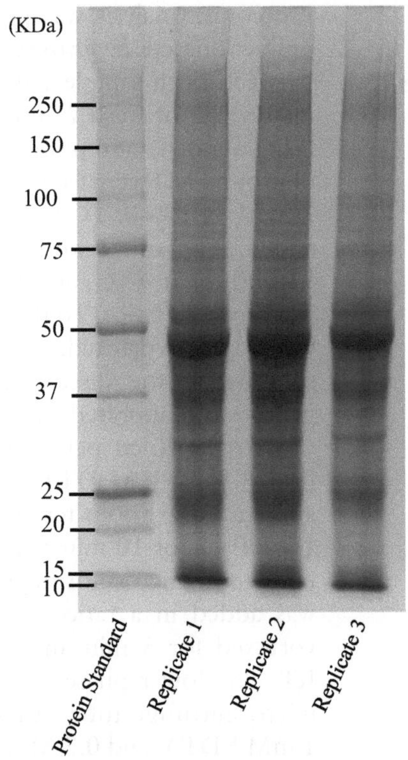

Fig. 4 SDS-PAGE gel obtained for three replicate protein extractions from *Eucalyptus grandis* leaf, which were used for subsequent fractionation and digestion

non-fractionated sample were analyzed by nano-LC-MS/MS, and then three replicates of sample fractionated by SDS-PAGE according to **step 3.1**, as shown in Fig. 4, were analyzed by nano-LC-MS/MS.

3.6.2 Mass Spectrometric Analysis of E. grandis *Leaf by nanoLC-MS/MS*

Eucalyptus grandis leaf protein tryptic digests (both SDS-PAGE fractionation and in-solution single shot) were analyzed on a Q-Exactive HF-X Orbitrap system connected to an UltiMate 3000 nanoLC system. Peptides were loaded on a 300 μm × 5 mm C18 trap column. Chromatographic separation was performed using a reversed-phase fused silica capillary column with an internal diameter of 75 μm and length of 30 cm, in-housed packed with ReproSil-Pur 120 C18-AQ resin, with a particle size of 3 μm. Samples were loaded in buffer A (2% v/v ACN, 0.1% v/v FA), and peptides were separated over a 90-min gradient with a flow rate of 300 nL/min at 35 °C. The gradient was developed using buffer B (99.9% v/v ACN, 0.1% v/v FA) as follows: 2% buffer (10 min), 2–35% buffer B (70 min), 35–95% buffer B flow (8 min), 95% buffer B (5 min), 95–2% buffer B (2 min), and 2% buffer B

(5 min). Spectral acquisition was performed in positive ion mode with a default charge of 2^+. The scan range was set to 350–1450 m/z (60,000 resolution, 3e6 AGC target, 50 ms maximum injection time). Data-dependent acquisition (DDA) method was used to acquire MS/MS spectra of the top ten most abundant ions (15,000 resolution, 1e5 automatic gain control target, 100 ms maximum injection time, and 1.4 *m/z* isolation window width).

3.6.3 Bioinformatic Analysis of *E. grandis* Leaf Mass Spectrometric Data

MaxQuant 2.2.0.0 software was used for peptide-to-spectrum matching (PSM) of the raw files obtained from nanoLC-MS/MS, as described in Subheading 3.5.3. The search on MaxQuant was performed against *E. grandis* proteome fasta file downloaded from NCBI (44, 428 proteins, November 2022).

3.6.4 *Eucalyptus grandis* Leaf Proteome Characterization Results

The results of the proteomics analysis of the *E. grandis* leaves, for both single-shot and SDS-PAGE fractionated samples, are shown in Table 3.

The results as presented in Table 3 and Fig. 5 show that SDS-PAGE fractionation more than doubled the number of reproducible proteins identified when compared to single-shot analysis.

Figure 6 illustrates the distribution of peptides identified over the fractions, with fractions 2, 3, and 4 showing a greater number of peptides identified, while fraction 6 contains the greatest number of proteins identified.

3.7 Conclusion

In this report, we have provided detailed instructions for how to perform SDS-PAGE gel-based protein fractionation and high-pH reversed-phase peptide fractionation as the first stage of shotgun proteomics analysis experiments. We have demonstrated this on two different samples, tuna meal waste and eucalyptus leaves, and shown that fractionation enables the identification of many more peptides and proteins from the same amount of starting material.

Table 3
Summary of protein identification data for single-shot and SDS-PAGE gel fractionation for *E. grandis* leaf

Row	*Eucalyptus grandis* leaf	Peptides R1[a]	Peptides R2	Peptides R3	Proteins R1	Proteins R2	Proteins R3	Peptide RSD[b] (%)	Protein RSD (%)	Reproducibly identified proteins
1	Single shot	11,429	18,000	10,136	1766	1428	1461	7.8	12.0	935
2	In-gel eight fractions	17,915	7259	6951	2268	2456	2282	2.2	4.5	2048

[a]R1,2,3 refer to biological replicates
[b]*RSD* relative standard deviation

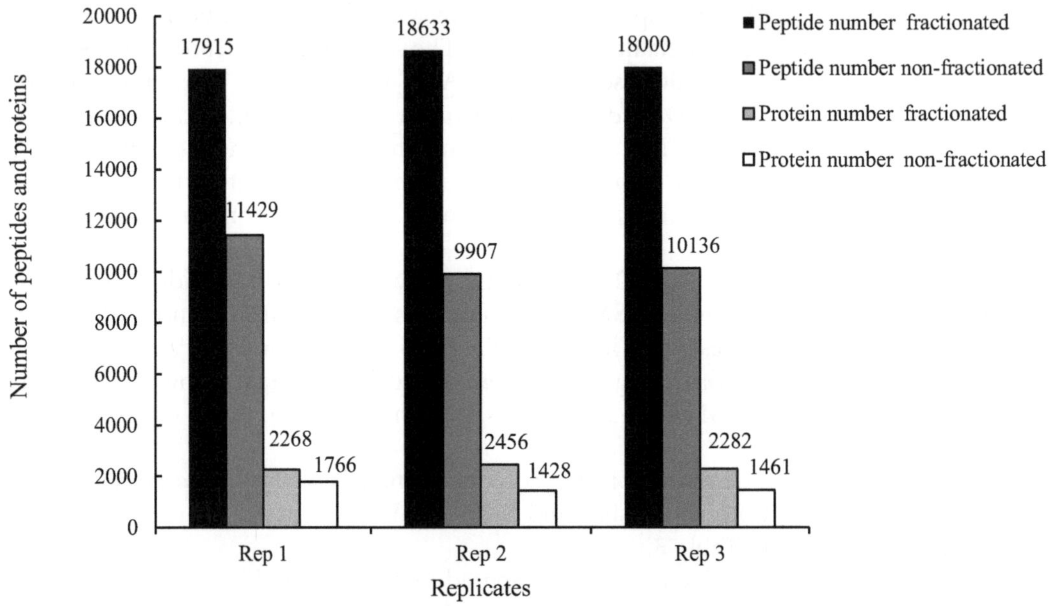

Fig. 5 Comparison between the number of peptides and proteins identified from non-fractionated and fractionated *E. grandis* leaf

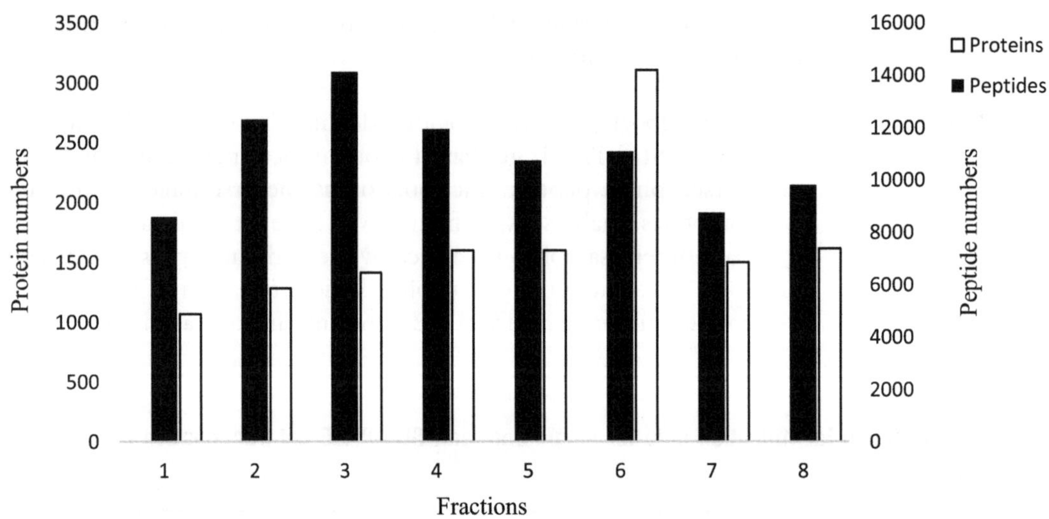

Fig. 6 Number of peptides and proteins identified in each fraction for *E. grandis* leaf

4 Notes

1. When making the SDS running buffer, it is recommended to prepare a higher concentration stock solution (e.g., 10×) and dilute to 1× when needed. These can be stored at room temperature for up to 6 months.

2. It is recommended to prepare a higher concentration stock solution of loading buffer (e.g., 5×) and dilute as necessary to 1×. We typically store the stock solution in 1-mL aliquots at − 20 °C, where it is stable for up to 12 months..
3. We usually make up 50 mL of 100 mM NH_4HCO_3 solution when starting in-gel digestion as it can be used to make other solutions required for the protocol. This should be made fresh.
4. Trypsin dissolved in acid is stable when frozen for up to 3 months, but trypsin diluted in NH_4HCO_3 solution should be used on the same day it is prepared.
5. 10% TCA in acetone solution can be prepared in advance and kept at 4 °C, and β-mercaptoethanol should be added just before use.
6. Prepare fresh DTT stock and add to these solutions just before use to avoid it decaying and becoming less efficient at reduction.
7. If using non-screw top microcentrifuge tubes, the caps will sometimes pop open during heating. We typically seal the cap with a small piece of parafilm to ensure the tubes stay closed.
8. Precast gels will usually come in a packet with some liquid to prevent the gel from drying out. It is important to wear gloves when handling these gels as residual unpolymerized acrylamide may be present, which is a potent neurotoxin and should be handled and discarded appropriately as hazardous waste. Precast gels will also have protective strips at the bottom which must be removed prior to use.
9. When loading gels, we typically ensure samples are evenly spaced as possible within the available wells to avoid inadvertent sample crossover. The protein marker is only loaded into a single well on the edge which helps to orient the gel later.
10. At 170 V this takes approximately an hour, so it is a good opportunity to go get lunch.
11. It is very easy to rip the gel during this process, so it is important to gently pry the cassette open. Placing the edge of the spatula in between the plastic sheets of the cassette and slowly rolling your wrist forward with light but constant force are a good way to open the cassette with minimal risk of accidentally tearing the gel.
12. When rinsing excess stain, it is helpful to use a piece of lint-free paper towel or tissue to rub the residual Coomassie stain from the sides of the container. Removing excess stain speeds up the destaining process.
13. Adding a small piece of absorbent lint-free paper towel helps to absorb the Coomassie stain and speeds up the destaining

process. You may need to change the destaining solution a few times depending on how dark the gel is.

14. During the excision process, the gel can dry out and become sticky which makes it difficult to work with. It is important to use a little bit of ultrapure water to keep the gel hydrated.
15. At this stage, the gel pieces should look white and shrivelled. That is ideal as it will help with ensuring that the reduction, alkylation, and trypsin solutions are properly absorbed into the gel during those steps.
16. It is important to keep the trypsin solution and the gel pieces cold during this stage as the low temperature inhibits the enzymatic activity of the trypsin and allows it to enter the gel matrix with minimal autodigestion.
17. The gel pieces may be sticky at this point and adhere to the pipette tip, so take care to avoid accidentally removing some when transferring liquids.
18. If peptides are already in solution, you can bring the volume up to 300 μL with 0.1% TFA.
19. The standard number of fractions for high-pH fractionation is 8 (5%, 7.5%, 10%, 12.5%, 15%, 17.5%, 20%, and 50% ACN). You can increase the number of fractions by adding more steps to the gradient. The calculations in Table 1 will make 1 mL of each elution solution. As only 300 μL is needed of each solution per sample, this volume is sufficient for three samples.
20. The spin columns that come in kits generally have a protective white cap at the end which should be removed prior to use. It is important to spin the columns when first using them to settle the resin and discard any remaining solution from manufacture and shipping.
21. The sample flowthrough should be retained in the event the peptides do not bind to the column. It may also contain very hydrophilic peptides that do not effectively bind to the column.
22. This flowthrough should be retained as a "wash fraction." It is not a true fraction but is kept for recovery purposes in case the samples do not bind to the column.
23. The membrane should place under the first line from the bottom of the tip. If packed too loosely, the membrane will move during centrifugation resulting in sample loss. Conversely, packing the membrane too tightly will cause a blockage and greatly reduce the flow rate of the tip (*see* Fig. 1). Each membrane "core" can load 20 μg of peptide, so a sample of 100 μg will require five cores to be packed into a single tip.

24. To centrifuge the tips, we cut a small hole into the cap of a 1.5-mL microcentrifuge tube wide enough in diameter that the tip can fit comfortably without pinching and with approximately 1 cm distance from the bottom of the tube (*see* Fig. 1). It is important to centrifuge the tips at a low speed. High-speed centrifugation can force the membrane out the end of the tip.
25. It is generally a good idea to retain the sample "flowthrough" in the event that the tip was not packed correctly or the peptides did not bind to the cores. If any sample remains above the membrane, centrifugation can be continued until all the sample has passed through the membrane.
26. For thorough collection of peptides, this step can be repeated once more.

Acknowledgments

The authors acknowledge support from Macquarie University in the form of Ph.D. scholarship funding and from the Australian Research Council in the form of funding for the Industrial Transformation Training Centre for Facilitating Advancement of Australian Bioactives.

References

1. States DJ, Omenn GS, Blackwell TW, Fermin D, Eng J, Speicher DW et al (2006) Challenges in deriving high-confidence protein identifications from data gathered by a HUPO plasma proteome collaborative study. Nat Biotechnol 24(3): 333–338
2. Malmström J, Lee H, Aebersold R (2007) Advances in proteomic workflows for systems biology. Curr Opin Biotechnol 18(4):378–384. https://doi.org/10.1016/j.copbio.2007.07.005
3. Deng L, Handler DCL, Multari DH, Haynes PA (2021) Comparison of protein and peptide fractionation approaches in protein identification and quantification from *Saccharomyces cerevisiae*. J Chromatogr B 1162:122453. https://doi.org/10.1016/j.jchromb.2020.122453
4. Mostovenko E, Hassan C, Rattke J, Deelder AM, van Veelen PA, Palmblad M (2013) Comparison of peptide and protein fractionation methods in proteomics. EuPA Open Proteom 1:30–37. https://doi.org/10.1016/j.euprot.2013.09.001
5. Chandramouli K, Qian P-Y (2009) Proteomics: challenges, techniques and possibilities to overcome biological sample complexity. Hum Genomics Proteomics 2009:239204
6. Fang Y, Robinson DP, Foster LJ (2010) Quantitative analysis of proteome coverage and recovery rates for upstream fractionation methods in proteomics. J Proteome Res 9(4):1902–1912. https://doi.org/10.1021/pr901063t
7. Mirzaei M, Soltani N, Sarhadi E, George IS, Neilson KA, Pascovici D et al (2014) Manipulating root water supply elicits major shifts in the shoot proteome. J Proteome Res 13(2): 517–526. https://doi.org/10.1021/pr400696u
8. Wu Y, Mirzaei M, Pascovici D, Chick JM, Atwell BJ, Haynes PA (2016) Quantitative proteomic analysis of two different rice varieties reveals that drought tolerance is correlated with reduced abundance of photosynthetic machinery and increased abundance of ClpD1 protease. J Proteome 143:73–82. https://doi.org/10.1016/j.jprot.2016.05.014

Chapter 4

Protein Sample Preparation for Bottom-Up, Label-Free Quantitative Proteomics of Adipose Tissue

Jane I. Khudyakov

Abstract

Adipose tissue (AT) is a complex, multifunctional endocrine organ that plays a significant role in animal evolution and human disease. Profiling of the proteome, or the set of proteins produced by a cell or tissue at a given time, can be used to explore the myriad functions of adipose tissue and understand its role in health and disease. The main challenges of adipose tissue proteomics include the high lipid and low protein content of the tissue and association of many proteins with lipid droplets. Here, we present a protocol for gel-free, label-free, bottom-up, relative quantitative proteomics of adipose tissue based on findings from the literature and our laboratory that yields reproducible protein and peptide identification rates while minimizing cost and processing time. This approach involves tissue homogenization, protein precipitation from homogenates, solubilization and denaturation of proteins in a buffer containing 5% sodium deoxycholate (SDC, an acid-insoluble detergent) and 5 mM tris(2-carboxyethyl)phosphine (TCEP, a reducing agent), alkylation with chloroacetamide, and in-solution tandem digestion with trypsin and Lys-C enzymes in the presence of 1% SDC. Acidification of peptides efficiently removes SDC prior to desalting and mass spectrometry. This method has been used successfully in our laboratory by both experienced researchers and those with limited technical backgrounds, including high school, undergraduate, and graduate students. We have identified >1500 proteins in adipose tissue of non-model mammals (e.g., blubber of marine mammals) spanning a dynamic range of 10^5 using this approach, including proteins of interest for comparative physiology such as adipokines, metabolic and antioxidant enzymes, lipid droplet proteins, metabolite transporters, and mitochondrial proteins, among others.

Key words Adipose tissue (AT), Label-free quantitative proteomics, Bottom-up proteomics, Shotgun proteomics, In-solution digest, Sodium deoxycholate (SDC)

1 Introduction

Adipose tissue (AT) is a complex, multifunctional endocrine organ that plays a significant role in animal evolution and disease [1]. Proteome profiling has been used to compare AT types and depots, examine changes in their proteomes during disease, and characterize their secretomes, primarily in biomedical systems [2]. AT proteomes have also been examined in domesticated animals, hibernating mammals, and marine mammals [3–8]. The challenges

Taufika Islam Williams (ed.), *Tissue Proteomics: Methods and Protocols*, Methods in Molecular Biology, vol. 2884,
https://doi.org/10.1007/978-1-0716-4298-6_4,

of AT proteomics include the tissue's high lipid and low protein content, association of many proteins with lipid droplets, and the potential masking of proteins of interest by highly abundant structural (extracellular matrix) proteins. The impact of a large dynamic range on protein identification has been shown in other matrices such as plasma and serum [9, 10]. Protein recovery from AT can be maximized by efficient tissue lysis, delipidation, and protein solubilization, for which numerous approaches have been developed and compared, each with advantages and disadvantages [11–16]. Practical considerations for selecting a protein sample preparation protocol include reproducibility, throughput, and cost, while technical considerations include the potential for protein loss during processing and compatibility of reagents with enzymatic digestion and liquid chromatography-mass spectrometry (LC-MS). These considerations will also inform whether quantitative proteomics experiments use a label-free or isobaric labeling approach [17, 18]. Isobaric labeling is more costly and requires additional fractionation to reduce sample complexity prior to LC-MS [18]. Here, we present a low-cost and reproducible protocol for gel-free, label-free, bottom-up quantitative proteomics of AT based on findings from the literature and our laboratory [4, 19, 20].

Protein sample preparation involves cell lysis, protein extraction, protein cleanup, reduction of disulfide bonds, alkylation of cysteine sulfhydryl groups, additional protein cleanup, and enzymatic digestion of proteins to peptides. As inefficient protein digest reduces peptide identification rates by nanoLC-MS, protocol optimization is mainly aimed at maximizing digestion efficiency [21]. Sample preparation begins with mechanical and/or chemical tissue lysis, usually in an alkaline (pH ~8) buffer, such as ammonium bicarbonate (ABC) or triethylammonium bicarbonate (TEAB), which stabilizes proteins. Ammonium-containing buffers are preferred over Tris and phosphate buffers as they reduce protein modification during denaturation and are volatile during electrospray ionization (ESI) [22]. The lysis buffer may contain chaotropic reagents (e.g., urea, guanidine hydrochloride), ionic or nonionic detergents (e.g., sodium dodecyl sulfate, SDS, sodium deoxycholate, SDC, CHAPS), and/or reducing agents [e.g., dithiothreitol (DTT) or Tris(2-carboxyethyl)phosphine (TCEP)] to improve protein solubilization and extraction from membranes [15]. Protease inhibitors may be added to the lysis buffer to inhibit the activity of native proteases during homogenization, which may cause nonspecific protein cleavage. However, the inhibitor cocktail must be selected carefully as it may interfere with trypsin digest [23]. In cases where sample quantity is limited and both proteins and nucleic acids must be extracted from the same tissue sample, the tissue may be lysed in a phenol-guanidine buffer such as TRIzol®, which may increase recovery of metabolic enzymes, lipid droplet proteins, mitochondrial proteins, and others of interest for physiological studies [19, 24]. Mechanical lysis is commonly performed by bead beating, homogenization, sonication, or a combination of these approaches, depending on the physical properties of the tissue. For example, bead beating and/or sonication may be necessary to lyse subcutaneous AT, which has a high content of structural fibers, whereas homogenization may be sufficient to lyse visceral AT, which is less stiff [25].

Following lysis, the homogenates may be delipidated (if not using TRIzol®) by organic phase extraction (e.g., using chloroform and methanol) or protein precipitation (e.g., with acetone) [26]. Either approach may lead to some protein loss. For example, membrane proteins may partition into multiple layers during methanol/chloroform extraction, leading to their loss, while basic and hydrophobic proteins may be lost during acetone precipitation [13, 27]. Optimization may be necessary to determine which delipidation method, if any, maximizes peptide identification for a particular sample type or protein target (e.g., membrane, lipid droplet, or exosome proteins). Further protein loss may occur if precipitated protein pellets are not adequately solubilized prior to downstream processing; solubilization may be improved using buffers containing thiourea, SDS, or SDC [14, 28]. In our experience, protein pellets from AT are significantly easier to solubilize than those from protein-rich tissues such as muscle, especially in ABC buffer containing 5% SDC. Protein denaturation and reduction are often conducted at elevated temperatures (typically 37 °C or 60 °C) to improve protein solubility. However, lower temperatures (<32 ° C) may be necessary when using urea as a denaturant because carbamylation of lysine residues, which can reduce protein digestion efficiency, is accelerated at higher temperatures [22, 29]. Therefore, protein sample preparation protocols that use 8 M urea for denaturation also require its dilution to <1 M to reduce interference with digestion [30]. To avoid these issues, our laboratory now uses 5% SDC instead of urea. Protein denaturation is followed by alkylation, most commonly using either iodoacetamide (IAA) or chloroacetamide (CAA). Recent work has shown that IAA causes offsite alkylation of multiple amino acids, especially if used following denaturation with DTT [16, 31]. Offsite modifications may be reduced, and peptide identification rates increased by using the combination of TCEP and CAA for denaturation and alkylation, respectively [16, 31]. TCEP is also preferred over DTT due to its higher solubility in water and resistance to oxidation [32]. Alkylation is quenched using TCEP or DTT, followed by dilution with ABC and total protein quantification (e.g., by using BCA assay) prior to digestion.

Protein digestion is one of the most critical steps of sample preparation for nanoLC-MS, and many studies have examined conditions that improve digestion efficiency and peptide identification rates [21, 33]. Trypsin is the most commonly used protease in proteomic studies. However, recent work has demonstrated that the combination of trypsin and Lys-C enzymes outperforms trypsin alone and that digestion efficiency is significantly enhanced when conducted in the presence of a detergent such as SDC [11, 14, 21, 30]. However, detergents used during sample preparation must be removed prior to nanoLC-MS due to their interference with reversed-phase high-performance liquid chromatography (HPLC) and ESI. This has led to the development of acid-labile detergents such as RapiGest that are hydrolyzed at acidic pH [33]. As many of these proprietary reagents are costly, SDC, which is insoluble at low pH, has emerged as the most popular detergent for protein sample preparation. Several studies, including those in AT, have shown that SDC yields higher peptide and protein identification rates and protein coverage compared to RapiGest and other proprietary acid labile detergents [11, 21]. Following enzymatic digest, SDC can be precipitated by adding trifluoroacetic acid (TFA) to pH < 2 and centrifugation [33]. Alternatively, SDC may be removed using phase transfer with ethyl acetate; however, it may lead to higher protein loss than acid precipitation [34]. Several studies have examined the potential for peptide loss due to coprecipitation with SDC [14, 34]. While findings have been inconsistent, an additional pellet wash step was shown to increase peptide recovery [34]. After protein digest and SDC removal, the final steps of peptide preparation include sample cleanup or desalting, most commonly using C-18 resin spin columns or tips, and quantification using fluorometric or colorimetric assays.

Multiple approaches have been developed in recent years to simplify and accelerate sample preparation for proteomics, reducing technical variability and increasing throughput. These approaches include filter-assisted sample preparation (FASP), suspension trapping (S-Trap), single-pot, solid-phase-enhanced sample preparation (SP3), Sample Preparation by Easy Extraction and Digestion (SPEED), and multiple commercially available kits [16, 35–38]. Several recent proteome analyses of AT have utilized the FASP approach (e.g., [39–41]). However, the SDC-based in-solution digest method, such as the one presented here, remains the most straightforward, cost-effective, and time-efficient approach that provides similar if not improved proteome coverage compared to FASP [16]. This method has been used successfully in our laboratory for label-free AT protein quantification by LC-MS by both experienced researchers and those with limited technical backgrounds, including high school, undergraduate, and graduate students. We have identified adipokines, metabolic and antioxidant enzymes, lipid droplet proteins, mitochondrial, endoplasmic reticulum, and Golgi proteins, various transport proteins, and components of the ribosomal machinery and extracellular matrix, among others, in mammalian AT using this approach (e.g., [4] and unpublished observations), with a dynamic range that spans five orders of magnitude.

2 Materials

Prepare all solutions using HPLC or LC-MS-grade water and MS-grade reagents. Prepare and store all reagents at room temperature (unless indicated otherwise). Wear lab coat and disposable gloves at all times to minimize contamination of reagents and samples by keratin. Organic solvents such as chloroform and acetonitrile and strong acids should be used in a fume hood. Follow all waste disposal regulations when disposing of waste materials.

2.1 Equipment

1. Heat block, thermomixer, or water bath capable of heating to 37 °C and 60 °C.
2. Refrigerated centrifuge.
3. Vortexer with microcentrifuge tube holder/rack/adaptor.
4. Bead mill or tissue homogenizer (e.g., NextAdvance Bullet Blender, Qiagen TissueLyser, OMNI Bead Ruptor, MPBio FastPrep) and tubes prefilled with zirconium or metal beads (≥ 1.5 mm).
5. Handheld homogenizer or sonicator.
6. Lyophilizer.

2.2 Reagents for Simultaneous Protein and RNA Extraction from AT

1. TRIzol® or QIAzol® reagent.
2. Chloroform.
3. Ethanol.
4. Isopropyl alcohol.
5. Guanidine Wash Solution: 0.3 M guanidine hydrochloride in 95% ethanol (stored at room temperature).

2.3 Stock Solutions and Reagents for Protein Sample Preparation

1. ABC Buffer: 50 mM ammonium bicarbonate (ABC); stored at room temperature for up to 1 month.
2. TCEP Stock: 500 mM tris(2-carboxyethyl)phosphine HCl (TCEP) in 50 mM ABC, made fresh.
3. CAA Stock: 200 mM chloroacetamide (CAA) in 50 mM ABC, made fresh.
4. Trypsin/Lys-C Stock: 1 mg/mL Trypsin/Lys-C mix in 50 mM acetic acid (or as specified by manufacturer); store aliquots at −80 °C for up to 1 year.
5. Sodium deoxycholate (SDC).
6. Acetone.
7. Acetonitrile (ACN).
8. Trifluoroacetic acid (TFA).
9. Formic acid (FA).

2.4 Solutions for Protein Denaturation, Reduction, and Digestion

1. SDC Denaturing Buffer: 5% (w/v) SDC, 5 mM TCEP, in 50 mM ABC, made fresh; requires heating at 37 °C for ~10 min to resuspend SDC.
2. Digest Dilution Buffer: 1% (w/v) SDC in 50 mM ABC.
3. SDC Pellet Wash Solution: 0.5% TFA (in water).
4. Peptide Resuspension Solution: 0.1% FA in LC-MS-grade water.

2.5 Kits

1. Total protein quantification kit, e.g., bicinchoninic acid (BCA) assay.
2. Quantitative fluorometric or colorimetric peptide assay kit.
3. Peptide desalting spin column kit.

3 Methods

Follow best practices for sample blinding and randomization to reduce technical bias during sample preparation and analysis by nanoLC-MS (*see* **Note 1**). Carry out all procedures at room temperature unless otherwise specified. Proteins may be kept on ice or at 4 °C (for up to 24 h) between processing steps to reduce the potential for nonspecific protein degradation and oxidation [42, 43]. Avoid multiple freeze-thaw cycles (*see* **Note 2**).

3.1 Protein Extraction from AT

Use one of the two methods described below (*see* Subheading 3.1.1 or 3.1.2) to lyse tissue and extract proteins. Subheading 3.1.1 can be used for simultaneous isolation of nucleic acids and proteins. Subheading 3.1.2 may require an additional sample delipidation step. The two different approaches may vary in proteome coverage (e.g., [19]).

3.1.1 AT Lysis in TRIzol® or QIAzol® (for Simultaneous Protein and RNA Extraction)

1. For additional details about this protocol, see [24] and the TRIzol® or QIAzol® manufacturer's protocol.
2. Homogenize <100 mg of fresh or frozen AT or cells in 1000 μL of TRIzol® or QIAzol® reagent until the homogenates have no visible particles remaining (*see* **Note 3**).
3. Centrifuge for 10 min at 12,000 × *g* and 4 °C to pellet insoluble cellular material (i.e., cell fragments and intact cells). Discard overlying lipid layer. Transfer homogenates to clean microcentrifuge tubes.
4. Add 200 μL of chloroform to homogenates, vortex for 30 s to mix, incubate for 5 min, and centrifuge for 15 min at 12,000× *g* and 4 °C to separate the aqueous and organic phases.
5. Remove the aqueous phase containing nucleic acids and purify RNA extracts as directed by the manufacturer; contaminating

DNA can be removed by enzymatic digest. Keep the organic phase on ice during RNA processing.

6. Add 300 μL of ethanol to the remaining organic phase and interphase, vortex to mix, incubate for 5 min, and centrifuge for 5 min at 2000× *g* and 4 °C to pellet DNA. The DNA pellets can be discarded or further processed according to the manufacturer's protocol.
7. Split supernatants in two 300–350 μL aliquots and place into two clean 1.5-mL microcentrifuge tubes. Alternatively, the total volume of the supernatant can be transferred to a larger (≥ 2.0-mL) tube.
8. Add 750 μL of isopropyl alcohol to each of the two supernatants (or 1.5 mL if using a larger tube), vortex, incubate for 10 min, and centrifuge for 10 min at 12,000× *g* and 4 °C to pellet the proteins. This produces duplicate protein pellets for each sample (*see* **Note 4**). Discard the supernatant.
9. Wash the protein pellets three times with 1.0-mL Guanidine Wash Solution, followed by one wash in 1.0-mL ethanol. Let the pellets air dry briefly (<5 min), but do not overdry (*see* **Note 5**).
10. The organic phase and interphase remaining after completing Subheading 3.1.1 **step 5** can be stored at 4 °C overnight, while the protein pellets obtained in Subheading 3.1.1 **step 8** can be stored in Guanidine Wash Solution at −80 °C for at least one year before further processing.

3.1.2 AT Lysis in SDC Denaturing Buffer

1. Homogenize <100 mg of fresh or frozen AT or cells in 1000 μL of SDC Denaturing Buffer (*see* **Note 3**).
2. Centrifuge for 10 min at 12,000× *g* and 4 °C to pellet insoluble cellular material (i.e., cell fragments and intact cells). Remove the overlying lipid layer and transfer the homogenates in 200 μL aliquots to clean, acetone-compatible, polypropylene microcentrifuge tubes on ice.
3. Optional delipidation step: Add 800 μL of acetone (precooled at −20 °C) to each 200 μL aliquot of homogenate. Vortex and incubate for at least 1 h at −20 °C.
4. Centrifuge for 10 min at 13,000–18,000× *g* and 4 °C to pellet the proteins (*see* **Note 4**).
5. Discard the supernatant and let the pellets dry briefly (<5 min) but do not overdry (*see* **Note 5**).

3.2 Protein Denaturation, Reduction, and Alkylation

1. Add 200 μL of SDC Denaturing Buffer (pre-warmed to 60 °C to dissolve SDC) to each microcentrifuge tube containing protein pellet(s) (*see* **Note 6**). Vortex, spin down, and heat at 60 °C to fully resuspend the proteins. Large protein pellets, those from fibrous AT, and those precipitated using

Subheading 3.1.1 may require several rounds of sonication (e.g., up to five cycles of 15-s sonication with 30-s incubation on ice between cycles [24]) to completely solubilize. Incomplete resuspension will cause protein loss.

2. Incubate proteins in SDC Denaturing Buffer for 1 h at 60 °C. Place microcentrifuge tubes on ice to cool.
3. Add 22 μL of 200 mM CAA to each microcentrifuge tube (20 mM final concentration) and incubate in the dark for 30 min.
4. Add 2.2 μL of 500 mM TCEP to each microcentrifuge tube (5 mM final concentration) to quench alkylation (*see* **Note 7**).
5. Add 776 μL of 50 mM ABC to each microcentrifuge tube to dilute the SDC concentration to 1%. Protein solutions can be frozen at −80 °C for up to a year before further processing (*see* **Note 2**).

3.3 Protein Digestion and SDC Precipitation

1. Use a protein quantification assay to estimate total protein concentration in the samples. If the protein samples are concentrated, aliquots may be diluted with 50 mM ABC for concentrations to fall within the range of the standard curve (usually <2 μg/μL) of the assay.
2. Dilute 100–200 μg of protein to 0.5 μg/μL final concentration in Digest Dilution Buffer (to maintain the SDC concentration at 1%). Add trypsin/Lys-C enzyme mix at a ratio of enzyme to protein specified by the manufacturer (typically 1:50 or 1:100, e.g., 1–2 μL enzyme to 100 μg protein [44]).
3. Incubate at 37 °C for 12–16 h with shaking, if a shaking heat block or water bath is available.
4. Add TFA to a final concentration of 1% (e.g., 5 μL for 500 μL of digest) and pH < 2 (verified using pH paper) to inactivate the enzymes and precipitate SDC. Vortex immediately to mix and centrifuge for 10 min at 12,000× *g* and 4 °C to pellet the SDC.
5. Transfer the supernatants to clean microcentrifuge tubes. Work quickly as the pellets will begin to resolubilize within several minutes after centrifuging.
6. Add 500 μL of SDC Pellet Wash Solution to the remaining SDC pellets and incubate with vortexing for 10 min to resuspend proteins that may have co-precipitated with SDC.
7. Centrifuge again for 10 min at 12,000× *g* and 4 °C. Transfer the supernatants to the same tubes as in Subheading 3.3, **step 5**.
8. Lyophilize samples until dry and store at −80 °C until further processing (*see* **Note 2**).

3.4 Peptide Desalting and Quantification

1. Resuspend tryptic peptides in the appropriate buffer specified for by the desalting kit protocol (e.g., 0.1% TFA). Incubate peptides with vortexing for 10–30 min to improve resuspension and reduce protein loss.
2. Proceed with desalting as specified by the manufacturer's protocol. Most desalting protocols use ACN to elute peptides from desalting columns or tips.
3. Lyophilize eluted peptides dry and store at −80 °C until further processing (*see* **Note 2**).
4. Resuspend peptides in 20–50 μL Peptide Resuspension Solution. Incubate while vortexing for 10–30 min.
5. Dilute a 5 μL aliquot of resuspended peptides 1:10 or 1:20 in 50 mM ABC and use for peptide quantification using a colorimetric or fluorometric kit. A fluorometric assay is generally more sensitive but is more expensive and requires access to a fluorometer.
6. Dilute peptides to 100–200 ng/μL final concentration with Peptide Resuspension Solution for nanoLC-MS (*see* **Note 8**).

3.5 nanoLC-MS

NanoLC-MS methodology varies significantly by instrumentation, data acquisition approach (targeted vs untargeted, data-dependent vs data-independent acquisition of mass spectra), protein quantification approach (absolute vs relative quantification, isobaric labeling vs label-free quantification), and sample type and complexity (reviewed in [45, 46]) and is beyond the scope of this sample preparation protocol. NanoLC gradients and mass spectrometer parameters should be empirically determined and optimized, if possible. Proteomics core facilities often use optimized instrument parameters that should be broadly applicable to different sample types.

3.6 MS Data Analysis

A number of approaches have been developed for bioinformatic analysis and visualization of MS data [47, 48] and are beyond the scope of this sample preparation protocol. Both proprietary and open-source platforms are available that can be used for peptide and protein identification and quantification. Consultation with a proteomics expert or statistician is highly recommended. In our laboratory, we typically use MaxQuant software for protein identification (by both database search and de novo sequencing) and label-free quantification [49, 50], Perseus software for parsing MaxQuant output files [51], and the R package LIMMA for identifying differentially abundant proteins [52]. We typically identify >1200 proteins in marine mammal blubber and find several to dozens of proteins that are differentially abundant in response to fasting, stress, and other physiological challenges [4, 20, and unpublished observations]; more proteins may be identified and quantified by optimizing LC-MS parameters and using a DIA approach.

4 Notes

1. To reduce technical bias, it is recommended to de-identify and randomize samples during processing and analysis by nanoLC-MS [53]. Technical variability caused by sample preparation may be minimized by reducing the numbers of researchers handling samples, sample blocks, processing steps, and freeze-thaw cycles (*see* **Note 2**).
2. Repeated freeze-thaw cycles (> 2) may impact protein stability, leading to artificial and unpredictable changes in protein abundance [42, 54]. Proteins may be stored for up to a year at −80 °C [43].
3. The homogenization and lysis procedure will vary depending on sample type and starting amount. Fibrous AT and large amounts of tissue may require bead beating or grinding under liquid nitrogen. Vortexing or a handheld homogenizer may be sufficient for lysing cultured adipocytes and small amounts of soft AT. This step may require optimization, but many examples are found in the literature (e.g., [19, 55–57]).
4. If starting with a small amount of tissue and precipitated protein pellets are very small, replicate pellets from the same sample can be combined into a single tube for further processing.
5. Limit protein pellet drying time as proteins may become more difficult to solubilize if overdried, especially if using Subheading 3.3, **step 1** [24, 58].
6. The amount of Denaturing Buffer used to resuspend protein pellets may need to be adjusted depending on the size of the pellet and desired protein concentration.
7. Failure to quench alkylation with CAA may result in over-alkylation, or alkylation of other amino acids in addition to cysteine, which may impact peptide identification rates [59].
8. Typically, 0.5–1 μg of peptides is injected for nanoLC-MS/MS. Three replicate injections of each sample are recommended for label-free quantification [17].

Acknowledgments

The author is grateful to B. Neely, C. Vierra, and the instructors and teaching assistants of the Cold Spring Harbor Laboratory 2017 Proteomics Course (especially G. Knudsen, D. Pappin, and K. Rivera) for advice on protein sample preparation and LC-MS/MS analyses and to R. Hekman and J. Deyarmin for assistance with method development.

References

1. Pond CM (2017) The evolution of mammalian adipose tissues. In: Symonds ME (ed) Adipose tissue biology. Springer International Publishing, Cham, pp 1–59
2. Kim EY, Kim WK, Oh K-J, Han BS, Lee SC, Bae K-H (2015) Recent advances in proteomic studies of adipose tissues and adipocytes. Int J Mol Sci 16(3):4581–4599
3. Grabek KR, Martin SL, Hindle AG (2015) Proteomics approaches shed new light on hibernation physiology. J Comp Physiol B 185(6):607–627. https://doi.org/10.1007/s00360-015-0905-9
4. Khudyakov JI, Holser RR, Vierra CA, Ly ST, Niel TK, Hasan BM et al (2022) Changes in apolipoprotein abundance dominate proteome responses to prolonged fasting in elephant seals. J Exp Biol 225(4). https://doi.org/10.1242/jeb.243572
5. Ballinger MA, Andrews MT (2018) Nature's fat-burning machine: brown adipose tissue in a hibernating mammal. J Exp Biol 221(Pt Suppl 1). https://doi.org/10.1242/jeb.162586
6. Almeida AM, Ali SA, Ceciliani F, Eckersall PD, Hernández-Castellano LE, Han R et al (2021) Domestic animal proteomics in the 21st century: a global retrospective and viewpoint analysis. J Proteome 241:104220. https://doi.org/10.1016/j.jprot.2021.104220
7. Tian J-j, Zhang J-m, Yu E-m, Sun J-h, Xia Y, Zhang K et al (2020) Identification and analysis of lipid droplet-related proteome in the adipose tissue of grass carp (Ctenopharyngodon idella) under fed and starved conditions. Comp Biochem Physiol Part D Genomics Proteomics 36:100710. https://doi.org/10.1016/j.cbd.2020.100710
8. Kershaw JL, Botting CH, Brownlow A, Hall AJ (2018) Not just fat: investigating the proteome of cetacean blubber tissue. Conserv Physiol 6(1):coy003. https://doi.org/10.1093/conphys/coy003
9. Sajic T, Hopfgartner G, Szanto I, Varesio E (2011) Comparison of three detergent-free protein extraction protocols for white adipose tissue. Anal Biochem 415(2):215–217. https://doi.org/10.1016/j.ab.2011.04.023
10. Tu C, Rudnick PA, Martinez MY, Cheek KL, Stein SE, Slebos RJC et al (2010) Depletion of abundant plasma proteins and limitations of plasma proteomics. J Proteome Res 9(10): 4982–4991. https://doi.org/10.1021/pr100646w
11. Pasing Y, Colnoe S, Hansen T (2017) Proteomics of hydrophobic samples: fast, robust and low-cost workflows for clinical approaches. Proteomics 17(6). https://doi.org/10.1002/pmic.201500462
12. O'Rourke MB, Januszewski AS, Sullivan DR, Lengyel I, Stewart AJ, Arya S et al (2023) Optimised plasma sample preparation and LC-MS analysis to support large-scale proteomic analysis of clinical trial specimens: Application to the Fenofibrate Intervention and Event Lowering in Diabetes (FIELD) trial. Proteomics. Clin Appl 17(3):2200106. https://doi.org/10.1002/prca.202200106
13. Moore SM, Hess SM, Jorgenson JW (2016) Extraction, enrichment, solubilization, and digestion techniques for membrane proteomics. J Proteome Res 15(4):1243–1252. https://doi.org/10.1021/acs.jproteome.5b01122
14. Scheerlinck E, Dhaenens M, Van Soom A, Peelman L, De Sutter P, Van Steendam K et al (2015) Minimizing technical variation during sample preparation prior to label-free quantitative mass spectrometry. Anal Biochem 490:14–19. https://doi.org/10.1016/j.ab.2015.08.018
15. Dapic I, Baljeu-Neuman L, Uwugiaren N, Kers J, Goodlett DR, Corthals GL (2019) Proteome analysis of tissues by mass spectrometry. Mass Spectrom Rev 38(4–5):403–441. https://doi.org/10.1002/mas.21598
16. Varnavides G, Madern M, Anrather D, Hartl N, Reiter W, Hartl M (2022) In search of a universal method: a comparative survey of bottom-up proteomics sample preparation methods. J Proteome Res 21(10):2397–2411. https://doi.org/10.1021/acs.jproteome.2c00265
17. Zhao L, Cong X, Zhai L, Hu H, Xu J-Y, Zhao W et al (2020) Comparative evaluation of label-free quantification strategies. J Proteome 215: 103669. https://doi.org/10.1016/j.jprot.2020.103669
18. Chen X, Sun Y, Zhang T, Shu L, Roepstorff P, Yang F (2021) Quantitative proteomics using isobaric labeling: a practical guide. Genomics Proteomics Bioinformatics 19(5):689–706. https://doi.org/10.1016/j.gpb.2021.08.012
19. Khudyakov JI, Deyarmin JS, Hekman RM, Pujade Busqueta L, Maan R, Mody MJ et al (2018) A sample preparation workflow for adipose tissue shotgun proteomics and

proteogenomics. Biol Open 7(11). https://doi.org/10.1242/bio.036731

20. Deyarmin J, Hekman R, Champagne C, McCormley M, Stephan A, Crocker D et al (2020) Blubber proteome response to repeated ACTH administration in a wild marine mammal. Comp Biochem Physiol Part D Genomics Proteomics 33:100644. https://doi.org/10.1016/j.cbd.2019.100644
21. León IR, Schwämmle V, Jensen ON, Sprenger RR (2013) Quantitative assessment of in-solution digestion efficiency identifies optimal protocols for unbiased protein analysis. Mol Cell Proteomics 12(10):2992–3005. https://doi.org/10.1074/mcp.M112.025585
22. Sun S, Zhou J-Y, Yang W, Zhang H (2014) Inhibition of protein carbamylation in urea solution using ammonium-containing buffers. Anal Biochem 446:76–81. https://doi.org/10.1016/j.ab.2013.10.024
23. Clifton J, Huang F, Rucevic M, Cao L, Hixson D, Josic D (2011) Protease inhibitors as possible pitfalls in proteomic analyses of complex biological samples. J Proteome 74(7):935–941. https://doi.org/10.1016/j.jprot.2011.02.010
24. Simões AES, Pereira DM, Amaral JD, Nunes AF, Gomes SE, Rodrigues PM et al (2013) Efficient recovery of proteins from multiple source samples after trizol® or trizol®LS RNA extraction and long-term storage. BMC Genomics 14(1):181. https://doi.org/10.1186/1471-2164-14-181
25. Fontanella CG, Toniolo I, Foletto M, Prevedello L, Carniel EL (2022) Mechanical behavior of subcutaneous and visceral abdominal adipose tissue in patients with obesity. PRO 10(9):1798
26. Fic E, Kedracka-Krok S, Jankowska U, Pirog A, Dziedzicka-Wasylewska M (2010) Comparison of protein precipitation methods for various rat brain structures prior to proteomic analysis. Electrophoresis 31(21):3573–3579. https://doi.org/10.1002/elps.201000197
27. Crowell AMJ, Wall MJ, Doucette AA (2013) Maximizing recovery of water-soluble proteins through acetone precipitation. Anal Chim Acta 796:48–54. https://doi.org/10.1016/j.aca.2013.08.005
28. Serra A, Zhu H, Gallart-Palau X, Park JE, Ho HH, Tam JP et al (2016) Plasma proteome coverage is increased by unique peptide recovery from sodium deoxycholate precipitate. Anal Bioanal Chem 408(7):1963–1973. https://doi.org/10.1007/s00216-016-9312-7
29. Kollipara L, Zahedi RP (2013) Protein carbamylation: in vivo modification or in vitro artefact? Proteomics 13(6):941–944. https://doi.org/10.1002/pmic.201200452
30. Hakobyan A, Schneider MB, Liesack W, Glatter T (2019) Efficient tandem LysC/trypsin digestion in detergent conditions. Proteomics 19(20):1900136. https://doi.org/10.1002/pmic.201900136
31. Müller T, Winter D (2017) Systematic evaluation of protein reduction and alkylation reveals massive unspecific side effects by iodine-containing reagents. Mol Cell Proteomics 16(7):1173–1187. https://doi.org/10.1074/mcp.M116.064048
32. Getz EB, Xiao M, Chakrabarty T, Cooke R, Selvin PR (1999) A comparison between the sulfhydryl reductants tris(2-carboxyethyl) phosphine and dithiothreitol for use in protein biochemistry. Anal Biochem 273(1):73–80. https://doi.org/10.1006/abio.1999.4203
33. Danko K, Lukasheva E, Zhukov VA, Zgoda V, Frolov A (2022) Detergent-assisted protein digestion—on the way to avoid the key bottleneck of shotgun bottom-up proteomics. Int J Mol Sci 23(22):13903
34. Lin Y, Liu Y, Li J, Zhao Y, He Q, Han W et al (2010) Evaluation and optimization of removal of an acid-insoluble surfactant for shotgun analysis of membrane proteome. Electrophoresis 31(16):2705–2713. https://doi.org/10.1002/elps.201000161
35. Doellinger J, Schneider A, Hoeller M, Lasch P (2020) Sample preparation by easy extraction and digestion (SPEED)—a universal, rapid, and detergent-free protocol for proteomics based on acid extraction. Mol Cell Proteomics 19(1):209–222. https://doi.org/10.1074/mcp.TIR119.001616
36. Hughes CS, Moggridge S, Müller T, Sorensen PH, Morin GB, Krijgsveld J (2019) Single-pot, solid-phase-enhanced sample preparation for proteomics experiments. Nat Protoc 14(1):68–85. https://doi.org/10.1038/s41596-018-0082-x
37. Wiśniewski JR, Zougman A, Nagaraj N, Mann M (2009) Universal sample preparation method for proteome analysis. Nat Methods 6(5):359–362. https://doi.org/10.1038/nmeth.1322
38. Zougman A, Selby PJ, Banks RE (2014) Suspension trapping (STrap) sample preparation

method for bottom-up proteomics analysis. Proteomics 14(9):1006. https://doi.org/10.1002/pmic.201300553

39. Gaspar RS, Delafiori J, Zuccoli G, Carregari VC, Prado TP, Morari J et al (2023) Exogenous succinate impacts mouse brown adipose tissue mitochondrial proteome and potentiates body mass reduction induced by liraglutide. Am J Physiol-Endocrinol Metab 324(3): E226–EE40. https://doi.org/10.1152/ajpendo.00231.2022
40. Ben Meir YA, Daddam JR, Kra G, Kamer H, Portnick Y, Levin Y et al (2022) Proteomic analysis of adipose tissue revealing differentially abundant proteins in highly efficient mid-lactating dairy cows. Sci Rep 12(1):9721. https://doi.org/10.1038/s41598-022-13964-x
41. Tang X, Li J, Zhao W-g, Sun H, Guo Z, Jing L et al (2019) Comprehensive map and functional annotation of the mouse white adipose tissue proteome. PeerJ 7:e7352. https://doi.org/10.7717/peerj.7352
42. Zimmerman LJ, Li M, Yarbrough WG, Slebos RJC, Liebler DC (2012) Global stability of plasma proteomes for mass spectrometry-based analyses. Mol Cell Proteomics 11(6). https://doi.org/10.1074/mcp.M111.014340
43. Dufresne J, Hoang T, Ajambo J, Florentinus-Mefailoski A, Bowden P, Marshall J (2017) Freeze-dried plasma proteins are stable at room temperature for at least 1 year. Clin Proteomics 14(1):35. https://doi.org/10.1186/s12014-017-9170-0
44. Glatter T, Ludwig C, Ahrné E, Aebersold R, Heck AJR, Schmidt A (2012) Large-scale quantitative assessment of different in-solution protein digestion protocols reveals superior cleavage efficiency of tandem Lys-C/trypsin proteolysis over trypsin digestion. J Proteome Res 11(11):5145–5156. https://doi.org/10.1021/pr300273g
45. Gillet LC, Leitner A, Aebersold R (2016) Mass spectrometry applied to bottom-up proteomics: entering the high-throughput era for hypothesis testing. Annu Rev Anal Chem 9(1):449–472. https://doi.org/10.1146/annurev-anchem-071015-041535
46. Rozanova S, Barkovits K, Nikolov M, Schmidt C, Urlaub H, Marcus K (2021) Quantitative mass spectrometry-based proteomics: An overview. In: Marcus K, Eisenacher M, Sitek B (eds) Quantitative methods in proteomics. Springer, New York, pp 85–116
47. Chen C, Hou J, Tanner JJ, Cheng J (2020) Bioinformatics methods for mass spectrometry-based proteomics data analysis. Int J Mol Sci 21(8):2873
48. Schessner JP, Voytik E, Bludau I (2022) A practical guide to interpreting and generating bottom-up proteomics data visualizations. Proteomics 22(8):2100103. https://doi.org/10.1002/pmic.202100103
49. Tyanova S, Temu T, Cox J (2016) The MaxQuant computational platform for mass spectrometry-based shotgun proteomics. Nat Protoc 11(12):2301–2319. https://doi.org/10.1038/nprot.2016.136
50. Cox J, Hein MY, Luber CA, Paron I, Nagaraj N, Mann M (2014) Accurate proteome-wide label-free quantification by delayed normalization and maximal peptide ratio extraction, termed MaxLFQ. Mol Cell Proteom 13(9):2513–2526. https://doi.org/10.1074/mcp.M113.031591
51. Tyanova S, Temu T, Sinitcyn P, Carlson A, Hein MY, Geiger T et al (2016) The Perseus computational platform for comprehensive analysis of (prote)omics data. Nat Methods 13(9):731–740. https://doi.org/10.1038/nmeth.3901
52. Ritchie ME, Phipson B, Wu D, Hu Y, Law CW, Shi W et al (2015) Limma powers differential expression analyses for RNA-sequencing and microarray studies. Nucleic Acids Res 43(7): e47. https://doi.org/10.1093/nar/gkv007
53. Nakayasu ES, Gritsenko M, Piehowski PD, Gao Y, Orton DJ, Schepmoes AA et al (2021) Tutorial: best practices and considerations for mass-spectrometry-based protein biomarker discovery and validation. Nat Protoc 16(8): 3737–3760. https://doi.org/10.1038/s41596-021-00566-6
54. Hassis ME, Niles RK, Braten MN, Albertolle ME, Ewa Witkowska H, Hubel CA et al (2015) Evaluating the effects of preanalytical variables on the stability of the human plasma proteome. Anal Biochem 478:14–22. https://doi.org/10.1016/j.ab.2015.03.003
55. Anderson AE, Wu I, Parrillo AJ, Wolf MT, Maestas DR, Graham I et al (2022) An immunologically active, adipose-derived extracellular matrix biomaterial for soft tissue reconstruc-

tion: concept to clinical trial. NPJ Regen Med 7(1):6. https://doi.org/10.1038/s41536-021-00197-1

56. Stroh AM, Lynch CE, Lester BE, Minchev K, Chambers TL, Montenegro CF et al (2021) Human adipose and skeletal muscle tissue DNA, RNA, and protein content. J Appl Physiol 131(4):1370–1379. https://doi.org/10.1152/japplphysiol.00343.2021
57. An YA, Scherer PE (2020) Mouse adipose tissue protein extraction. Bio-protocol 10(11): e3631. https://doi.org/10.21769/BioProtoc.3631
58. Feist P, Hummon AB (2015) Proteomic challenges: sample preparation techniques for microgram-quantity protein analysis from biological samples. Int J Mol Sci 16(2): 3537–3563. https://doi.org/10.3390/ijms16023537
59. Guo M, Weng G, Yin D, Hu X, Han J, Du Y et al (2015) Identification of the over alkylation sites of a protein by IAM in MALDI-TOF/TOF tandem mass spectrometry. RSC Adv 5(125):103662–103668. https://doi.org/10.1039/C5RA18595E

Chapter 5

Sample Preparation and Processing for Quick, Universal, and Insightful Microbial Proteomics

Clément Lozano and Jean Armengaud

Abstract

Next-generation shotgun proteomics is one of the most valuable tools for gaining insight into the function of organisms. By providing a list of peptides and abundance information, proteomics enables the identification of proteins, their quantities, posttranslational modifications, and localization. The most refined shotgun proteomics workflow involves protein extraction, trypsin digestion, ultrahigh-performance liquid chromatography coupled to high-resolution tandem mass spectrometry, and confident assignment of resulting spectra to peptide sequences. In this study, we present a versatile, time- and cost-efficient experimental workflow for protein extraction, digestion, and analysis that can be applied to any type of microorganism. Our experimental procedure exhibits superior sensitivity compared to gel-based protocols and can be used for comparative microbial proteomics to highlight key players that explain phenotypic differences between conditions or for proteotyping new microbial isolates for taxonomic purposes.

Key words Microbial proteomics, Taxonomical proteotyping, Versatile proteomic workflow, Tandem mass spectrometry

1 Introduction

Proteomics delivers an accurate molecular picture of an organism by depicting its proteins and their abundance. Doing so, proteomics allows for the investigation of microorganisms, permitting the study of their functional response toward multiple environmental conditions [1], antibiotic treatment [2, 3], and pollutant exposure [4, 5], or establishes a descriptive proteome analysis [6]. It also allows exploratory investigations for diagnostic purposes as biomarkers can be highlighted on the basis of their abundance and specificity [7, 8]. Proteomics can also be used to probe the interactions between microorganisms or between microorganisms and their host as the methodology can be applied to complex samples. Indeed, metaproteomics has been shown to be an efficient procedure to highlight key players and their functions [9, 10].

Taufika Islam Williams (ed.), *Tissue Proteomics: Methods and Protocols*, Methods in Molecular Biology, vol. 2884, https://doi.org/10.1007/978-1-0716-4298-6_5,

Shotgun proteomics involves protein extraction, proteolysis with trypsin (or another suitable protease), and resolution of the resulting peptides via a nano-flow liquid chromatographic (LC) system coupled to a high-resolution tandem mass spectrometer (MS). This instrument provides molecular weight information for the peptides, fragments the most abundant ones into smaller ions by targeting peptide bonds, and delivers the molecular weight of these fragments. To identify peptides and proteins, the in silico digestion of organism-specific protein sequences from a protein database (*see* **Note 1**) provides all the possible theoretical masses of tryptic peptides, as well as the expected MS/MS spectra, which are deduced from the established MS/MS fragmentation rules for digested peptides. Matching the experimental MS/MS spectra and the precise molecular weight of the precursors to these theoretical data enables confident peptide identification. Subsequently, peptide mapping following parsimony rules leads to the confident protein identification.

Label-free peptide relative quantification can be conducted by extracting the signal intensity of a specific m/z value—known as extracted ion chromatogram (XIC) quantification [11]—or via spectral counting by summing the peptide to spectrum match (PSM) occurrence for a given peptide [12]. Protein abundance can be obtained by cumulating quantification values of the corresponding peptides or by using the most abundant as a reference peptide [13].

Sample preparation has been documented as a critical step in proteomic analyses. Many options—with pros and cons—are available and have been discussed, from the extraction [14, 15] to the digestion of proteins [16] and the fractionation of the resulting peptides [17]. We have previously described a robust protocol for microbial proteomics based on denaturing electrophoresis of proteins and in-gel proteolysis that provides ultrapure peptides [18]. However, the protocol is rather time-consuming. Here, we describe a quick and versatile shotgun proteomic workflow suitable for the study of any microorganisms (*see* **Note 2**). Protein extraction involves bead beating using an optimized bead mixture as previously described [19], and trypsin proteolysis is carried out via an improved microplate-adapted SP3 protocol [20], amenable to high-throughput experiments. The protocol has previously showed greater sensitivity compared to in gel and S-trap digestion [16] and has been successfully applied to the proteotyping of microorganisms [21] and the characterization of pathogenic viruses [7].

2 Materials

Reagents for mass spectrometry use must be analytical grade. Solution should be prepared with ultrapure water.

2.1 Harvesting Microorganism Proteins

1. Microorganism culture (*see* **Note 3**).
2. Centrifuge.

2.2 Protein Extraction

1. 1× lithium dodecyl sulfate solution (LDS). Composition of the 1× solution is 106 mM Tris base(hydroxymethyl)-aminomethane (Tris) HCl, 141 mM Tris base, 2% LDS, 10% glycerol, 0.51 mM ethylenediaminetetraacetic acid (EDTA), and 5% β-mercaptoethanol (BME).
2. Heating block.
3. 2-mL screw-cap microcentrifuge tubes.
4. Bead beating mixture: equal amount of 0.1-mm silica beads, 0.1-mm glass beads, and 0.5-mm glass beads (can be ordered from Bertin Technologies) (*see* **Note 4**).
5. Sonication bath such as ultrasonic cleaner USC 300 T (VWR).
6. Bead beater, e.g., Precellys Evolution instrument (Bertin Technologies).
7. Centrifuge adapted for 2-mL microcentrifuge tubes.

2.3 Reduction, Alkylation, and Digestion of Proteins

1. Reduction solution: 30 mM dithiothreitol (DTT) in a 50 mM ammonium bicarbonate (ABC) solution.
2. Alkylation solution: 105 mM iodoacetamide (IAA) in 50 mM ABC (*see* **Note 5**). Store in the dark.
3. Ultrapure water.
4. Magnetic beads solution: 1:1 mixture of "hydrophilic beads" SpeedBeads magnetic carboxylate modified particles (reference: GE45152105050250, Sigma-Aldrich) and "hydrophobic beads" SpeedBeads magnetic carboxylate modified particles (reference: GE65152105050250, Sigma-Aldrich) in water at 50 μg/μL (*see* **Note 6**).
5. Acetonitrile (ACN).
6. Ethanol: 70%.
7. Trifluoroacetic acid (TFA) 5% in water.
8. Enzyme solution: Reconstitute lyophilized sequencing-grade trypsin to a final concentration of 0.1 μg/μL in 50 mM acetic acid (*see* **Note 7**).
9. Digestion solution: 1 μL of enzyme solution in 30 μL 50 mM ABC.

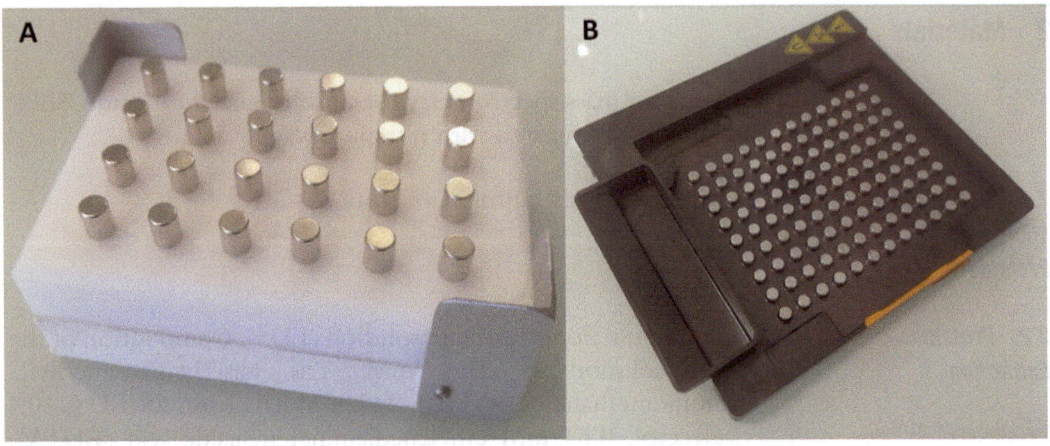

Fig. 1 Plate stand with 4 × 6 magnets with custom-made plate holder (Invitrogen) (**a**) and all-in-one magnetic plate holder (Tecan) (**b**)

10. 96-well microplate.
11. Magnetic plate holder, e.g., Magnetic Stand-96 (ref AM10027 Invitrogen) (*see* Fig. 1 for examples of plate holder suitable for this protocol).
12. Heating block.

2.4 UHPLC-MS/MS

1. Ultra-high performance liquid chromatography system.
2. High-resolution tandem mass spectrometer.

2.5 Peptide and Protein Identification

1. Search engine for LC-MS/MS raw data interpretation.

3 Methods

The procedures described below should be conducted at room temperature unless specified otherwise. Microbial culture should be conducted in sterile conditions. In order to avoid contaminants such as keratins, wool clothing should be avoided, and the use of a lab coat is mandatory. The experimental workflow from the sample processing to the functionally characterized proteome is described in Fig. 2.

3.1 Harvesting Microbial Proteins

Transfer microbial biomass from a Petri dish to a 1.5-mL microcentrifuge tube, estimate the weight of material, and add 60 μL LDS 1× buffer per mg of biomass. Alternatively, a liquid culture can be centrifuged for 5 min at 3000 *g*, the pellet weighted, and cell pellet resuspended in 60 μL LDS 1× buffer per mg of biomass (*see* **Note 8**).

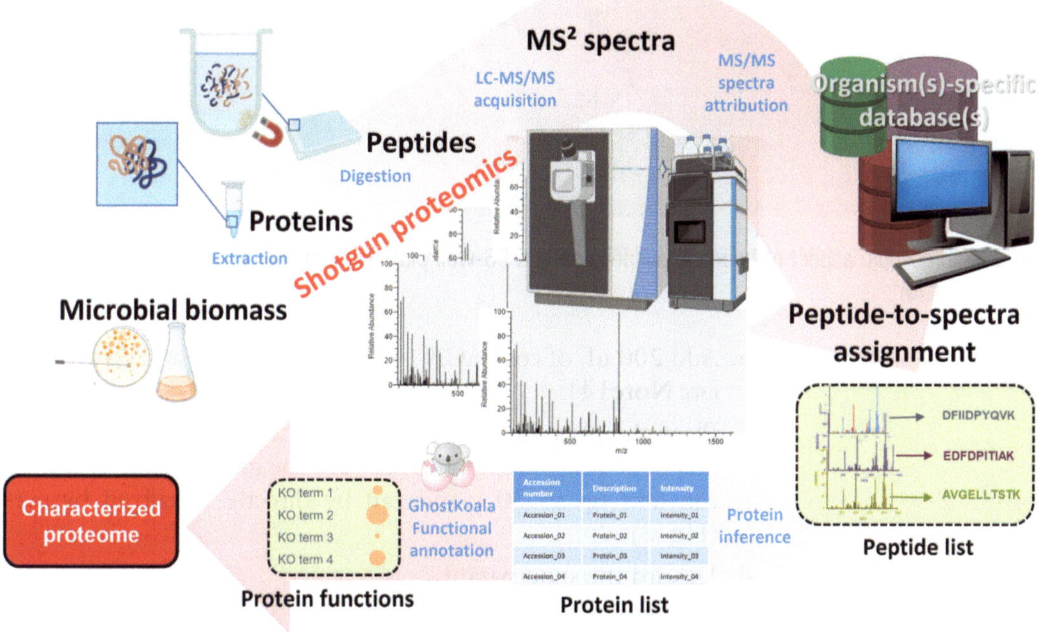

Fig. 2 Schematic representation of the experimental and data interpretation workflow

3.2 Protein Extraction

1. Heat at 99 °C for 5 min using a heat block (*see* **Note 9**).
2. Sonicate for 5 min in an ultrasonic bath at 45 kHz and 200 W (*see* **Note 10**).
3. Transfer the sample from the microcentrifuge tube to a 2-mL screw-cap microtube adapted for bead beating and containing the bead mixture described above.
4. Perform bead beating at 10,000 rpm for ten cycles of 30 s, with 30 s of pause between each cycle.
5. Centrifuge the tube at 16,000 *g* for 1 min.
6. Transfer the supernatant to a clean microcentrifuge tube (*see* **Note 11**).

3.3 Reduction, Alkylation, and Digestion of Proteins

1. Transfer 20 μL of the protein solution into a well of a 96-well plate.
2. Add 4 μL of DTT (30 mM) to get a final concentration of 5 mM. Incubate for 10 min at 50 °C (*see* **Note 12**).
3. Place the plate on ice for 2 min so that the temperature decreases.
4. Add 4 μL of IAA (105 mM) for a final concentration of 15 mM. Incubate for 10 min at room temperature in the dark (*see* **Note 13**).
5. Add 4 μL of the magnetic beads mixture (i.e., 200 μg of beads).

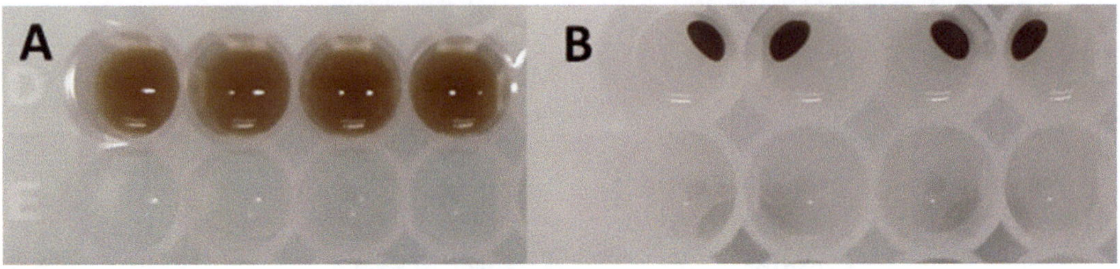

Fig. 3 Macroscopic aspect of beads in suspension in a 96-well plate (**a**) and with the plate on the magnetic plate holder (**b**)

6. Add 200 µL of cold ACN (4 °C) and pipet up and down twice (*see* **Note14**).
7. Place the plate on a magnetic plate holder and wait until the magnetic beads are retained toward the magnet (this operation can take up to 30 s and can be directly visualized by eye) (Fig. 3).
8. Discard the supernatant.
9. Add 200 µL of cold ethanol 70% (4 °C) and pipet up and down twice while keeping the plate on the magnetic plate holder (*see* **Note 15**).
10. Repeat **steps** 7 and **8**.
11. Remove the supernatant.
12. Add 180 µL of ACN.
13. Remove the supernatant.
14. Remove the plate from the magnetic plate holder.
15. Wait until the ACN fully evaporates (beads should become clearer). This operation can be faster if done under a fume hood with the plate lid on.
16. Resuspend the beads with 30 µL of fresh proteolytic digestion solution (*see* **Note 16**).
17. Incubate for 30 min at 50 °C (*see* **Note 17**).
18. Place the plate on the magnetic plate holder.
19. Transfer 25 µL of the peptide digest into an autosampler vial.
20. Acidify by adding 2.75 µL of 5% TFA (final concentration of 0.5%) (*see* **Note 18**).

3.4 UHPLC-MS/MS

Settings are described for a high-resolution orbitrap tandem mass spectrometer coupled to an HPLC system (*see* **Note 19**), equipped with reversed-phase C18 µ-precolumn (5 µm, 100 Å, 300 µm i.d. × 5 mm) followed by a C18 column (2 µm, 100 Å, 75 µm i.d. × 50 cm) (*see* **Note 20**).

1. Load 200 ng of the acidified peptide mixture (*see* **Note 21**) to be separated over a 30-min linear gradient from 5% to 25% solvent B (ACN) and 5 min from 25% to 40% at a 250 nL/min flow rate (*see* **Note 22**).
2. Record full-scan mass spectra from 350 to 1800 *m/z* and MS/MS on the 20 most abundant precursor ions (minimum signal required at 5e4, possible charge states: 2+ and 3+) and apply a 10-s dynamic exclusion window. Isolate the secondary ions within a 2 m/z window (*see* **Note 23**).
3. Convert the resulting .raw file in .mgf format for subsequent identification (*see* **Note 24**).

3.5 Peptide and Protein Identification

1. Search MS/MS spectra against the database corresponding to your microbial species (*see* **Note 25**) using the following parameters: 2 (maximum number of missed cleavages), 5 ppm (mass tolerance for the parent ion), 0.02 Da (mass tolerance for the product ions), carbamidomethylated cysteine residues (fixed modification), oxidized methionine residues (variable modification), and asparagine and glutamine deamidation (variable modification) (*see* **Note 26**).

3.6 Spectral Count-Based Peptide and Protein Quantification

1. Generate a peptide and protein list from a search engine with a target FDR at 1% (*see* **Note 27**).
2. Using a pivot table, calculate the spectral counts (*see* **Note 28**) for every proteins and their respective peptides. At this stage, removing multiple spectra assignation by keeping the spectra showing the greatest peptide and protein scores is recommended.
3. In order to have a quantitation variable more representative of the protein biomass, calculate the normalized spectral abundance factor (NSAF) of each protein as defined in McIlwain et al. (2012) [12] by dividing the spectral counts per the number of residues of each polypeptide. Alternatively, estimate the normalized spectral abundance factor (NSAF) of each protein as defined in Christie-Oleza et al. (2012) [1] by dividing the spectral counts per the molecular weight (expressed in kDa) of each polypeptide.

3.7 Protein Functional Annotation

1. Generate a .fasta file compiling all identified proteins.
2. Provide this .fasta file to GhostKoala for KEGG orthology (KO) annotation (www.kegg.jp/ghostkoala) (*see* **Note 29**).
3. At this stage, GhostKoala provides the number of proteins per KO term. Weight the functional attribution by the NSAF of each protein for each sample (*see* Fig. 4 for an example of functional analysis).

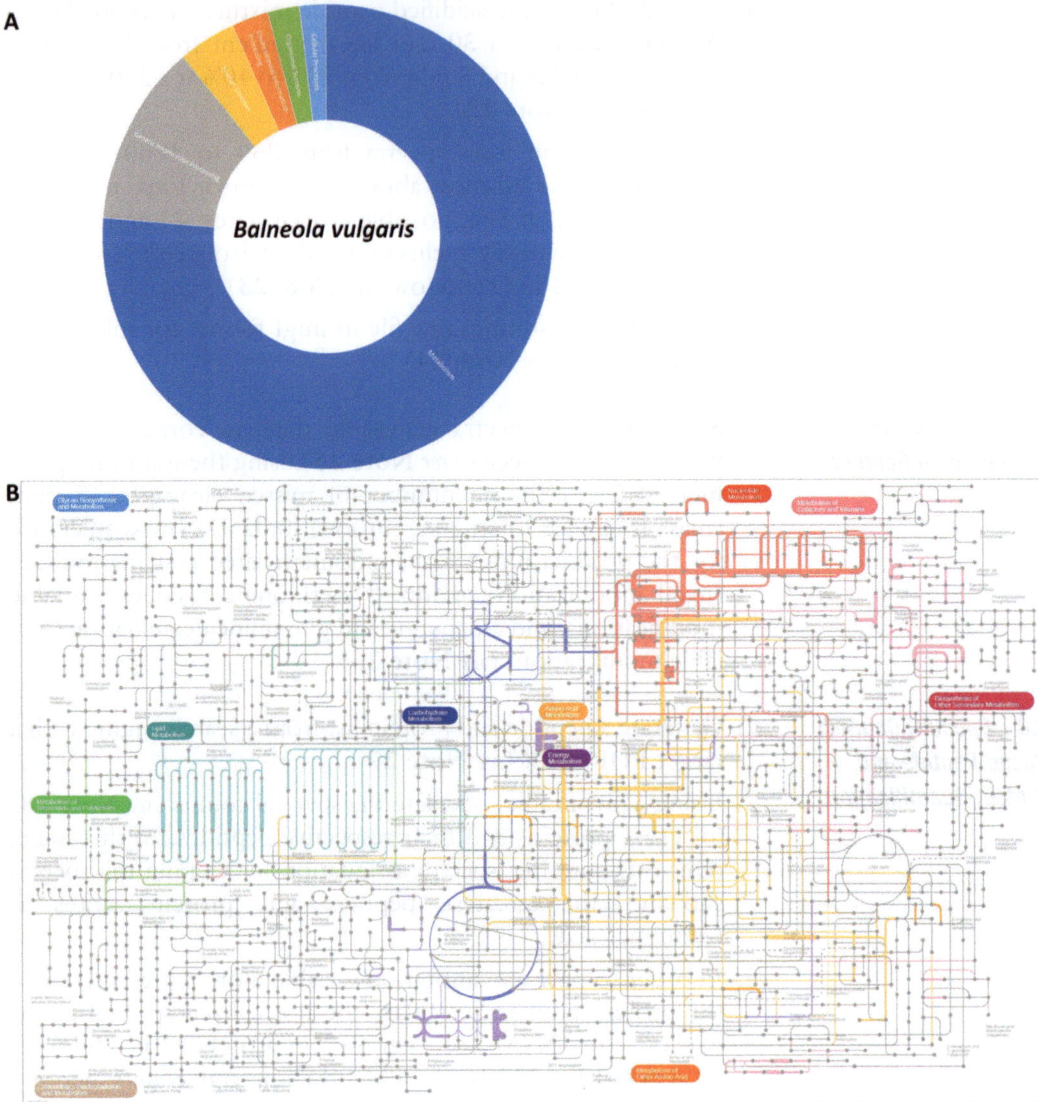

Fig. 4 KEGG functional characterization of the proteomes of *Balneola vulgaris* weighted by the %NSAF of each protein (**a**) and metabolism landscape from iPath3 (**b**). Line width is proportional to the %NSAF observed in each pathway

4 Notes

1. The database constitution is major in proteomics. The more representative it is of the sample, the better the MS/MS spectra will be assigned. If contaminants are known, or likely to be present in the sample, a contaminant database can also be

added, in order to increase attribution percentage and decrease false-positive rates.

2. Although other protocols could be considered, this shotgun proteomics workflow based on SP3 digestion coupled to a short liquid chromatography gradient is well suited to characterizing relatively simple microbial proteomes.
3. Biological biomass from colonies harvested from solid culture or biofilms, or cell pellets from liquid culture, is appropriate for protein extraction. Alternatively, this protocol can be conducted from proteins collected from culture supernatants if establishing the exoproteome is the objective.
4. We recommend the use of this bead mixture, previously optimized for increasing the yield of lysis of bacterial and eukaryotic microorganisms [16].
5. Although we recommend extemporaneous preparation of IAA and DTT solutions, aliquots can be stored for up to 2 weeks at −20 °C.
6. As recommended by the original author of the SP3 method, we recommend the use of these references for conducting the procedure [20]. Alternatively, glass beads can be used as described in Johnston et al. [22].
7. We recommend following supplier instructions for enzyme reconstitution. However, acidic solutions are advised to avoid trypsin autolysis. Reconstituted enzyme can be aliquoted and stored at −20 °C for several months.
8. This protocol works for any type of biomass regardless of its provenance (Petri dishes, liquid culture, etc.). While we provided a volume per biomass, weighing biomass can be an experimental constraint. If so, we recommend a 3:1 buffer to biomass volume ratio. A total of 10 mg of biomass is considered a comfortable amount to characterize a microbial proteome.
9. Heating the sample to 99 °C degrades recalcitrant membranes and accelerates protein denaturation.
10. Sonication is an additional step which ensures optimal protein extraction by breaking down the cell wall. Sonication also degrades residual DNA and facilitates the dissolution of solids.
11. Microcentrifuge tube of 0.5 or 1.5 mL can be used. If low amount of microbial biomass (from 10 to 50 mg) is processed, the use of low-binding tubes and tips is recommended in order to avoid material loss via adsorption.
12. Dithiothreitol (DTT) is widely used to reduce disulfide bonds from cystein residues, permitting protein unfolding and

solubilization. Alternatively, other reducing agents such as tris (2-carboxyethyl)phosphine (TCEP) can be used.

13. IAA allows alkylation of free thiol groups on cysteines. The covalent bonding of the alkyl groups stabilizes the protein structure, allowing easy access by the protease. As IAA is photosensitive, it is advisable to carry out this step in the dark. Alternatively, other alkylating agents such as chloroacetaldehyde (CAA) can also be used.
14. Acetone induces protein precipitation and aggregation on beads. The use of cold ACN is recommended as it will decrease protein solubility, hence increasing protein aggregation.
15. Ethanol is used to wash residual unwanted cellular material not removed by ACN alone. Alternatively, methanol or isopropanol could be used at this stage.
16. If the initial protein amount is not a limiting factor, the volume of enzyme solution can be increased to work with greater volumes of peptides.
17. Longer digestion time up to 2 h at 50 °C can be conducted to reduce miscleavages. Longer incubation time at this temperature can induce liquid evaporation. Alternatively, digestion at 30 °C from 1 h to 3 h can be conducted or overnight at RT.
18. Acidification by adding TFA stops the digestion and favorizes positive charges amenable to C18 capture on offline desalting systems or trapping columns. It is recommended to perform peptide quantification prior to acidification as TFA can be incompatible with fluorimetric assays.
19. While any high-resolution LC-MS/MS setup is suitable, this protocol has been conducted in our lab with an Exploris 480 (Thermo Fisher) mass spectrometer equipped with an EASY-Spray™ source coupled to a Vanquish Neo nano-HPLC system (Thermo Fisher).
20. The chromatography settings used to optimize this workflow is based on a reversed-phase Acclaim PepMap100 C18 μ-precolumn (5 μm, 100 Å, 300 μm i.d. × 5 mm, Thermo Fisher) followed by an EASY-Spray C18 column (2 μm, 100 Å, 75 μm i.d. × 50 cm, Thermo Fisher). We recommend desalting samples offline if using a system without trapping column. Gradient from 30 to 120 min can be used with a 50-cm column, while we recommend shorter column for gradient below 30 min and vice versa.
21. 200 ng was noted as a good amount of peptides to avoid trapping column saturation with hydrophobic peptides and column degradation with the setup abovementioned. Injection of greater quantities is possible, but desalting with C18 spin tips/column is then recommended. Injection of lower

quantities as low as 40 ng is also possible if sample quantity is a limiting factor. Injection of too low amount would be at the cost of proteome coverage. Peptide quantification can be performed via fluorimetric assays such as the Pierce™ Quantitative Peptide Assays.

22. The length of the gradient should be adjusted depending on the project objectives. Depending on the sample complexity, the proteome exhaustivity required, or the time dedicated to the analysis, one could require longer active separation gradients such as 60 min or 90 min from 5% to 25% acetonitrile. In order to increase the analysis exhaustivity, separation gradients up to 120 min can be considered if the studied proteome is of high complexity.

23. The mass spectrometry settings described above are generic and applicable to a wide range of proteomic applications. However, parameters can be optimized depending on the goal of the study. For instance, dynamic exclusion time could be increased to avoid multiple sampling of the same peptide, permitting the increase of unique peptides scanned during the analysis and the detection of low abundant proteins. Depending on the objective of the study, dynamic exclusion and ion intensity fragmentation threshold should be adapted. Analyses in data-independent acquisition (DIA) mode can be considered to improve proteome coverage.

24. Raw to mgf file conversion can be performed via Proteome Discoverer (Thermo Fisher) or MSConvert (ProteoWizard) softwares.

25. Databases at the .fasta format can be obtained from public repositories such as NCBI or UniProt.

26. The search for other posttranslational modifications (PTMs) can be of interest. Acetylation at the N-terminus of proteins for which the initial methionine has been removed may occur in a wide range of bacteria. Variable modifications such as phosphorylation, acetylation, methylation, ubiquitination, and glycosylation should be adjusted according to the biological context and study objectives.

27. The protein list used for functional annotation can be generated with any suitable proteomics software (e.g. MASCOT, Proteome Discoverer, MaxQuant, Skyline, Peaks, etc.). The FDR is classicaly set at 1% but could be set at a different value depending on the project goal.

28. Depending on the experiment goal, one could use extracted ion chromatograms (XIC) for quantification. XIC-based quantification could be recommended for a more accurate low to mid abundance LFQ. To do so, the use of MaxQuant could be an alternative to MASCOT, especially for comparative label-

free quantification (LFQ), for which we advise the use of MaxLFQ [23]. For DIA-based LFQ, DIA-NN is a reliable option.

29. Several alternatives for functional analysis tools are described in a protocol previously published in Grenga et al. [24].

References

1. Christie-oleza JA, Fernandez B, Nogales B, Bosch R, Armengaud J (2011) Proteomic insights into the lifestyle of an environmentally relevant marine bacterium. ISME J 6(1): 124–135. https://doi.org/10.1038/ismej.2011.86
2. Sivaranjani M, Leskinen K, Aravindraja C (2019) Deciphering the antibacterial mode of action of alpha-mangostin on staphylococcus epidermidis RP62A through an integrated transcriptomic and proteomic approach. *Front Microbiol* 10(February):1–16. https://doi.org/10.3389/fmicb.2019.00150
3. Thangamani S et al (2016) Antibacterial activity and mechanism of action of auranofin against multi-drug resistant bacterial pathogens. *Sci Rep* 6(October 2015):1–13. https://doi.org/10.1038/srep22571
4. Lozano C, Lee C, Wattiez R, Lebaron P, Matallana-Surget S (2021) Unraveling the molecular effects of oxybenzone on the proteome of an environmentally relevant marine bacterium. Sci Total Environ 793:148431. https://doi.org/10.1016/j.scitotenv.2021.148431
5. Pinel-Cabello M et al (2021) Delineation of cellular stages and identification of key proteins for reduction and biotransformation of Se (IV) by Stenotrophomonas bentonitica BII-R7. J Hazard Mater 418:126150. https://doi.org/10.1016/j.jhazmat.2021.126150
6. Zivanovic Y et al (2009) Genome analysis and genome-wide proteomics of Thermococcus gammatolerans, the most radioresistant organism known amongst the Archaea. *Genome Biol* 10(6). https://doi.org/10.1186/gb-2009-10-6-r70
7. Lozano C, Grenga L, Gallais F, Miotello G, Bellanger L, Armengaud J (2022) Mass spectrometry detection of Monkeypox virus: comprehensive coverage for ranking the most responsive peptide markers. Proteomics, August:e2200253. https://doi.org/10.1002/pmic.202200253
8. Gouveia D et al (2020) Proteotyping SARS-CoV-2 virus from nasopharyngeal swabs: a proof-of-concept focused on a 3 min mass spectrometry window. J Proteome Res 19(11):4407–4416. https://doi.org/10.1021/acs.jproteome.0c00535
9. Armengaud J (2023) Metaproteomics to understand how microbiota function: the crystal ball predicts a promising future. Environ Microbiol 25(1):115–125. https://doi.org/10.1111/1462-2920.16238
10. Salvato F, Hettich RL, Kleiner M (2021) Five key aspects of metaproteomics as a tool to understand functional interactions in host-associated microbiomes. PLoS Pathog 17(2): 1–10. https://doi.org/10.1371/JOURNAL.PPAT.1009245
11. Belouah I, Blein-Nicolas M, Balliau T, Gibon Y, Zivy M, Colombié S (2019) Peptide filtering differently affects the performances of XIC-based quantification methods. *J Proteomics* 193(October 2018):131–141. https://doi.org/10.1016/j.jprot.2018.10.003
12. McIlwain S, Mathews M, Bereman MS, Rubel EW, MacCoss MJ, Noble WS (2012) Estimating relative abundances of proteins from shotgun proteomics data. BMC Bioinform 13:308. https://doi.org/10.1186/1471-2105-13-308
13. Matzke MM et al (2013) A comparative analysis of computational approaches to relative protein quantification using peptide peak intensities in label-free LC-MS proteomics experiments. Proteomics 13(3–4):493–503. https://doi.org/10.1002/pmic.201200269
14. Chen S, Rillig MC, Wang W (2009) Improving soil protein extraction for metaproteome analysis and glomalin-related soil protein detection. Proteomics 9(21):4970–4973. https://doi.org/10.1002/pmic.200900251
15. Baniasad M et al (2022) Optimization of proteomics sample preparation for identification of host and bacterial proteins in mouse feces. Anal Bioanal Chem 414(7):2317–2331. https://doi.org/10.1007/s00216-022-03885-z
16. Hayoun K, Gouveia D, Grenga L, Pible O, Armengaud J, Alpha-bazin B (2019) Evaluation of sample preparation methods for fast proteotyping of microorganisms by tandem

mass spectrometry. Front Microbiol 10 (September):1–13. https://doi.org/10.3389/fmicb.2019.01985

17. Manadas B, Mendes VM, English J, Dunn MJ (2010) Peptide fractionation in proteomics approaches. Expert Rev Proteomics 7(5): 655–663. https://doi.org/10.1586/epr.10.46
18. Hartmann EM, Allain F, Gaillard JC, Pible O, Armengaud J (2014) Taking the shortcut for high-throughput shotgun proteomic analysis of bacteria. Methods Mol Biol 1197:275–285. https://doi.org/10.1007/978-1-4939-1261-2_16
19. Hayoun K, Gaillard J, Pible O, Alpha-bazin B, Armengaud J (2020) High-throughput proteotyping of bacterial isolates by double barrel chromatography-tandem mass spectrometry based on microplate paramagnetic beads and phylopeptidomics. J Proteomics 226 (February):103887. https://doi.org/10.1016/j.jprot.2020.103887
20. Hughes CS, Moggridge S, Müller T, Sorensen PH, Morin GB, Krijgsveld J (2019) Single-pot, solid-phase-enhanced sample preparation for proteomics experiments. Nat Protoc 14(1): 68–85. https://doi.org/10.1038/s41596-018-0082-x
21. Lozano C, Kielbasa M, Gaillard J-C, Miotello G, Pible O, Armengaud J (2022) Identification and characterization of marine microorganisms by tandem mass spectrometry proteotyping. Microorganisms 10(4):719. https://doi.org/10.3390/microorganisms10040719
22. Johnston HE et al (2022) Solvent precipitation SP3 (SP4) enhances recovery for proteomics sample preparation without magnetic beads. Anal Chem 94(29):10320–10328. https://doi.org/10.1021/acs.analchem.1c04200
23. Cox J, Hein MY, Luber CA, Paron I, Nagaraj N, Mann M (2014) Accurate proteome-wide label-free quantification by delayed normalization and maximal peptide ratio extraction, termed MaxLFQ. Mol Cell Proteomics 13(9):2513–2526. https://doi.org/10.1074/mcp.M113.031591
24. Grenga L, Gouveia D, Armengaud J (2022) Profiling SARS-CoV-2 infection by high-throughput shotgun proteomics. Methods Mol Biol 2452:167–182. https://doi.org/10.1007/978-1-0716-2111-0_11

Chapter 6

Proteomic Analysis of Single Hairs

Nicole M. McLoughlin, Romy E. Keane, Rebecca J. Tidy, Zachary C. Goecker, Robert H. Rice, Joel P. A. Gummer, Ashley M. Spicer, and Glendon J. Parker

Abstract

Hair is a ubiquitous and robust mammalian tissue with biological, clinical, forensic, social, and economic significance. The hair shaft proteome reflects both structural proteins, dominated by cuticular intermediate filament keratins and associated proteins, and proteins involved in the final cellular processes of terminally differentiating corneocytes prior to cornification. These distinct biological processes involve cell maintenance, biosynthesis, senescence, and xenobiotic response. Because growth occurs rapidly and predictably, there is also temporal organization. The hair shaft proteome also contains genetic and phylogenetic information in the amino acid sequence. Chemically the shaft is highly robust, the result of a highly structured protein matrix with abundant isopeptide and disulfide intermolecular bonds. Sample processing is therefore a challenge that requires robust chemistries but also minimizes the introduction of additional chemical modifications. This protocol depends on the combination of sodium dodecanoate and high levels of reductant to denature the matrix. The result is a proteome that is both readily accessible and can provide biological information to geneticists, developmental biologists, toxicologists, human and wildlife forensic scientists, scientists in the cosmetics industry, and wildlife ecologists.

Key words Hair, Hair shaft protein, Cuticular corneocytes, Cornification, Shotgun mass spectrometry, Tissue proteomics, Forensic proteomics, Species identification

1 Introduction

Hair is a ubiquitous and robust mammalian tissue with biological, clinical, forensic, and economic significance. The hair shaft contains the remains of multiple different cell types [1, 2]. Because of this, the hair shaft is relevant for investigators as a biological model of differentiation and of a range of biological conditions [3], including senescence [4, 5], genetic mutations [6], and environmental exposure [7]. The end result is a proteome that is both readily accessible and can provide a wealth of biological information to geneticists, developmental biologists, wildlife biologists,

Taufika Islam Williams (ed.), *Tissue Proteomics: Methods and Protocols*, Methods in Molecular Biology, vol. 2884, https://doi.org/10.1007/978-1-0716-4298-6_6,

toxicologists, forensic scientists, as well as scientists in the cosmetics industry [3].

Proteomic processing of hair shafts is a challenge due to the highly cross-linked nature of the keratin matrix [8]. Type-I and type-II keratin proteins are the basis of a highly organized hierarchical structure of intermediate filaments, maintained by extensive hydrogen bonding and reinforced by intra- and intermolecular covalent bonds [9–11]. These cross-links fall into two categories: isopeptide bonds, the result of transglutaminase activity, and disulfide bonds, the result of disulfide bond formation between adjacent cysteines. The latter is reinforced by an extensive network of cementing keratin-associated proteins [12]. This highly structured matrix accounts for the physical robustness and tensile properties of the hair shaft [13].

Unlike most tissues, the hair shaft is not readily solubilized by standard proteolytic methods, particularly the outer cuticle [8]. Normally, full solubilization and peptide extraction are a precondition for maximum sensitivity in proteomic mass spectrometry. However, the conditions required to achieve this can introduce peptide modifications that reduce the concentration of native tryptic peptides and reduce the information gained from downstream bioinformatic analyses [14–16]. This method addresses these concerns in two ways: it incorporates gentler chemistries, such as lower temperatures and shorter incubation times, and it specifically targets disulfide bonds by increases in the level of reductant in the extraction and digestion buffers, bringing the final dithiothreitol (DTT) concentration to 100 mM. This protocol uses a strong detergent, sodium dodecanoate, that is insoluble at or below pH 5 but, unlike the chemically similar and popular sodium dodecyl sulfate, is uncharged and can be extracted from the sample using repeated phase extraction with ethyl acetate. Data yields over 2000 peptides typically occur in samples that see a substantial swelling, often to greater than 1 mm in diameter, and softening of the hair shaft during the extraction phase. These shafts soften to the extent that a magnetic micro stir bar can disrupt the shaft and therefore create a suspension of keratinized matrix [8].

Other hair shaft protocols make different choices to optimize peptide yields and balance the demands of solubilization and downstream compatibility. While in our hands the above method gave the best results and was time efficient [8], other approaches have been developed that use different strategies to maximize peptide yields and achieve equivalent results [17]. These include physical disruption through milling of the hair shaft to maximize the surface area-to-volume ratio [2], the use of sodium dodecyl sulfate-polyacrylamide gel electrophoresis (SDS-PAGE) to resolve and fractionate hair shaft proteins for greater depth of coverage [18], the use of urea/thiourea instead of detergent in the "Shindai method" [17, 20], or the use of additional proteolytic enzymes with different specificities. These protocols incorporate other priorities, such as time, cost, or depth of coverage [2, 17–19]. Strategies that use harsher chemistry may result in complete solubilization with shorter incubation times, but they also tend to have poorer information outcomes [19].

2 Materials

1. Distilled water or water purified by reverse osmosis to be used to generate *solid-phase extraction (SPE)-purified water* (*see* **Note 1**).
2. *1 M ammonium bicarbonate (ABC) in SPE-purified water* (*see* **Note 2**).
3. *Wash solution* containing 2% sodium dodecanoate (SD): 50 mM ABC working solution (*see* **Note 3**).
4. *1 M dithiothreitol (DTT) in SPE-purified water* (*see* **Note 4**).
5. *0.5 M iodoacetamide (IAA) in SPE-purified water* (*see* **Note 5**).
6. *Ammonium hydroxide (NH_4OH, 3% w/w) in SPE-purified water (see* **Note 6**).
7. *Ethyl acetate.* Mass spectrometry grade.
8. *Trifluoroacetic acid (TFA).* Mass spectrometry grade, in 1-mL ampules.

3 Methods

3.1 Solution Preparation, Decontamination, and Quality Control

In solution preparation, chemical precision may conflict with the need to maintain a maximum level of cleanliness to minimize environmental contamination. The protocols below ensure a maximum level of cleanliness. Human skin proteins are in the air, and on all laboratory surfaces, including the interior surfaces of disposables and instrumentation, as well as within reagents. Practices that reduce this contamination risk include the use of standard laboratory personal protective equipment (PPE), a protected workspace such as a polymerase chain reaction (PCR) workstation, SPE of aqueous reagents (Subheading 3.2.1), and the use of mass spectrometry (MS)-grade reagents for the preparation of stocks and solutions (*see* **Note** 7). Quality control of environmental contamination can occur through inclusion of reagent blanks in all sample sequences. The inclusion of prewash and post-wash chromatography conditioning runs can also be used to monitor and measure sample-to-sample carryover on the instrument.

3.2 Proteomic Digestion of Single Hairs

3.2.1 Preparation of SPE-Purified Water

1. Disassemble a 10-mL syringe barrel from the plunger and attach a Luer Lock SPE (i.e., C18) column onto the syringe barrel.
2. Condition the column by passing about 5–10 mL of methanol through it. Pour the methanol into the syringe barrel with the attached SPE filter and reattach the syringe plunger. Discard the methanol wash.
3. Ensure that air is not inserted into the SPE C18 matrix. Pre-condition the SPE column by running water through the pre-conditioned column. The water should be the cleanest available, either distilled or purified by reverse osmosis.
4. Discard the first 2 mL before collecting SPE-water into unused pristine, sealable glassware, or pristine polypropylene vials.
5. Store the eluted water at room temperature (*see* **Note 1**). Use this SPE-purified water to generate all aqueous solutions for the protocol.

3.2.2 Hair Shaft Washing

1. Cut a 2-cm (20-mm) section of a single hair shaft. Reserve any remaining hair for additional analyses.
2. In a 1.5-mL tube, rinse hair with 500 μL of the wash solution (2% SD—50 mM ABC). Vortex well. Discard solution and repeat with fresh wash solution two more times, for a total of three washes.
3. Remove and allow hairs to dry. Cut into 2-mm pieces using a razor blade or scissors and forceps (*see* **Notes 8** and **9**). Place all sections into a fresh 1.5-mL low-binding polypropylene microcentrifuge vial.

3.2.3 Protein Reduction

1. Add 100 μL of wash solution (2% SD—50 mM ABC) to each sample tube.
2. Add 10 μL 1 M DTT to each sample vial.
3. Add a single (1) micro stir bar to each sample vial.
4. Centrifuge vials briefly with a pulse to collect the contents at the bottom of the vial and ensure the sample hair is fully submerged in the solution.
5. Stir samples on stir plate on low-medium speed for 6 h. Ensure hair stays fully submerged in the solution. Manually submerge the sections as required with a clean pipette tip.

3.2.4 Protein Alkylation and Detergent Extraction

1. Add 40 μL of 0.5 M IAA to each sample tube.
2. Continue stirring for another 45 min. Ensure the samples are protected from light.
3. Add 2 μL TFA to each sample tube (to adjust pH to ~3). Aliquot 2–5 μL onto a pH strip to confirm pH.

4. Add 175 µL ethyl acetate to each sample. Vortex well.
5. Centrifuge the samples at maximum speed for 3 min to decrease the volume of the interphase.
6. Remove the ethyl acetate (top phase) by pipetting and then discarding. If any of the lower phase is drawn into the pipette tip, gently dispense any lower-phase material back into the tube. This will be visible in the tip since there is an obvious interphase line between the two layers.
7. Repeat the extraction two more times (three extractions total). Take care to pipette out as much as possible of the upper ethyl acetate phase on the last extraction.
8. Add 6.3 µL concentrated SPE-filtered ammonium hydroxide (2.8–3.0% w/w) and 6.3 µL 1 M ABC to each sample tube. Gently mix (to adjust pH to ~8.5). Aliquot 2–5 µL onto a pH strip to confirm pH.
9. Store the samples overnight or as long as needed at ≤ −20 °C.

3.2.5 Sample Digestion

1. Remove samples from the freezer to thaw.
2. Resolubilize trypsin (20 µg) with 20 µL SPE water from aliquoted microcentrifuge tubes. Gently mix with pipetting.
3. Add the entire 20 µL of trypsin to sample (one vial per sample). This is equivalent to a 1:5 ratio of protease to protein. Lower ratios of protease to protein (1:10 or 1:20) may work as well (*see* **Note 10**).
4. Stir the samples on a magnetic stir plate on a low-medium speed for 6 h (*see* **Note 11**). Ensure hair remains fully covered by the solution. Push down as needed (*see* **Note 12**).
5. After the digestion is completed, remove the stir bars using an external magnet.
6. Centrifuge the samples at maximum speed (at least 13,000 *g*) for 15 min in a benchtop centrifuge at 4 °C.
7. Pipette each supernatant into new tubes with centrifugal filters.
8. Centrifuge the samples at maximum speed (at least 13,000 *g*) for 5 min at 4 °C.
9. Aliquot 60 µL of each sample into a fresh pre-labeled low--binding polypropylene microcentrifuge vial for storage at −20 °C until nanoLC-MS analysis (*see* **Notes 13–15**).

3.3 Sample Analysis by nanoLC-MS

Proteomic analysis by shotgun mass spectrometry should be conducted according to the standard methods and protocols of liquid chromatography and mass spectrometry manufacturers [21]. A raw data file is produced and then analyzed for peptide spectra matching using any one of a range of established bioinformatic peptide spectra matching algorithms.

3.4 Bioinformatic Analysis

In setting the search parameters, a human reference proteome should be used, such as the curated UniProt human reference protein database (proteome = UP000005640, www.uniprot.org) or the equivalent protein FASTA file downloaded from the National Center of Bioinformatic Information based on human genome assembly GRCh38.p14 (www.ncbi.nlm.nih.gov/datasets/taxonomy/9606/). Since many hair proteins overlap with the CRAPome [22], this bioinformatic filter should not be used. In conducting a search, the peptide and fragment parameters should reflect those of the instrument used. For example, on a Thermo Scientific Orbitrap Exploris 480 Mass Spectrometer, a 10 ppm and 0.02 Da threshold is appropriate for primary and secondary (i.e., fragmentation) masses, respectively. In addition, many peptides from keratin-associated proteins may have over ten cysteines per peptide. If the search algorithm does not have a more permissive search phase for posttranslational modification (PTM) number and type per peptide, such as with PEAKs™ [23] or X! TANDEM [24], highly modified peptides will be lost from the analysis, and a maximum number of variable PTM assignments and missed cleavage number per peptide should be used, depending on the capacity of the software to tolerate a larger search space. Otherwise, default criteria may be used. As a minimum, the first stage should include variable deamidation of asparagine and glutamine (+1 NQ) and oxidized methionine (+16 M). Typically, most practitioners assume that all cysteines are modified with carbamidomethylation (+57 C). This may not be the case in hair because of the high number of cysteines. If it is important to identify the maximum number of peptides, then variable carbamidomethylation should be assumed. As with any peptide spectral matching (PSM) run, standard internal quality control thresholds based on peptide sequence assignments using decoy databases should be used. Typically, this is a 1% false discovery rate for peptides, but a lower threshold of 0.1% can be used if higher stringency is necessary.

4 Notes

1. To minimize contamination, only pour from dry chemical stocks into a pristine 50-mL vial. Ideally attempt to weigh the reagent by pouring from the vial, only using a spatula if necessary. Use pristine weigh paper on the analytical balance. Care should be given to regular cleaning of the workspace with water and 70% ethanol, as well as ensuring protection of reagents and disposables from environmental dust. Unused vials, pipette tips, and reagents should be covered at all times.

2. Add 949 mg of ABC to a 15-mL polypropylene vial and fill to the 12 mL mark with SPE-purified water. Vortex to dissolve and pass the entire solution through a preconditioned SPE column with filter. Discard the first 2 mL of the wash before collecting the remaining 10 mL by dispensing the stock into 10 × 1-mL aliquots in 1.5-mL low-binding microcentrifuge vials. Store at ≤ −20 °C.
3. Weigh 166.0 mg of ABC, fill to the 42 mL mark with SPE-purified water, and vortex to dissolve. Pass the entire solution through a preconditioned SPE cartridge column with filter. Discard the first 2 mL of the wash before collecting the remaining 40 mL in a new 50-mL polypropylene vial. Add 800 mg of SD to the tube. Seal the cap with parafilm before mixing by vortexing and placing in an incubator at 70 °C for a few minutes to solubilize and gently mix. The addition of SD will slightly lower the concentration of ABC, below 50 mM. Warming the 2% SD/50 mM ABC detergent solution in an incubator for a few minutes may be needed to allow the sodium dodecanoate to fully dissolve and go into solution. The parafilm on the cap of the falcon tube will help prevent leakage.
4. Weigh 1851.0 mg DTT and transfer into a 15-mL falcon tube. Fill to the 12 mL mark with SPE-purified water. Vortex to dissolve. Pass the entire solution through a preconditioned SPE column with filter. Discard the first 2 mL of the wash before collecting the remaining 10 mL by aliquoting them into roughly 10 × 1-mL aliquots in 1.5-mL pristine low-binding microcentrifuge vials. Store at ≤ −20 °C.
5. Weigh 1110 mg IAA and transfer to a 15-mL falcon tube. Fill to the 12 mL mark with SPE-purified water. Vortex to dissolve. Pass the entire solution through a preconditioned SPE column with filter. Discard the first 2 mL of effluent before collecting the remaining 10 mL by dispensing roughly 1 mL into 10 × 1.5-mL low-binding microcentrifuge vials. Store at ≤ −20 °C. Iodoacetamide is light sensitive. It is best to work quickly and efficiently when handling this reagent and always keep stock solutions under cover and out of the exposure of light as much as possible when using. If necessary, cover incubators and equipment in foil.
6. Add 0.3 g of solid ammonium hydroxide to a 15-mL polypropylene vial and add SPE-purified water to a total of 12 mL. Mix or vortex to dissolve. Pass the entire solution through a preconditioned SPE column with filter. Discard the first 2 mL of effluent before collecting the remaining 10 mL by dispensing roughly 1 mL into 10 × 1.5-mL low-binding polypropylene microcentrifuge vials. Store at 4 °C.

7. Once solvent has been added to the syringe, take care not to push air through the column. If this does occur, the air can be removed by washing with water and repeating the MeOH preconditioning step. Only do this step if it is difficult to force the solution through the SPE column. Because this step employs a C18 solid-phase resin cartridge, impurities cannot be removed from aqueous stocks or solutions containing organics or detergents/surfactants.
8. Cutting hairs into small pieces can be challenging. It is best to work on a dark surface when handling light-colored hairs and a light surface when working with dark-colored hairs.
9. Depending on humidity, hair pieces may attract static electricity and "jump" around while being cut.
10. The trypsin amount used in this protocol is high; 20 μg of trypsin is roughly a protease-to-protein ratio of 1:5. We found that the typical ratio of 1:50 as recommended in fully soluble protein mixtures was not as efficient for a partially insoluble digest.
11. The rounded/oval-shaped micro stir bars are more efficient at mixing and disrupting the softened hair shaft than the square-edged ones.
12. Samples may stir best when placed in a tube rack that is two inches above the stir plate and set to low-medium speed.
13. If working with less soluble keratin-based samples (such as skin and thicker hair shafts), then allow the samples to stir longer than 6 h to maximize the digestion process.
14. We find that a subfraction of human hairs shafts (less than 5%) are intractable, failing to denature in the presence of even harsher chemistries. A systematic analysis of these samples is difficult due to the small sample size. Anecdotally we observe a higher likelihood of this phenomenon in female African Americans who often include heat treatments in their personal grooming. These may increase the prevalence of cross-linking through Maillard chemistries.
15. Use of microcentrifuge filters (0.22 or 0.45 μm) is necessary to remove particulates that may interfere with downstream chromatography. These filters should be employed when remixing frozen aliquots of trypsin digests since precipitation may occur during the freeze/thaw cycle.

Acknowledgments

This work is or has been supported by the National Institute of Justice (2015-DN-BX-K065 and 15PNIJ-22-GG-03566-SLFO),

California Department for Fish and Wildlife (CDFW, P2195005, P2295005, P2395001), and the West Australian Government (ChemCentre). The authors thank members of the Parker, Rice, CDFW, and ChemCentre laboratories for assistance and advice in protocol development and manuscript preparation.

References

1. Parker GJ, McKiernan HE, Legg KM, Goecker ZC (2021) Forensic proteomics. Forensic Sci Int Genet. https://doi.org/10.1016/j.fsigen.2021.102529
2. Parker GJ et al (2016) Demonstration of protein-based human identification using the hair shaft proteome. PLoS One 11(9): e0160653. https://doi.org/10.1371/journal.pone.0160653
3. Adeola HA et al (2018) Emerging diagnostic and therapeutic potentials of human hair proteomics. Proteomics Clin Appl 12(2): 1700048. https://doi.org/10.1002/prca.201700048
4. Franklin RN et al (2020) Proteomic genotyping: using mass spectrometry to infer SNP genotypes in pigmented and non-pigmented hair. Forensic Sci Int 310:110200. https://doi.org/10.1016/j.forsciint.2020.110200
5. Rosenberg AM et al (2021) Quantitative mapping of human hair greying and reversal in relation to life stress. elife 10. https://doi.org/10.7554/eLife.67437
6. Ahmed A, Almohanna H, Griggs J, Tosti A (2019) Genetic hair disorders: a review. Dermatol Ther 9(3):421–448. https://doi.org/10.1007/s13555-019-0313-2
7. Naudin G et al (2019) Human pollution exposure correlates with accelerated ultrastructural degradation of hair fibers. Proc Natl Acad Sci 116(37):18410–18415. https://doi.org/10.1073/pnas.1904082116
8. Goecker ZC et al (2020) Optimal processing for proteomic genotyping of single human hairs. Forensic Sci Int Genet 47(102314). https://doi.org/10.1016/j.fsigen.2020.102314
9. Bragulla HH, Homberger DG (2009) Structure and functions of keratin proteins in simple, stratified, keratinized and cornified epithelia. J Anat 214(4):516–559. https://doi.org/10.1111/j.1469-7580.2009.01066.x
10. Yang FC, Zhang Y, Rheinstadter MC (2014) The structure of people's hair. PeerJ 2:e619. https://doi.org/10.7717/peerj.619
11. Vermeire P-J et al (2021) Molecular interactions driving intermediate filament assembly. Cells 10(9):2457
12. Litman T, Stein WD (2023) Ancient lineages of the keratin-associated protein (KRTAP) genes and their co-option in the evolution of the hair follicle. BMC Ecol Evol 23(1):7. https://doi.org/10.1186/s12862-023-02107-z
13. Wolfram LJ (2003) Human hair: a unique physicochemical composite. J Am Acad Dermatol 48(6 Suppl):S106–S114. https://doi.org/10.1067/mjd.2003.276
14. Rice RH (2011) Proteomic analysis of hair shaft and nail plate. J Cosmet Sci 62:229–236
15. Laatsch CN et al (2014) Human hair shaft proteomic profiling: individual differences, site specificity and cuticle analysis. PeerJ 2: e506. https://doi.org/10.7717/peerj.506
16. Lee YJ, Rice RH, Lee YM (2006) Proteome analysis of human hair shaft: from protein identification to posttranslational modification. Mol Cell Proteomics 5(5):789–800
17. Catlin LA et al (2019) Demonstration of a mitochondrial DNA-compatible workflow for genetically variant peptide identification from human hair samples. Forensic Sci Int Genet 43: 102148. https://doi.org/10.1016/j.fsigen.2019.102148
18. Zhang Z et al (2019) Sensitive method for the confident identification of genetically variant peptides in human hair keratin. J Forensic Sci. https://doi.org/10.1111/1556-4029.14229
19. Adav SS et al (2018) Studies on the proteome of human hair – identification of histones and Deamidated keratins. Sci Rep 8(1):1599. https://doi.org/10.1038/s41598-018-20041-9
20. Nakamura A, Arimoto M, Takeuchi K, Fujii T (2002) A rapid extraction procedure of human hair proteins and identification of phosphorylated species. Biol Pharm Bull 25(5):569–572
21. Wang X, Zhang B (2014) Integrating genomic, transcriptomic, and interactome data to improve peptide and protein identification in shotgun proteomics. J Proteome Res 13(6): 2715–2723. https://doi.org/10.1021/pr500194t

22. Mellacheruvu D et al (2013) The CRAPome: a contaminant repository for affinity purification-mass spectrometry data. Nat Methods 10(8):730–736. https://doi.org/10.1038/nmeth.2557
23. Zhang J et al (2012) PEAKS DB: de novo sequencing assisted database search for sensitive and accurate peptide identification. Mol Cell Proteomics 11(4):M111 010587. https://doi.org/10.1074/mcp.M111.010587
24. Fenyo D, Eriksson J, Beavis R (2010) Mass spectrometric protein identification using the global proteome machine. Methods Mol Biol 673:189–202. https://doi.org/10.1007/978-1-60761-842-3_11

Chapter 7

Shotgun Proteomics Protocol for Insects

Madhushri S. Varunjikar, Ikram Belghit, Eystein Oveland, Magnus Palmblad, and Josef D. Rasinger

Abstract

Shotgun proteomics can be applied to identify and study insect species in diverse research areas such as agriculture, forensics, biodiversity conservation, and food safety. In this chapter, we have provided a detailed protocol for shotgun proteomics analytical methods involving enzymatic digestion of insect proteins using trypsin, separation using high-performance liquid chromatography, and detection of separated peptides using high-resolution mass spectrometry. The protocol also covers the utilization of bioinformatics software for protein identification and spectral library building, proposing both proteomic database-dependent and independent methods. This chapter provides a valuable foundation for applying insect proteomics by discussing a step-by-step protocol.

Key words Shotgun proteomics, Insects, Spectral library matching, Protein identification

1 Introduction

Proteomics involves the study of proteins, which serve as the fundamental building blocks of cellular structures within an organism. By examining the proteins present in a specific tissue of an organism at a given time point, proteomics offers insights into cellular, biological, and molecular processes [1]. Through a shotgun proteomic approach, it becomes possible to identify protein profiles and quantify proteins within tissues or cells [2]. In the shotgun proteomic approach, proteins are enzymatically digested after extraction from various samples. This process typically involves the use of trypsin, which cleaves the protein backbone, specifically at the C-terminal of arginine and lysine residues. The digested peptides from the samples are separated using reversed-phase high-performance liquid chromatography (HPLC). After the separation, using data-dependent acquisition, the peptides are typically detected using Orbitrap, ion trap, or time-of-flight (TOF) mass analyzers [3]. Acquired MS2 spectra, i.e., tandem mass spectra

Taufika Islam Williams (ed.), *Tissue Proteomics: Methods and Protocols*, Methods in Molecular Biology, vol. 2884,
https://doi.org/10.1007/978-1-0716-4298-6_7,

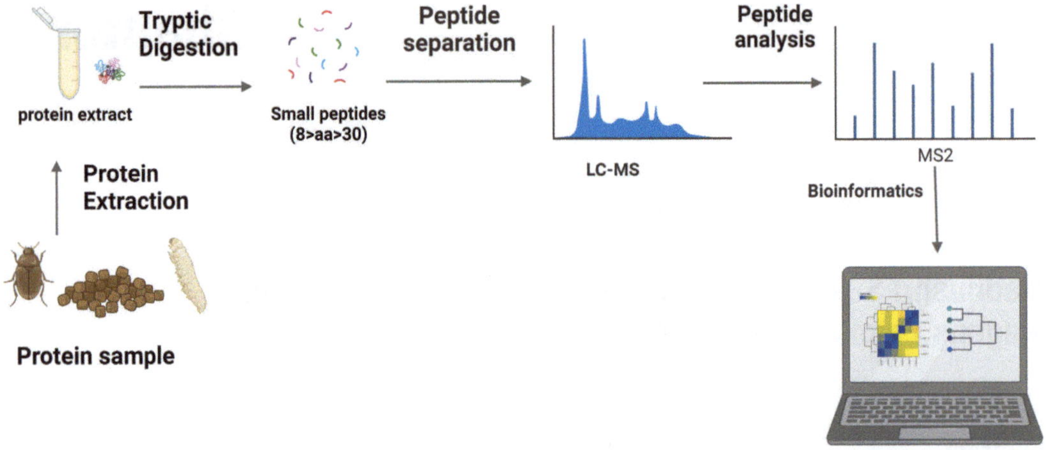

Fig. 1 Shotgun proteomic workflow used for insect proteomics. Protein samples from insects were processed and enzymatically digested with trypsin enzyme using filter-aided sample preparation (FASP). The digested samples were then separated using high-performance liquid chromatography tandem mass spectrometry (HPLC-MS/MS) to acquire proteomic data. (Modified from and reproduced with permission of © Varunjikar 2023 [18]. Created with BioRender.com)

generated by the mass analyzer, are searched against protein databases or spectral libraries to identify peptides and proteins. This chapter describes how proteomics protocols can be implemented for the identification of insect species, the study of insect proteins, and the detection of allergens in food-grade insect samples. In our previous works, we have implemented the proteomics workflows presented here for insects, including black soldier fly (*Hermetia illucens*), yellow mealworm (*Tenebrio molitor*), lesser mealworm (*Alphitobius diaperinus*), house cricket (*Acheta domesticus*), and morio worm (*Zophobas morio)* [3–5], as shown in Fig. 1. For non-model species such as insects other than fruit fly (*Drosophila melanogaster)*, the sizes of protein sequence databases are commonly small, posing challenges to proteomics reference database-dependent workflows [6]. In addition to traditional proteome database-centered search methods, we have adopted database-independent approaches. These include spectral library-based analysis and direct spectral comparison to analyze mass spectra, alongside the use of traditional proteome database search methods. The advantage of the spectral library-based method is that it does not require a priori knowledge of genome or protein sequences, thereby benefiting studies on non-model species such as insects [7]. The bioinformatics workflow described here prioritizes open-source software over commercial alternatives, ensuring a reproducible workflow that is easy to implement across diverse laboratories.

2 Materials

2.1 Insect Sample Collection

Targeted insect species can be collected from the natural habitat, cell culture from insects, or insects reared in laboratory conditions on substrates of study [4] or obtained from insect-producing companies [5]. Store insect samples at −80 °C before processing to avoid any degradation of protein samples.

2.2 Protein and Peptide Extractions

Prepare all solutions in HPLC-grade water and use analytical-grade chemicals. Buffer quantities mentioned here are suitable for approximately 24 samples. In this protocol, we have exemplified vendors for different equipment and instrumentation. However, similar equipment and instrumentation from other reputable vendors can also be used, and the choice of supplier may depend on availability and user preference.

1. Lysis buffer: 0.1 M Tris–HCl, 4% SDS, pH 7.6. Add 2.5 mL of 1 M Tris–HCl, pH 7.6, to a cylinder or glass beaker. Weigh 1-g sodium dodecyl sulfate (SDS) and transfer it to a cylinder. Add water to make up 25 mL of solution. Dissolve SDS entirely into the solution.
2. Protein Assay Reagent: Combine 6-mL Protein Assay Reagent Pierce™ 660 with 300 mg of the ionic reagent (*see* **Note 1**).
3. Urea solution: 8 M urea, pH 8.5, and 0.1 M Tris–HCl. Weigh 24 g urea to a glass beaker. Add 5 mL of 1 M Tris–HCl and make up the volume by adding 26.5 mL of water.
4. Dithiothreitol (DTT) stock solution: 1 M DTT, pH 7.6. Weigh 1540 mg DTT and add to 10 ml 0.1 M Tris–HCl, pH 7.6.
5. Urea DTT solution: 8 M urea, 0.1 M Tris–HCl, and 10 mM DTT. Prepare 13-mL urea DTT solution 8 M urea/0.1 M Tris–HCl solution by adding 30 μL of DDT.
6. Iodoacetamide (IAA) solution: 50 mM iodoacetamide and 8 M urea, pH 8.5. Weigh 23.125 mg of IAA in a glass tube and dissolve in 8 M urea solution (*see* **Note 2**).
7. Ammonium bicarbonate solution: 50 mM ammonium bicarbonate. Weigh 39.5 mg of ammonium bicarbonate in a glass beaker and add 20 mL of water to make 1 M stock solution. From stock solution, prepare 50 mM solution.
8. Trypsin solution: Trypsin, 50 mM ammonium bicarbonate, 0.1 M $CaCl_2$. Using sequencing grade trypsin, dissolve 20 μg trypsin in 100 μL trypsin resuspension buffer. Add 1183 μL 50 mM ammonium bicarbonate solution and 13 μL of 0.1 M $CaCl_2$ solution.
9. Sodium chloride (NaCl): 0.5 M NaCl.

10. C_{18} activation solution: Mix 25 mL of methanol with 25 mL of water.
11. Equilibration solution: Add 2.5 mL of acetonitrile and 250 μL of trifluoroacetic acid into a glass beaker in 47.25 mL of water.
12. Sample buffer: Mix 10 mL of acetonitrile, 1 mL of trifluoroacetic acid, and 39 mL of water.
13. Elution buffer: Mix 35-mL acetonitrile and 15-mL water.
14. Sample Grinding kit One Plus (GE Healthcare Life Science, Piscataway, USA).
15. Prediluted Protein Assay Standard BSA solutions (Thermo Fisher Scientific).
16. Protein Assay Reagent Pierce™ 660 (Thermo Fisher Scientific).
17. 96-well plate.
18. Pierce™ Ionic Detergent Combability Reagent (Thermo Fisher Scientific).
19. Microcon 30-kDa Centrifugal Filter (Merck-Millipore).
20. 1.5-mL protein low-binding tubes (Thermo Fisher Scientific).
21. Pierce™ C_{18} Spin Columns (Thermo Fisher Scientific).
22. Vortex for tubes (650 rpm).
23. Centrifuge for 1.5-mL tubes.
24. 37 °C heating cabinet.
25. NanoDrop spectrophotometer for peptide quantification.

2.3 High-Performance Liquid Chromatography-High Resolution Mass Spectrometry (HPLC/HR-MS)

1. HeLa Protein Digest Standard Pierce™ (Thermo Scientific, San Jose, CA).
2. Mobile phase A: 0.1% formic acid in water, vol/vol. Mix 500 μL formic acid with 500-mL water. To prevent bubbles in the system, degas the mobile phase in an ultrasonic bath.
3. Mobile phase B: 0.1% formic acid in acetonitrile, vol/vol. Mix 500 μL formic acid with 500-mL acetonitrile (ACN). To prevent bubbles in the system, degas the mobile phase in an ultrasonic bath.
4. HeLa protein standard: Dissolve 20 μg of Pierce™ HeLa Protein Digest Standard in solution with 0.1% formic acid in water/acetonitrile (98:2, vol/vol).
5. An HPLC system equipped with an autosampler, binary gradient pump, and column compartment with a thermostat (Vanquish Horizon binary HPLC, Thermo Scientific).
6. High-resolution mass spectrometry system Q Exactive Orbitrap (Thermo Scientific) equipped with an electrospray source

(similar equipment and instrumentation from other reputable vendors can also be used).

7. Analytical column: 2.2 μm Acclaim Vanquish C18, 2.1 × 250 mm from Thermo Scientific (similar equipment from other reputable vendors can also be used).
8. Ultrasonic bath.
9. Software for data acquisition and data preprocessing: TraceFinder™, Xcalibur™, and Chromeleon™.

2.4 Bioinformatics Software

1. Tandem mass spectra comparison software compareMS2 [8].
2. msConvert (version: 3.0., ProteoWizard) [9].
3. SpectraST (version 5.0) [10].
4. Trans-Proteomics Pipeline (TPP) (version 5.2.0) [11].

3 Methods

3.1 Protein and Peptide Extraction

Using laboratory personal protective equipment (PPE), such as safety glasses, a laboratory coat, and disposable gloves, is mandatory for all protein and peptide extraction experiments [7]. Notably, keratin contamination from laboratory personnel is a common issue in protein samples, necessitating careful precautions [12]. Adhering to good laboratory practices involves consistently wearing gloves, refraining from touching the body with gloved hands, avoiding leaning over samples, and other relevant measures. To minimize the risk of side reactions and contamination, using fresh reagents is advised [7]. All the samples should be kept on ice to prevent potential degradation and nonspecific cleavage of proteins.

3.1.1 Protein Extraction

For protein extraction, protein samples (frozen insects, tissue, or freeze-dried powder) are lysed using lysis buffer to extract proteins (Fig. 1).

1. Prepare a grinding tube by vertexing to loosen the beads. If the beads are dry, add water to resuspend.
2. Centrifuge grinding tubes for 1 min at 15,000× *g*.
3. Remove excess water from all tubes.
4. Add 200–300 μL lysis buffer.
5. Weigh 100 mg or 50 mg—as shown in Table 1, depending on the nature of the samples—into a test tube of the One Plus Grinding kit and add 1-mL lysis buffer.
6. Homogenize with a pestle for around 1 min.
7. Add more lysis buffer to make a final 1000 or 200 μL volume.
8. Incubate all tubes at 95 °C on a heating block for 5 min.

Table 1
Approximate protein output and required volume of lysis buffer for different sample types

Sample	Input (mg)	Output (μg protein/μL)	Volume μL (lysing buffer)
Tissue	100	5–15	1000
Protein powder	50	4–30	1000
Cell pellet	Pellet	1–5	200

9. Centrifuge at 15,000× *g* for 10 min at room temperature.
10. Transfer the supernatant into a new tube. Avoid precipitates and fat.
11. Store samples at −80 °C until the following procedure.

3.1.2 Determination of Protein Concentration

1. Prepare adequate Protein Assay Reagent with the ionic detergent compatibility reagent (*see* **Note 1**).
2. In a 96-well plate, add 10 μL of each of 7 standards in duplicate (Prediluted Protein Assay Standard BSA set).
3. Add 10 μL of each sample in duplicate by diluting with lysis buffer. Adjust the assay for samples with low or high protein yield by diluting with lysis buffer (e.g., 3 μL of sample diluted with 7 μL lysis buffer).
4. Add 10 μL of lysis buffer in a well for the negative control.
5. Add 150 μL of Protein Assay Reagent to all wells with the ionic detergent compatibility reagent.
6. Add 150 μL of Assay Reagent with the ionic reagent as a buffer blank to a well.
7. Incubate the plate for 5 min at room temperature.
8. Measure the color intensity using a spectrophotometer measuring absorbance at 660 nm.
9. Calculate protein concentration using absorbance values and standard measurements.

3.1.3 Filter-Aided Sample Preparation (FASP) for Peptide Extraction

The FASP protocol is used for tryptic digestion of peptides from protein mixtures [13]. It involves several steps to extract, digest, and clean up proteins from a sample.

1. Calculate the volume (μL) of a 200 μg protein input for each sample (*see* **Note 3**).
2. Calculate the volume of lysis buffer required to achieve a standardized total input volume for all samples, such as 30 μL, and add urea solution to the protein sample (*see* **Note 3**).

Day 1

1. To dissolve the SDS, thaw frozen samples and incubate them at 37 °C for 5 min. Afterwards, spin down the samples.
2. Add DTT solution until a concentration of 0.1 M is achieved. For example, add 1 μL of 1 M DTT to a 10 μL sample.
3. Heat the samples at 95 °C for 5 min.
4. Check the Microcon-30 centrifuge filter by adding 400 μL of urea solution to each filter tube, one per sample. Centrifuge samples at 11,000× *g* for 5 min (*see* **Note 4**).
5. Centrifuge the filter for 10–16 min (*see* **Notes 4** and **5**).
6. For a protein sample volume of 10 μL, mix with 100 μL of urea DTT solution aiming for an input of 40 μg protein (*see* **Note 3**).
7. Add the sample mixture to the centrifuge filter and incubate for 1 h at room temperature (*see* **Note 6**).
8. Mix the samples for 1 min at 650 rpm using a shaker incubator. Centrifuge at 11,000× *g* for 10–20 min (*see* **Notes 5** and 7).
9. Add 200 μL of urea DTT solution.
10. Mix for 1 min at 650 rpm using a shaker incubator.
11. Centrifuge samples at 11,000× *g* for 8–22 min (*see* **Note 8**).
12. Repeat **steps 8** and **9** for two washes (*see* **Note 9**).
13. Add 100 μL of 50 mM IAA solution (*see* **Note 2**).
14. Mix for 1 min at 650 rpm using a shaker incubator.
15. Incubate the samples in a dark environment for 20 min at room temperature.
16. Centrifuge the samples at 11,000× *g* for 8–20 min.
17. Add 200 μL of urea solution to the samples.
18. Mix the samples for 1 min at 650 rpm. Centrifuge at 11,000× *g* for 10–20 min.
19. Repeat the preceding two steps two more times, totaling three washes.
20. Discard all the liquid passed through the filter (*see* **Note 9**).
21. Add 100 μL of a 50 mM ammonium bicarbonate solution.
22. Mix the samples for 1 min at 650 rpm. Centrifuge at 11,000× *g* for 4–10 min, considering variations in the duration based on sample types.
23. Repeat the previous two steps twice, completing three washes with ammonium bicarbonate solution. Do not remove the liquid that has passed through the filter to minimize evaporation in the subsequent step.
24. Add 50 μL of trypsin solution to the samples.
25. Mix the samples for 1 min at 650 rpm.

26. Place parafilm around the lid to prevent evaporation.
27. Incubate the samples at 37 °C for 16 h.

Day 2

28. Transfer the filter into a new tube.
29. Centrifuge at 11,000× *g* for 2–12 min, considering variations in the duration based on sample types.
30. Add 40 μL of ammonium bicarbonate solution.
31. Mix for 1 min at 650 rpm using a shaker incubator. Centrifuge at 11,000× *g* for 2–12 min.
32. Repeat the previous two steps twice, resulting in three washes with the ammonium bicarbonate solution.
33. Add 50 μL of 0.5 M NaCl solution.
34. Mix for 1 min at 650 rpm using a shaker incubator. Centrifuge at 11, 000× *g* for 10–12 min. All peptides are released from the filters after this stage.
35. Transfer the peptide solutions into separate, low-binding protein tubes.
36. Measure the concentration of peptides using a NanoDrop spectrophotometer. At this stage we expect around 80–90% protein-to-peptide conversion.

3.1.4 C_{18} Spin Column Cleaning of Peptides

C_{18} protocol is widely used for sample cleanup, facilitating the removal of salts, solvents, and protecting groups that could otherwise interfere with mass spectrometric or spectrophotometric analysis of peptides. By utilizing reversed-phase chromatography, C_{18} protocol ensures high resolution for closely sized protein components. However, it is possible to bypass the C_{18} protocol by performing trap-and-elute nano-LC-MS/MS [12].

1. Measure the volume of the sample in each tube and make a note of it.
2. Mix the sample with C_{18} sample buffer in a ratio of three parts sample to one part buffer. For example, if you have 180 μL of sample, add 60 μL of sample buffer.
3. Insert the C_{18} column into a 1.5-mL microcentrifuge tube. Tap the C_{18} column gently on the table to ensure that the matrix settles to the bottom of the tube. Then remove the bottom (red) cap and insert the column into the tube.
4. Add 200 μL of activation buffer to the column to clean the walls and wet the matrix.
5. Centrifuge the tube at 1500× *g* for 1 min. Remove the eluate from the tube.
6. Repeat the previous two steps for a total of two washes.
7. Add 200 μL of equilibration solution to the column.

8. Centrifuge columns at 1500× g for 1 min. Remove the eluate from the tube. Repeat this step for a total of two equilibrations.
9. Transfer the column to a new 1.5-mL microcentrifuge tube.
10. Add the sample and sample buffer with a maximum volume of 150 μL.
11. Centrifuge the microcentrifuge tube at 1500× g for 1 min. To ensure optimal peptide binding to the column, re-add the eluate and centrifuge using the same conditions. Remove the eluate from the tube and save it until the complete sample is processed for troubleshooting. If there is any remaining sample or sample buffer, add it to the column and repeat the procedure for this load.
12. Introduce 200 μL of C_{18} equilibration solution.
13. Centrifuge the microcentrifuge tube at 1500× g for 1 min. Remove the eluate from the tube.
14. Repeat the previous two steps three times, resulting in four washes.
15. Place the column into a 1.5-mL (protein low-binding) microcentrifuge tube.
16. Add 20 μL of elution buffer.
17. Centrifuge the tube at 1500× g for 1 min.
18. Repeat the previous two steps with a fresh 20 μL of elution buffer, resulting in 40 μL of eluted peptides.
19. Measure the peptide concentration at 280 nm using NanoDrop.
20. Perform vacuum centrifugation at 35 °C for approximately 15–45 h to completely dry peptides.
21. The samples can now be frozen at −80 °C as completely evaporated preparations.

3.2 High-Performance Liquid Chromatography-High Resolution Mass Spectrometry (HPLC/HR-MS)

Digested peptide samples are separated using a high-performance liquid chromatograph (HPLC) using a reversed-phase column, and eluted peptides are detected using high-resolution mass spectrometry (HR-MS). We have exemplified vendors for HPLC and MS instrumentation due to the availability of these instruments in our laboratory. However, similar equipment and instrumentation from other reputable vendors can also be used, and the choice of supplier may depend on availability and user preference.

3.2.1 HPLC for Peptide Separation

Before starting the HPLC, *see* **Notes 10**, **11**, and **12**. The HPLC separation of the peptides is conducted using the Vanquish Horizon HPLC system with an Acclaim Vanquish C_{18} analytical column (*see* **Note 10**).

Table 2
The gradient conditions for 80 min of HPLC separation are given here, where 0.1% (vol/vol) formic acid in high-purity water as mobile phase A and 0.1% formic acid (vol/vol) in acetonitrile as mobile phase B

Condition	Time (min)	Mobile phase A, %	Mobile phase B, %
Peptide elution	0.0	98	2
	62.0	65	35
Cleaning	62.1	5	95
	67.0	5	95
Equilibration	67.1	98	2
	80.0	98	2

1. Set the column temperature at 50 °C and the flow to 400 μL/min. Check in the instrument and select the correct bottles as mobile phase A of 0.1% (vol/vol) formic acid in high-purity water (18.2 MΩ × cm) and 0.1% formic acid (vol/vol) in acetonitrile as mobile phase B (*see* **Note 12**). The gradient conditions for 80 min of HPLC separation are given in Table 2.
2. Create a sequence and run it using Xcalibur™ software.
3. Choose a clear and understandable file name for the sample. For instance, 2021-04-20-yellow_mealworm.sdl, i.e., the file name should specify the date of analyses and insect species.
4. Inject solvent first without any matrix: 0.1% formic acid in water; water and acetonitrile 98:2, vol/vol.

Adjust the gradient length for peptide separation depending on the need of the research. For the identification of insect species, smaller gradients can provide sufficient information [3]. However, longer gradients are recommended to identify a maximum number of proteins [14].

3.2.2 Mass Spectrometry Analyses

Peptides eluted during the HPLC analyses are sent to high-resolution mass spectrometers such as Q-Exactive Orbitrap mass spectrometer systems.

1. Set up tune parameters on the mass spectrometer (reference parameters in Table 3).
2. For untargeted proteomics, measure MS2 in data-dependent full-MS/MS2 acquisition mode (full-MS/ddMS^2 TopN) and set parameters as given in Table 4.

3.2.3 Sample Injection into HPLC/HR-MS System

Dissolve insect-derived peptide samples (dried after C_{18} processing in Subheading 3.1.4) in a solvent that matches the initial gradient conditions of the HPLC system, i.e., 0.1% formic acid in water/acetonitrile (98:2, vol/vol).

Table 3
Tune parameters for Q-Exactive Orbitrap mass spectrometer systems are given. These parameters were tested for five insect species [3]

Parameter	Value
Source type	Electrospray ionization source (ESI)
ESI polarity	Positive
ESI spray voltage	3.5 kV
Sheath gas flow rate	40 AU
Auxiliary gas flow rate	10 AU
Capillary temperature	320 °C
Probe heater temperature	400 °C
S-lens RF level	50 AU

Table 4
Parameters for collecting MS2 in data-dependent full-MS/MS2 acquisition mode (full-MS/ddMS^2 TopN) are given along with the values

Parameter	Value
Polarity	Positive
Scan range (full-MS)	350–2000 m/z
Maximum injection time (full-MS)	50 ms
Resolution (full-MS)	70,000 (FWHM at 200 m/z)
Resolution (ddMS^2)	35,000 (FWHM at 200 m/z)
AGC target (full-MS)	3.0×10^6
AGC target (ddMS^2)	3.0×10^6
Loop count (ddMS^2)	10
Dynamic exclusion time (ddMS^2)	30 s
Normalized collision energy (ddMS^2)	32 NCE
TopN value (precursor count, ddMS^2)	10
Fixed first mass (ddMS^2)	ON, 140 m/z
Isolation window (ddMS^2)	1.2 m/z
Precursor intensity threshold (ddMS^2)	2.0×10^4

3.2.4 Sample Analyses on HPLC/HR-MS

1. Structure the sequence so that the initial and final injections comprise pure solvent without any matrix (blank) using TraceFinder™, Xcalibur™, and Chromeleon™ (*see* **Note 13**).

2. Inject the HeLa cell protein digest solution (quality control or QC) before running the sample sequence.
3. Ensure that the system is functioning optimally for consistent and reliable results.
4. Evaluate instrumental performance by analyzing HeLa sample spectra using the bioinformatics approach described in Subheading 3.3.1.
5. If the instrument's performance is acceptable, analyze the remaining samples and collect tandem mass spectra (*see* **Note 14**).

3.3 Bioinformatic Analyses

Protein sequences and proteomes are abundant and readily accessible for model organisms. However, when it comes to non-model species, the protein sequence databases are relatively limited in size, which presents challenges to existing workflows [6]. For example, more protein sequences are available for the fruit fly (*Drosophila melanogaster*) than commercially essential insect species such as black soldier fly larvae (*Hermetia illucens*) and other species, as shown in Table 5. One approach to mitigate the scarcity of data is to perform protein identification through homology searches in conventional database searches using the proteome of closely related species such as fruit fly (*Drosophila melanogaster*). Typically, the proteome of a model species or even an entire class, such as Arthropoda, can be used to identify many proteins in insect samples

Table 5
Comparison of reviewed and unreviewed sequences from UniProt KB database in model organism fruit fly (*Drosophila melanogaster*) and non-model insect species such as black soldier fly (*Hermetia illucens*), yellow mealworm (*Tenebrio molitor*), lesser mealworm (*Alphitobius diaperinus*), house cricket (*Acheta domesticus*), and morio worm (*Zophobas morio).* More reviewed sequences were present in model insect species compared to non-model insect species

Organisms (insects)	Scientific name	Number reviewed sequences from UniProt KB	Number unreviewed sequences from UniProt KB
Fruit fly	*Drosophila melanogaster*	4926	364,541
Black soldier fly	*Hermetia illucens*	1	17,698
Yellow mealworm	*Tenebrio molitor*	31	19,356
Lesser mealworm	*Alphitobius diaperinus*	0	25
House cricket	*Acheta domesticus*	235	466
Morio worm	*Zophobas morio*	0	6

Modified from and reproduced with permission of © Varunjikar 2023 [18]

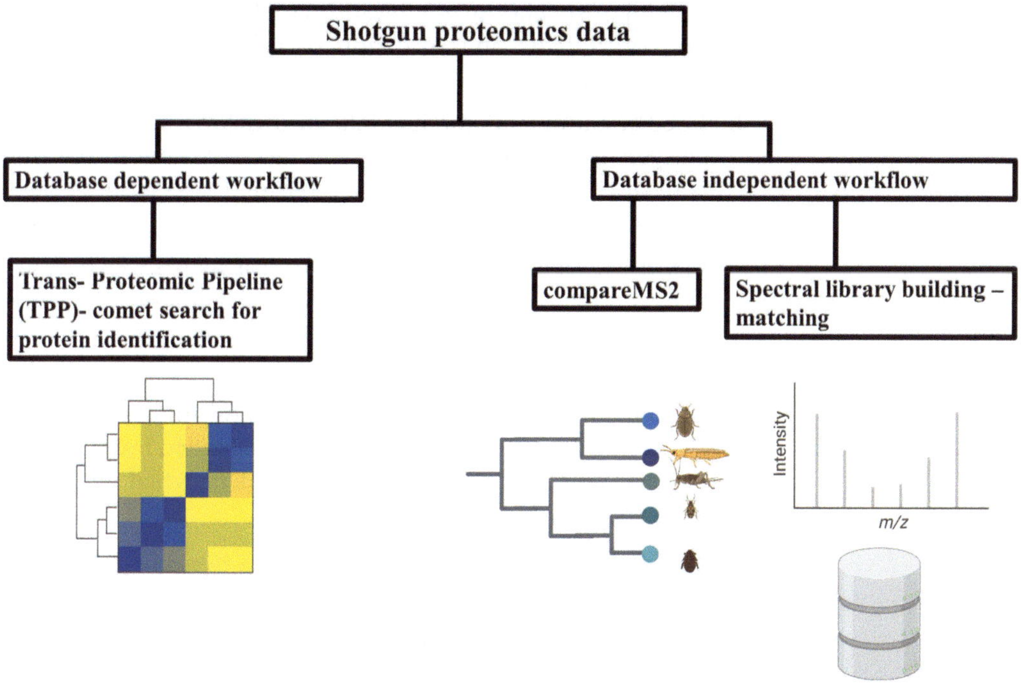

Fig. 2 Proteomic bioinformatics workflows for insect proteomics. If sequence information is available for species, protein identification can be performed, and if the protein sequence information is not available, it is possible to implement compareMS2 and spectra library-based approach. (Modified from and reproduced with permission of © Varunjikar 2023 [18]. Created with BioRender.com)

[3, 14]. An alternative approach to proteomics analysis involves utilizing database-independent methods, as shown in Fig. 2. For instance, direct spectral comparison of tandem mass spectra using tools like compareMS2 or employing a spectral library-based approach can be implemented. These methods provide valuable options for analyzing proteomics data without relying on traditional database searches [3–5].

3.3.1 Protein Identification

1. Obtain .raw files from the mass spectrometer and convert the .raw files to the .mzML format using msConvert (version: 3.0., ProteoWizard) [9].
2. Download the protein sequence database for insect species of interest using the NCBI or UniProt database.
3. Match tandem mass spectra (.mzML format) against protein sequences (.fasta format) (*see* **Note 15**) using the open-source Comet search engine implemented in the Trans-Proteomics Pipeline (TPP) (version 5.2.0) [15].
4. In a Comet search, set the precursor mass tolerances to 4.5 ppm and fragment ion tolerances to 20 ppm and choose trypsin as the digestive enzyme, allowing for two nonenzymatic

termini. Also, choose modifications such as carbamidomethylation of cysteine and oxidation of methionine as fixed and variable modifications, respectively.

5. Use the output of search results or .pepXML files for further analyses using PeptideProphet and ProteinProphet, applying a false discovery rate (FDR) threshold of 1% to ensure reliable identification of peptides and proteins [16].
6. Post-process your output using statistical analyses such as analysis of variance (ANOVA) or principal component analyses (PCA) based on differentially expressed proteins across different experimental groups.

Implement a similar bioinformatic workflow for detecting allergens wherein allergen sequences can be downloaded from NCBI or UniProt (**step 2** in Subheading 3.3.1) and matched against the collected spectra [3].

1. Use mzML files converted in **step 1** in Subheading 3.3.1.
2. Choose targeted allergens for insect species using (www.allergen.org) and download allergen sequences from NCBI or UniProt to create an allergen database.
3. Match of tandem mass spectra (.mzML format) against protein sequences (.fasta format) using the Comet search engine implemented in Trans-Proteomics Pipeline (TPP) (version 5.2.0).
4. Process .pepxml files generated by the Comet search further using PeptideProphet and ProteinProphet, applying a false discovery rate (FDR) threshold of 1% to ensure reliable identification of peptides and proteins.
5. Use the resultant list of proteins and peptides to determine known allergens in the sample.

3.3.2 Direct Spectral Comparison (compareMS2) and Spectral Library (SpectraST)

For molecular phylogenetic studies, compareMS2 tools are implemented [17].

1. Convert spectra files into .mgf format using msConvert (version: 3.0., ProteoWizard) [9].
2. Use compareMS2 tool GUI [8] for generating molecular phylogeny of insect species.

To identify species using spectral libraries, for each species in the set, generate spectral libraries using tandem mass spectra in . mzML or files generated after the Comet search pepXML files [7]. Create spectral libraries from mass spectrometry data of known insect species (reference libraries) and subsequently use them to identify of data obtained from unknown samples.

1. Using SpetraST (version 5.0), create spectral libraries using tandem mass spectra .mzML files generated from reference

samples or files generated after the Comet search .pepXML files [3]. For detailed information about creating libraries for unidentified spectra, follow the protocol by Onder et al. [7].

2. Match unknown samples to the created library and count matching spectra with a dot product above 0.8, representing a valid spectra match. The spectra library against which the highest number of matching spectra (dot product above 0.8) are acquired can be used to determine the identity of samples [3].

4 Notes

1. When determining protein concentration, it is essential to consider specific criteria when using the Pierce™ 660 protein assay kit. For instance, adjustments should be made for high concentrations of various compounds. The lysis buffer in this method contains a high SDS content, requiring the addition of an ionic detergent compatibility reagent.
2. In the FASP protocol, IAA solution is light sensitive and must be prepared freshly in the dark.
3. The filter capacity is stated as 0.2–200 μg (maximum 300 μg) in the original protocol [13]. Adjust the protein volume as necessary in the FASP protocol. However, if the volume exceeds 25 μL, additional urea solution with DTT (more than 100 μL) must be added during **step 5** of the protocol on day 1 to compensate for the higher SDS input. This step reduces the SDS concentration to 0.5% (weight/volume), making it compatible with the filter. If the sample volume exceeds 25 μL, add more urea solution accordingly. For larger volumes, such as 30 μL (maximum 50 μL), mix with 200 μL of urea solution (maximum 350 μL). Dilution of SDS is essential as Microcon Centrifugal Filters are sensitive to higher SDS concentrations.
4. At the start of the FASP protocol, check the Microcon Centrifugal Filter; approximately one-third of the volume should pass through the filter, and if a more considerable amount has passed after **step 4** in Subheading 3.2.2. FASP Day 1, first centrifugation, replace the filter with a new one.
5. In the FASP protocol, ensure that the centrifuge speed does not exceed 11,000× *g* when utilizing the FASP filter.
6. Always perform centrifugation in the FASP protocol with the filter cover securely closed and at room temperature.
7. Prevent the FASP filter from drying out during centrifugation, and discontinue centrifugation once approximately 50 μL of residue remains in the filter as the urea solution might crystalize and block the filter.

8. Centrifugation time for the FASP protocol varies depending on the sample. Insect samples rich in fat might take longer to pass through the FASP filters.
9. Remove the solution passed through filters when collection tubes fill beyond recommended levels during urea or urea DTT solution washes.
10. It is essential to purge the HPLC pump at a high flow rate (e.g., 5 mL/min) in two scenarios: when the instrument has been inactive or when the previous mobile phase composition differs from the following setup. This step helps eliminate any air trapped in the pump head and ensures that any residual mobile phase from previous runs is effectively flushed out.
11. It is crucial to ensure compatibility between the syringe wash solution and the injection phase of the sample. This ensures that the syringe wash solution does not interfere with the sample and maintains the integrity of the analysis.
12. Ensure that the remaining volume of solution in bottles A and B is adequate for the duration of the analytical run. It is crucial to verify enough solutions to complete the analysis without interruption. Verify that the volume of the rear seal wash solution is sufficient for the entire analytical run. This will contribute to a smooth operation of the instrument and not generate a need for additional refill of solutions during the analysis.
13. Before running a sample injection sequence on the mass spectrometer, it is essential to verify the validity of accurate mass calibration of the instrument. It is crucial to wait 20–30 min before starting the process. This waiting time is necessary as the mass spectrometer requires a warm-up after loading the tune file, enabling solutions to evaporate effectively. If the mass spectrometer is not sufficiently heated, there is a risk of contaminating the capillary within the MS.
14. Maintaining the system in optimal condition and performing proper calibration is essential.
15. In the case of insect species, the data can be matched against protein sequences from the specific species of interest. However, if an adequate number of sequences are unavailable, an alternative is to align the data against protein sequences from the *Arthropoda* species or the proteome of a fruit fly.

Acknowledgments

The present work was financed by the Norwegian Research Council (NFR) project "ClimSeaFood" (15804) and the "MultiOmicsTools" project (Ministry of Trade, Industry and Fisheries,

15470, IMR). We are thankful to Dr. Ingus Perkons (EU-FORA fellow) for helping in drafting parts of the in-house mass spectrometry workflow. All figures here are modified or generated using Bio render (www.biorender.com).

Declaration of Interest The author Dr. J.D. Rasinger currently is employed with the European Food Safety Authority (EFSA) in the Food Ingredients and Packaging Unit (FIP). However, the present chapter is published under the sole responsibility of the author Dr. J.D. Rasinger and may not be considered as an EFSA scientific output. The positions and opinions presented in this article are those of the author alone and are not intended to/do not necessarily represent the views/any official position of EFSA. The author's main contribution to this chapter was made before joining EFSA when still employed at the Institute of Marine Research.

References

1. Olsen JV, Ong SE, Mann M (2004) Trypsin cleaves exclusively C-terminal to arginine and lysine residues. Mol Cell Proteomics 3:608–614
2. Duivesteijn SJ (2018) Advancing zebrafish models in proteomics. https://hdl.handle.net/1887/66790
3. Varunjikar MS, Belghit I, Gjerde J et al (2022) Shotgun proteomics approaches for authentication, biological analyses, and allergen detection in feed and food-grade insect species. Food Control 137:108888
4. Belghit I, Varunjikar M, Lecrenier M-CC et al (2021) Future feed control—tracing banned bovine material in insect meal. Food Control 128:108183
5. Belghit I, Lock EJ, Fumière O et al (2019) Species-specific discrimination of insect meals for aquafeeds by direct comparison of tandem mass spectra. Animals (Basel) 9(5):222
6. Heck M, Neely BA (2020) Proteomics in non-model organisms: a new analytical frontier. J Proteome Res 19:3595–3606
7. Önder Ö, Shao W, Lam H et al (2014) Tracking the sources of blood meals of parasitic arthropods using shotgun proteomics and unidentified tandem mass spectral libraries. Nat Protoc 9:842–850
8. Marissen R, Varunjikar MS, Laros JFJ et al (2023) compareMS2 2.0: an improved software for comparing tandem mass spectrometry datasets. J Proteome Res 22:514–519
9. Kessner D, Chambers M, Burke R et al (2008) ProteoWizard: Open source software for rapid proteomics tools development. Bioinformatics 24:2534–2536
10. Lam H, Deutsch EW, Eddes JS et al (2008) Building consensus spectral libraries for peptide identification in proteomics. Nat Methods 5:873–875
11. Deutsch EW, Mendoza L, Shteynberg D et al (2015) Trans-proteomic pipeline, a standardized data processing pipeline for large-scale reproducible proteomics informatics. Proteomics Clin Appl 9:745–754
12. Bai H, Collins LB, André MR et al (2023) A bottom-up proteomics workflow for a system containing multiple organisms. Rapid Commun Mass Spectrom 38(Suppl 1):e9431
13. Wiśniewski JR (2016) Quantitative evaluation of filter aided sample preparation (FASP) and multienzyme digestion FASP protocols. Anal Chem 88:5438–5443
14. Francis F, Mazzucchelli G, Baiwir D et al (2020) Proteomics based approach for edible insect fingerprinting in novel food: differential efficiency according to selected model species. Food Control 112:107135
15. Deutsch EW, Lam H, Aebersold R (2008) Data analysis and bioinformatics tools for tandem mass spectrometry in proteomics. Physiol Genomics 33:18–25
16. Keller A, Nesvizhskii AI, Kolker E et al (2002) Empirical statistical model to estimate the accuracy of peptide identifications made by

MS/MS and database search. Anal Chem 74: 5383–5392

17. Palmblad M, Deelder AM (2012) Molecular phylogenetics by direct comparison of tandem mass spectra. Rapid Commun Mass Spectrom 26:728–732
18. Varunjikar MS (2023) Proteomic tools for food and feed authentication. University of Bergen. https://hdl.handle.net/11250/3053481

Chapter 8

Insect Brain Proteomics: A Case Study of *Periplaneta americana*

Jayaprada R. Chunduri and Siddharth P. Sagar

Abstract

Insects are known invertebrate species with economic, ecological, pathological, and medicinal value, as well as closely associated with human populations. Entomophagy and entomotherapy are future promising prospects largely attributable to the abundant availability, high protein content, and climatic sustainability of insects. In particular, the insect brain is an important system with a secluded, compact, and protective exoskeleton. It is immunologically privileged and capable of producing a robust immune response against pathogens. It is also a source of materials that initiate key activity throughout the body. Proteomic interrogation of *Periplaneta americana* enables understanding the role of this insect in the fields of food and pharmacology. Proximate analyses of *P. americana* highlights its richness in proteins. Here we perform a simple proteomic analysis to study the brain proteome of *P. americana*. The processes applied during the study include gel-based isolation and separation of proteins, followed by NanoLC-MS (Orbitrap) analyses and bioinformatic interrogation of the data. The results demonstrated that this insect proteome comprises antimicrobial proteins, allergens, and proteins required for metabolic processes.

Key words Insect protein, Immune response, Isolation and separation, nanoLC-MS, Bioinformatics

1 Introduction

Insects play a key role as decomposers, pollinators (to maintain ecological balance), and alternate food sources and also in medicinal applications. They are a viable food source in many parts of the world due to the high protein content (35–65% of total dry matter) and rich source of essential amino acids [1–3]. However, insect consumption raises parallel problems as most of the insects are pathogen vectors, exposed to insecticides, and may comprise diverse allergens.

Entomotherapy is an alternate therapeutic option practiced in many parts of the world. Developing and developed countries including India, Mexico, Korea, China, Spain, Brazil, Argentina, Ecuador, and various African countries [4] consider insects to have medicinal value. Folk medicine applies insect secretions, ground-up

Taufika Islam Williams (ed.), *Tissue Proteomics: Methods and Protocols*, Methods in Molecular Biology, vol. 2884, https://doi.org/10.1007/978-1-0716-4298-6_8,

bodies, or *Pseudomyrmex* ant venom [5] for treatment purposes. The northeastern states of India (Arunachal Pradesh, Meghalaya, Nagaland, and Chhattisgarh) use insects in traditional Indian medicine to heal wounds and treat dog bites, asthma, liver disorders, urinogenital disorders, ulcers, fevers, tuberculosis, and nervous breakdown. Insects belonging to Diptera, Lepidoptera, Coleoptera, Hymenoptera, and Orthoptera taxonomic orders are applied for medicinal purposes [6]. Insect natural products of these organisms have been found to exhibit antimicrobial, antifungal, antiviral, anticancer, antioxidant, anti-inflammatory, and immunomodulatory effects [4]. Although insect species outnumber plant species by almost 16 times, scientific investigations on insects are 7000 times fewer than those of plants [7]. Proteomics studies of these organisms can be of great value and importance in the current scenario of drug development.

Insects can also be a cause of concern as vectors that spread diseases and pests [8]. Environmental (biotic and abiotic) factors and inhospitable conditions can impact insect metabolic activities and induce strong immune systems. The cellular and humoral components such as antimicrobial peptides (AMPs) thus produced are of interest [9, 10]. Insect AMPs comprise defensin, cecropin, attacin, lebocin, and other proline-rich peptides including gloverin and moricin. Structural-functional relationships as well as the potential applications of insect AMPs from Dipteran, Lepidopteran, Isopteran, Coleopteran, and Hymenopteran orders are well studied [11].

Previous studies on the American cockroach indicated proximate composition of crude protein (8.72 ± 1.3%), crude fat (17.64 ± 1.9%), along with carbohydrates and richness of major and minor minerals (Ca, Mg, Fw, K, P, Na, Mn, Cu, Zn) required for human health [12, 13]. American cockroach powder is considered as potential ruminant feed [14] and also exhibited pharmacological attributes of inhibitory or controlling activities against pain, viral infections, tumors, and inflammation, as well as improving immunity and tissue repair [15]. Antibacterial compounds extracted and isolated from cockroaches were found to exhibit homology with isoquinoline, thiazine groups, chromene derivatives, imidazoles, sulfonamides, furanones, flavanones, and analogs containing pyrroles. These compounds are known to exhibit broad-spectrum antimicrobial, anti-inflammatory, antitumor, and pain relief properties [16]. Three antibacterial proteins isolated from the hemolymph of *Blatta orientalis* exhibited potential antibacterial activity against *E. coli*, *S. aureus*, and *Salmonella typhi* [17].

Several enzymatic peptides have potential as antibiotics against drug-resistant pathogenic bacteria [18]. Some insect proteins contain enzymatic agents or enzybiotics. The EnzyBase database is a comprehensive repository for enzybiotics [19] that helps to identify proteins (enzymes) with anti-microbicidal potential. Peptides from the giant silk moth (*Hyalophora cecropia*) pupal hemolymph challenged the survival of live bacteria [20]. More than 150 antibacterial peptides have been found from diverse insect species. DLP2 and DLP4 have been identified in *Hermetia allicins* as effective in controlling methicillin-resistant *Staphylococcus aureus* (MRSA) [21]. Studies have been conducted on antimicrobial peptides from locusts, American cockroaches, grasshoppers, and fruit flies [22–25]. Protein molecules that perform multiple functions are known as moonlighting proteins [26, 27]. Such proteins [i.e., transferrin and glyceraldehyde 3-phosphate dehydrogenase (GAPDH)] may be effectively used in cancer treatment [28].

Isolation, identification, and characterization of animal protein are a multistep process. The exact methodology is selected depending on the availability of material and type of tissue. The approach of gel-based techniques (1D PAGE) was considered over solution-based approaches such as FASP or SP3 followed by mass spectrometry (Q-orbitrap), bioinformatics, and the antibacterial assessment of isolated proteins (Fig. 1). The reason behind selection of the gel-based approach was with an intent to minimize the cost and availability of equipment and chemicals in a developing country as well as to avoid structural damage and maintain the integrity of protein activity.

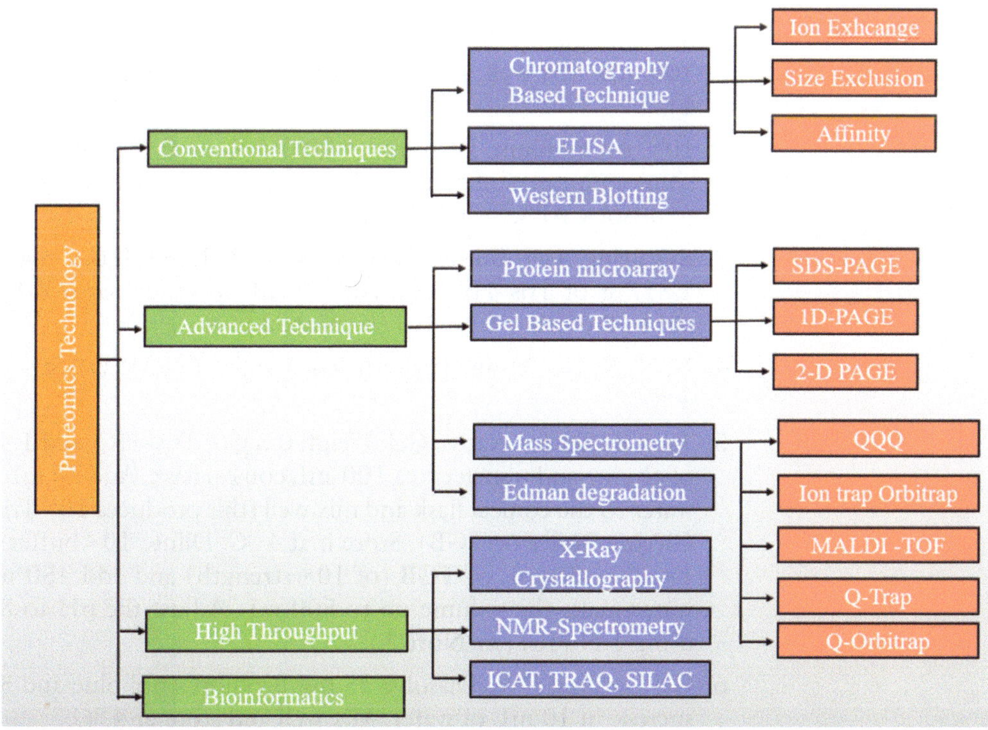

Fig. 1 Proteomics analysis methods

2 Materials

Sterile distilled water was used in the entire protocol. All solutions were prepared using high-performance liquid chromatography (HPLC)-grade solvents.

2.1 Protein Extraction

1. Phosphate-Buffered Saline (PBS): Weigh 0.8 g of sodium chloride, 0.144 g of disodium phosphate, 0.024 g of potassium dihydrogen phosphate, and 0.02 g of potassium chloride and transfer them to a 250-mL glass cylinder containing 100 mL of water. Mix well and adjust pH to 7.4 with 1 N sodium hydroxide. Store at room temperature.
2. Biuret Reagent: Weigh 0.15 g of copper sulfate and 0.6 g of potassium sodium tartrate and transfer them to a 250-mL conical flask. To this add 50 mL of water and allow it to stabilize (Solution A). Dissolve 8 g of NaOH in 100-mL water to prepare 2 M NaOH solution (Solution B). Add 37.5 mL of 2 M sodium hydroxide (Solution B) to 50 mL of Solution A in the conical flask and mix well. Bring the total volume up to 100 mL with water.
3. Glass beads.

2.2 Native Gel

1. 40% Acrylamide/Bisacrylamide Solution (40% BAA): Weigh 37.5 g of acrylamide and 1 g of bisacrylamide. Transfer to a 500-mL glass cylinder and add water to a final volume of 100 mL. Mix well, plug, and store at 4 °C until use (*see* **Note 1**).
2. 10% Ammonium Persulfate (APS) Solution: Weigh 1 g of ammonium persulfate and add to 10 mL of water. Stir gently to dissolve APS. Store it in fridge until further use (*see* **Note 1**).
3. Resolving Gel Buffer (1.5 M Tris–HCl pH 8.6.): Weigh 1.817 g of Tris–HCl and add 10 mL of water to it. Adjust pH to 8.6 using 1 N HCl (*see* **Note 1**).
4. N, N, N, N′-Tetramethyl-ethylenediamine (TEMED): Store at 4 °C.
5. Tank Buffer for Native Gel: Weigh 0.3 g of Tris–HCl and 1.4 g of glycine and transfer to a 100-mL conical flask. Add 50 mL of water to the conical flask and mix well (this produces 10× Tris/Glycine Buffer or TGB). Store it at 4 °C. Dilute 10× buffer to 1×. Take 50 mL of TGB (of 10× strength) and add 450 mL water make the volume up to 500 mL. Adjust the pH to 8.6 using 1 N HCl (*see* **Note 1**).
6. Gel Loading Dye: Dissolve 25 mg Bromophenol blue and 5 g sucrose in 10 mL of water. Mix well and store at 4 °C.

7. Staining Solution: Add 45 mL of methanol, 10 mL of glacial acetic acid, and 45 mL of water in a 250-mL glass measuring cylinder. Weigh 0.25 g of Coomassie Brilliant Blue R250 and add to the above mixture, mixing well. Store in an airtight glass container at room temperature.
8. Destaining Solution: Measure 45 mL of methanol, 10 mL of glacial acetic acid, and 45 mL of water using a graduated cylinder. Mix well and store at room temperature in an airtight glass container.
9. Protein Elution Buffer (0.25 M Tris–HCl, pH 6.8): Dissolve 0.3 g of Tris–HCl in 10 mL of water and adjust pH with 1 N HCl (*see* **Note 1**).

2.3 Antimicrobial Sensitivity Test

1. Muller Hilton (MH) Agar: Weigh 150 g of beef infusion, 0.75 g of starch, 8.75 g of casein, and 8.5 g of agar and mix well. Transfer it to a 1000-mL capacity conical flask. Add 500 mL of water, mix well, and adjust pH to 7.3 ± 0.2. Steam sterilize at 15 psi for 15 min and store at room temperature (*see* **Note 2**).
2. Sterile Whatman Filter Paper No. 10 discs of 5-mm diameter (*see* **Note 3**).
3. Sterile Saline: Mix 0.85 g of NaCl in 100 mL of water and steam sterilize it at 15 psi for 15 min.

2.4 Proteomics Sample Preparation for Liquid Chromatography-Mass Spectrometry (LC-MS)

A. *Protein Extraction from Tissue Lysate.*

1. Protein Extraction Buffer (for protein extraction from tissue): Take 7 M (4.29 g) of urea, 2 M (1.4 g) of thiourea, and 1 M (0.799 g) of ammonium bicarbonate (ABC) and transfer to 10 mL of water. Mix it well (*see* **Note 1**).

B. *In-Gel protein extraction.*

Use all analytical-grade chemicals for the preparation of reagents (*see* **Note 4**).

1. Gel Dehydrating Solution: A mixture of acetonitrile (ACN) and 50 mM of ABC in the ratio of 2:1 (50 mM ABC can be prepared by adding 0.4 g of ABC in 10 mL of water).
2. Gel Rehydrating Solution: 10 mM of dithiothreitol (DTT). Weigh 1.54 g of DTT and dissolve in 10 mL of water.
3. Digestion Buffer (25 mM ABC): Weigh 0.2 g of ABC and add to 10 mL of water. Vortex.
4. Trypsin: Dissolve 20 μg of trypsin in 1 mL of 25 mM of ABC.
5. 50 mM Iodoacetamide (IAA): Dissolve 10 mg of IAA in 1 mL of 100 mM ABC in a microcentrifuge tube.

6. 0.1% Trifluoroacetic Acid (TFA): Add 100 μL of TFA to 99.9 mL of water. Mix well.
7. Extraction Solutions:
   1. 50% ACN in 0.1% TFA: 1 mL ACN + 200 μL 1% TFA + 800 μL water.
   2. 60% ACN in 0.1% TFA: 1.2 mL ACN + 200 μL 1% TFA + 600 μL water.
   3. 80% ACN in 0.1% TFA: 1.6 mL ACN + 200 μL 1% TFA + 200 μL water.
8. Elution Buffer for Zip Tip: (50% ACN in 0.1% TFA). Mix 300 μL ACN in 240 μL water and 60 μL of 1% TFA (in water).

2.5 Liquid Chromatography and Mass Spectrometry

1. Solvent A: 0.1% formic acid (FA) in water.
2. Solvent B: Prepare a solution of acetonitrile (ACN) and 0.1% formic acid in the ratio of 80:20 (*see* **Note 4**).

2.6 Bioinformatics

1. Structure modelling.
   1. SWISS MODEL: https://swissmodel.expasy.org/interactive.
   2. RaptorX: http://raptorx6.uchicago.edu/ContactMap/.
2. Protein Docking.
   1. PDBsum: https://www.ebi.ac.uk/thornton-srv/databases/pdbsum/Generate.html.
   2. HawkDock: http://cadd.zju.edu.cn/hawkdock/.
3. Protein Characterization.
   1. Antimicrobial Peptide—collection of antimicrobial proteins (CAMP): http://www.camp.bicnirrh.res.in/predict/.
   2. Antiviral Predictor Meta-iAVP - http://codes.bio/meta-iavp/.
   3. Anticancerous—iACP - http://lin.uestc.edu.cn/server/iACP.

3 Methods

All procedures can be carried out at room temperature, under sterile conditions (unless otherwise stated). The water used in every procedure is sterile distilled water.

3.1 Protein Extraction for Antibacterial Sensitivity Test (AST)

1. Immobilize the adult cockroach by suffocating in 1% detergent water for 5 min. Wash the cockroach with water twice to remove the soap solution. Wipe it with sterile tissue paper. Immerse the cockroach in 70% methanol solution for surface sterilization. Rinse with water and wipe.
2. Fix the cockroach on a wax plate and dissect the head. Remove the sclerites carefully, extract the brain tissue, and immediately transfer it to a microcentrifuge tube containing 1 mL of PBS solution.
   1. Add 7–8 glass beads and homogenize the tissue by vortexing at 250 rpm for 10 min.
   2. Centrifuge the specimen homogenate at 8000 rpm for 15 min. Collect the supernatant in another tube labeled "BT" (brain tissue lysate).
   3. Add 500 μL of PBS to the tissue homogenate pellet. Vortex the microcentrifuge tube again at 250 rpm for 10 min, followed by centrifugation at 8000 rpm for 15 min. Collect the supernatant and add to the microcentrifuge tube labeled "BT."
   4. Repeat above steps thrice to concentrate the protein.
3. Verify the presence of protein in the supernatant by testing 10 μL of lysate with 30 μL of Biuret Reagent and check for color change. The change of color to purple indicates the presence of protein.

3.2 Native Gel

1. Prepare a gel by mixing 4.4 mL of water, 3 mL of 40% of BAA, 2.5 mL of Tris–HCl buffer (pH 8.6), 100 μL of APS, 10 μL of TEMED [29] and cast the gel in a casting cassette. Add a 7–10 well gel comb, careful to not introduce air bubbles. Allow the gel to set at room temperature (*see* **Note 5**).
2. Sample preparation: Take 70 μL of protein extract from the tissue lysate and mix it with 30 μL of Gel Loading Dye. Expose the sample to 50 °C for 5–10 s (*see* **Note 6**).
3. Add 25 μL of sample to each well. Add tank buffer in both upper and lower tanks.
4. Connect the electrophoresis setup with the negative electrode toward the upper tank (well) and the positive electrode on the opposite end (lower tank). Separate proteins at a constant 50 mV until the loading dye runs down to 3/4th of the gel height [29] (*see* **Note 7**).
5. Carefully dismantle the cassette and remove gel. Immediately stain it with staining solution for 15 min and then and destain for 30 min (Fig. 2).

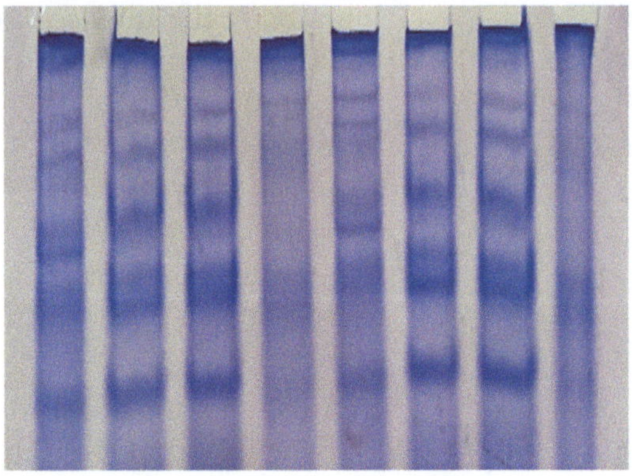

Fig. 2 Protein isolated from *Periplaneta* brain lysate sample

6. Cut out bands of protein with a blade and transfer each band to a separate microcentrifuge tube containing 1 mL of elution buffer. Centrifuge at 12,000 rpm for 20 min. Remove the supernatant and repeat the step thrice (add 1 mL of elution buffer to the bands, mix well, and then centrifuge) [30].
7. Check for the presence of protein using the Biuret Reagent. Change in color to purple indicates the presence of protein in the sample.

3.3 Antimicrobial Sensitivity Test: Disc Method

1. Pour 20 mL of MH agar into a sterile, glass petri dish. Allow it to solidify at room temperature (*see* **Note 8**).
2. Take a 0.1 mL of 24-h cultured bacteria and add 10 mL of sterile saline in a sterile test tube. Check the absorbance at 600 nm with a spectrophotometer and set the culture suspension's absorbance to 0.1.
3. Swab 0.1 mL of bacterial suspension onto a solidified MH agar plate.
4. Prepare 10 μL of protein sample discs of 5-mm diameter with Whatman's Filter Paper and place carefully on the swabbed plate. Incubate the plates at 37 °C for 24 h. Check the antimicrobial activity of protein by observing the clearance zone around the disc (zone of inhibition) (Fig. 3) [31].

3.4 Proteomic Sample Preparation for Liquid Chromatography-Mass Spectrometry (LC-MS)

Proteomics of brain lysate was performed for protein profiling. Both in-solution and in-gel digestion were considered in sample preparation for proteomics studies by LC-MS.

1. In-solution digestion (ISD) on brain tissue lysate.
    1. Mince and homogenize the brain tissue collected from five adult cockroaches.

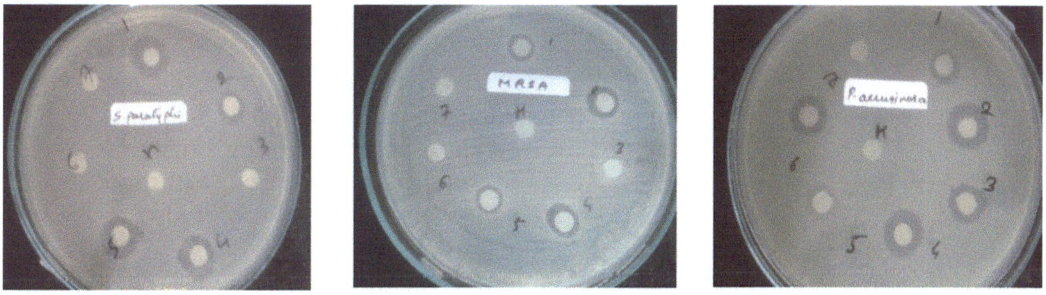

A B C

Fig. 3 Antibacterial activity of insect proteins on test bacteria: (**a**) *Salmonella paratyphi*, (**b**) Methicillin resistant *Staphylococcus aureus*, (**c**) *Pseudomonas aeruginosa*

   2. Transfer the homogenate into a 2-mL microcentrifuge tube containing 160 μL of freshly prepared protein extraction buffer.
   3. Heat at 95 °C for 20 min and then at 60 °C for 1.5 h. Digest the protein using reconstituted proteomics-grade trypsin [enzyme/protein ratio of 1:20 (w/w)] for 8–10 h at 37 °C [32].

2. In-gel protein digestion and extraction.
   1. Rinse two sets of 500 μL capacity snap cap microcentrifuge tubes prior to use with 2× 500 μL methanol and 500 μL water. Cut out the protein bands with a razor and place each band into a separate microcentrifuge tube.
   2. Rinsing: Rinse the gel pieces with 100 μL of 25 mM ABC solution.
   3. Washing and Dehydration: Dehydrate gel slices with 100 μL of Solution A (2:1 mixture of ACN/50 mM ABC) for 5 min. Remove the supernatant and add 100 μL of 25 mM ABC for 5 min. Repeat this step twice. The last washing should always be with Solution A for 5 min.
   4. Reduction/Alkylation (optional: perform this step if protein contains disulfide bonds): After the previous step (2–3 cycles of dehydration/rehydration), rehydrate the gel slice in 50 μL of 10 mM DTT. Incubate the gel slices at 56 °C for 1 h. Remove DTT and rinse the gel slices with 100 μL of 25 mM ABC. Add 50 μL of 100 mM IAA. Incubate the gel slices at room temperature in the dark for 30 min. Remove IAA and rinse the gel slices with 100 μl ABC.
   5. Dehydration: Dehydrate the gel slices with 100 μL of Solution A for 5 min. Remove the supernatant and add 100 μL of 25 mM ABC for 25 min (optional: repeat dehydration and rehydration one additional time). Remove the final supernatant and dry the gel slices in a speed vacuum (5–15 min) or heat dry (temperature note more than 40 ° C) for 3–5 min.

6. Digestion: Rehydrate the gel slice with trypsin, (about 10–20 μL) at 0 °C on ice. Incubate the gel slice on ice for 20–30 min until trypsin is absorbed. After the gel slices are completely rehydrated, add just enough (~50 μL) 25 mM ABC to cover gel slices in the microcentrifuge tubes. Incubate at 37 °C overnight (16 h). Periodically check the gel slices to make sure they are still just covered with 25 mM ABC. Add more if needed.
7. Extraction: Remove the supernatant and vortex the gel slices with 100 μL extraction buffer (50% ACN and 0.1% TFA) for 10 min and pool into a single microcentrifuge tube. Once again add 100 μL extraction solution (60% ACN and 0.1% TFA), sonicate for 1 min (2-s pulse, 1-s gap, 25% amplitude), vortex for 10 min, and add the supernatant to the previously pooled sample. Repeat the extraction step with 80% ACN containing 0.1% TFA, vortex for 10 min, and collect the supernatant. Dry the samples with a speed vacuum until the volume is reduced to 10–20 μL. This solution is used for LC-MS analysis after ZIP Tip cleanup [33].
8. Zip Tip C18 Procedure (*see* **Note 9**): Aspirate 10 μL of wetting solution into the tip and dispense to waste. Repeat this step thrice. Bind peptides to the Zip Tip pipette tip by aspirating and dispensing over 3–7 cycles for simple mixtures and up to 10 cycles for complex samples. Aspirate 0.1% TFA and dispense to waste. Repeat the wash five times (*see* **Note 10**). Dispense 3–5 μL of elution solution into a clean 0.5-mL microcentrifuge tube.
9. Using a standard pipette tip, aspirate and dispense eluant through the Zip Tip at least 8–10 times without introducing air. Sample recovery can be improved by increasing the elution volume to 10 μL (though at the expense of concentration).

3.5 Liquid Chromatography and Mass Spectrometry

1. Chromatography: The Thermo EASY nanoLC-1200 instrument is used for chromatographic separation in this method, though any nanoLC instrument would suffice. Maintain the sample flow at 4.0 μL/min. The total volume in pre-column equilibration must be 10 μL and that in analytical column equilibration must be 3 μL. Maintain the total mobile phase flow at 300 nL/min for 60 min. The nanoLC gradient will ramp from 5% MPB at the beginning to 10% MPB after 5 min, 50% MPB after 50 min, and finally 95% MPB after 55 min and 60 min [33].
2. Mass Spectrometry: A Thermo Scientific Q-Exactive Plus Orbitrap MS instrument is used for protein identification, though any high-resolution mass spectrometer (HRMS) with tandem

MS (MS2) capabilities would be appropriate. Run the samples for 60 min with a scan range of 350 *m/z* to 2000 *m/z* and a constant flow rate of 300 nL/min [33].

3. Analyze the proteomic data using a Thermo Scientific Proteome Discoverer (PD) 2.2 or other proteomic software. Consider the peak with highest intensity and confidence. X correlation values >2.0 usually indicate good reliability. Use PDB and UniProt databases for peptide/protein identification. Proteins associated with the cockroach brain lysate were identified based on the PDB/insect database (Table 1) [34].

3.6 Bioinformatics

1. SWISS MODEL and RaptorX: Check the accession number of identified proteins (from gel bands) and copy the FASTA sequence of the proteins. Paste the sequence into the slot provided on the SWISS MODEL server. Select the template with QMEAN4 score value close to 1. Select the command to model the respective protein template. Download the PDB file of the selected model. If QMEAN4 values are not close to 1, opt for RaptorX for homology modelling in which uGDT value should be more than 350 (Figs. 4 and 5) [35].
2. Protein Docking: Submit the PDB files of ligand (insect protein) and receptor (membrane protein of bacteria) to the HawkDock server for protein-protein interaction analysis. Check for the best interaction between both the proteins with highest scores among the given top ten models. Check the binding energy of the best model by selecting the command for MM/GBSA analysis. The lower the binding energy, the greater the reliability of the result. Download the PDB file (Fig. 6) [36, 37].
3. Submit the PDB file of docked protein interactions to the PDBsum server (with a valid email address). Click on the link provided by the server and check for detailed analyses of protein-protein interactions. The number of salt bridges, hydrogen bonds, disulfide bonds, and nonbonded interactions between the interactive regions addresses the relationship between the amino acids. Consider hydrogen bonds > salt bridges > nonbonded interactions in assessing binding (Fig. 7) [38].
4. Submit the FASTA sequence of an identified protein to the CAMP server, Meta-iAVP server, and iACP server. The analyses performed offer the extent of a protein's antimicrobial, antiviral, and anticancerous efficacy [39, 40].
5. The CAMP server also helps in the identification of most likely antimicrobial peptides using four different algorithms, i.e., support vector machine (SVM) classifier, random forest (RA) classifier, artificial neural network (ANN) classifier, and discriminant analysis (DA) classifier [41].

Table 1
Proteins identified from the brain tissue lysate of *Periplaneta americana* by high-resolution nanoLC-MS

Sr. no.	Accession	Description	Moonlight	M.W (kDa)
1	A1E2I6	Ubiquitin	Yes	8.5
2	Q95W47	Histone H3	Yes	11.8
3	Q1M0X8	MPA13 allergen	Yes	14.7
4	H6SWV2	Calmodulin	Yes	16.8
5	G8XWV3	Myosin light chain variant 1	Yes	17
6	D0UND9	Alpha-spectrin (fragment)	Yes	17.5
7	G8XWU4	Glutathione S transferase class delta variant 1	Yes	24.6
8	A0A2P0XIH0	Triosephosphate isomerase	Yes	26.8
9	A0A2P0XIZ5	Putative Per a allergen	Yes	28.3
10	A0A2P0XIX7	Putative Per a 7 allergen variant	Yes	32
11	A0A2P0XIF1	Putative Per a 7 isoallergen	Yes	32.4
12	Q2LIY7	Electron transfer flavoprotein subunit alpha	Yes	35.2
13	A0A2P0XIG5	Glyceraldehyde-3-phosphate dehydrogenase	Yes	35.4
14	A0A2P0XJ03	Putative Per a allergen	Yes	36
15	A0A2P0XJ12	Putative Per a 2-allergen variant	Yes	38.2
16	A0A2P0XIZ4	Fructose-bisphosphate aldolase	Yes	39.5
17	A1KY39	Arginine kinase	Yes	39.7
18	O02460	Elongation factor 1-alpha (fragment)	Yes	39.8
19	A0A2R4CIH4	Actin	Yes	41.8
20	A0A2P0XIF0	Putative Per a allergen	Yes	47.2
21	Q1WK32	Adipokinetic hormone receptor	Yes	51.9
22	A0A2P0XIF5	Putative Per a allergen	Yes	56
23	A0A2P0XIF9	Putative Per a allergen	Yes	71.1
24	A0A2P0XJ04	Putative Per a allergen	Yes	72.6
25	A0A2P0XIG9	Putative Per a 3 allergen variant	Yes	72.9
26	Q94643	Allergen (fragment)	Yes	75.5
27	H2F490	Transferrin	Yes	78.4
28	Q25641	Allergen Cr-PI	Yes	81.1
29	Q6JU91	Elongation factor-2 (fragment)	Yes	81.5
30	D3JUE9	Major allergen Cr-PI	Yes	82.1

(continued)

Table 1
(continued)

Sr. no.	Accession	Description	Moonlight	M.W (kDa)
31	A0A2P0XJ16	Putative Per a allergen	Yes	102.9
32	Q9BPS0	Vitellogenin-2	Yes	215.8
33	A0A2P0XIE8	Putative Per a 6 isoallergen	Yes	17.2
34	D0UMV2	SH2 domain-binding protein (fragment)	Yes	19.9
35	D0UNA0	Alpha-spectrin (fragment)	Yes	22.2
36	A0A2P0XIY7	Peptidyl-prolyl cis-trans isomerase	Yes	22.9
37	D0UMY8	Glucose-6-phosphate isomerase (fragment)	Yes	24.8
38	A0A059WIM	Aminopeptidase (fragment)	Yes	25.3
39	H6S3P0	Cytochrome c oxidase subunit 2	Yes	26.2
40	D0UKL7	Ala-tRNA synthetase (fragment)	Yes	26.2
41	D0J9I6	30S ribosomal protein S3	Yes	27
42	D0UKH2	RNA helicase (fragment)	Yes	27.6
43	D0J8R4	Malonyl CoA-acyl carrier protein transacylase	Yes	31.8
44	D0J9A1	Prephenate dehydrogenase	Yes	31.9
45	D0J945	Transketolase, N-terminal subunit	Yes	32.1
46	P0DSM7	Tropomyosin Per a 7.0102	Yes	32.8
47	D0J9L9	4-Hydroxy-tetrahydrodipicolinate synthase	Yes	32.8
48	Q9Y071	Homeobox protein engrailed-like	Yes	37.2
49	D8X0H9	Formyl-tetrahydrofolate synthetase (fragment)	Yes	37.8
50	D0J9G7	DNA-directed RNA polymerase subunit alpha	Yes	38
51	D8X0N4	Formyl-tetrahydrofolate synthetase (fragment)	Yes	38
52	D8X0I5	Formyl-tetrahydrofolate synthetase (fragment)	Yes	38
53	Q5MBV5	Parcxpwnx02	Yes	38
54	D8X0L9	Formyl-tetrahydrofolate synthetase (fragment)	Yes	38.2
55	A0A0U5AH5	Melatonin receptor	Yes	39.3
56	A0A0K1ZFV1	Putative green opsin	Yes	42.1
57	A0A0K1ZFW	Putative UV opsin	Yes	42.3
58	D0J8N6	Cystathionine gamma-synthase	Yes	42.5
59	C6F3V9	Cytochrome b	Yes	42.9
60	D0J9Q4	Elongation factor Tu	Yes	43.8

(continued)

Table 1
(continued)

Sr. no.	Accession	Description	Moonlight	M.W (kDa)
61	D0J8L0	Dihydrolipoyllysine-residue succinyl transferase component of 2-oxoglutarate dehydrogenase complex	Yes	45.7
62	Q9Y070	Homeobox protein engrailed-like	Yes	47.1
63	D0J999	S-adenosyl methionine synthase	Yes	47.1
64	D0J8Y7	Threonine-tRNA ligase	Yes	48.2
65	D0J9N5	Putative type II NADH dehydrogenase	Yes	49.3
66	D0J8U8	Threonine synthase	Yes	50.2
67	D0J980	GTPase Der	Yes	50.8
68	D0J8M4	Molybdopterin oxidoreductase membrane subunit	Yes	53.4
69	D0J9K9	Membrane protein insertase YidC	Yes	71.5
70	D0J9E8	Ribonuclease R	Yes	76.1
71	D3YP11	Per a 3 allergen	Yes	81.1
72	D0JA03	Chaperone protein ClpB	Yes	100.6
73	D0J9Z5	PDDEXK_1 domain-containing protein	Yes	106.9
74	D0J9U7	Leucine-tRNA ligase	Yes	111.3
75	D0J8W1	Phosphoribosylformylglycinamidine synthase	Yes	140.4
76	D0J9P8	DNA-directed RNA polymerase subunit beta	Yes	144
77	Q9U8M0	Vitellogenin-1	Yes	212.6
78	P84118	Bursicon (fragments) burs	No	8.4
79	A0A2P0XJ09	Putative Per a allergen	No	11.9
80	A0A0U2JZ21	CRF-DH (fragment)	No	13
81	O17447	p10	No	15
82	I1XB30	C-2 lysozyme	No	15
83	A0A2P0XIY8	Putative Per a 6 isoallergen	No	16.9
84	A0A2P0XIE3	Putative Per a 6 allergen variant	No	17.1
85	A0A481SQK2	Per a 16 allergen	No	21.5
86	A0A2P0XIZ1	Putative Per a 8 allergen variant	No	22.5
87	A0A2P0XJ18	Putative Per a allergen	No	23.6
88	Q9TZR6	Major allergen Per a 1.0101 (fragment)	No	26.2
89	A0A481SN44	ATP synthase subunit beta	No	38.2

(continued)

Table 1
(continued)

Sr. no.	Accession	Description	Moonlight	M.W (kDa)
90	D3JUE8	Major allergen Cr-PII	No	44.6
91	Q9XZ71	Troponin T	No	45.9
92	D0J8P1	ATP synthase subunit alpha	No	58.2
93	D0J8Z2	Arginine-tRNA ligase	No	68.6
94	B9W4N7	Hemocyanin subunit type 1	No	78.6
95	D0J8I9	Putative penicillin-binding protein 2	No	72.5
96	D0J8H1	Translation initiation factor IF-2	No	101.3

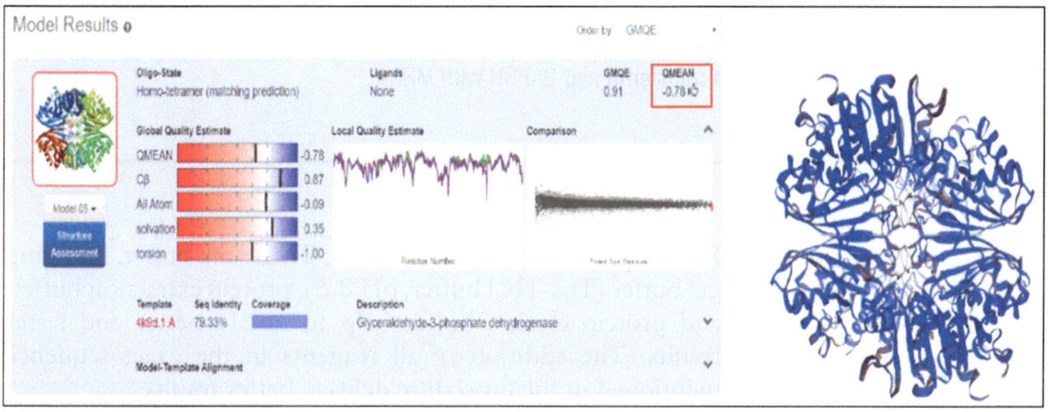

Fig. 4 Structure of GAPDH using SWISS MODEL

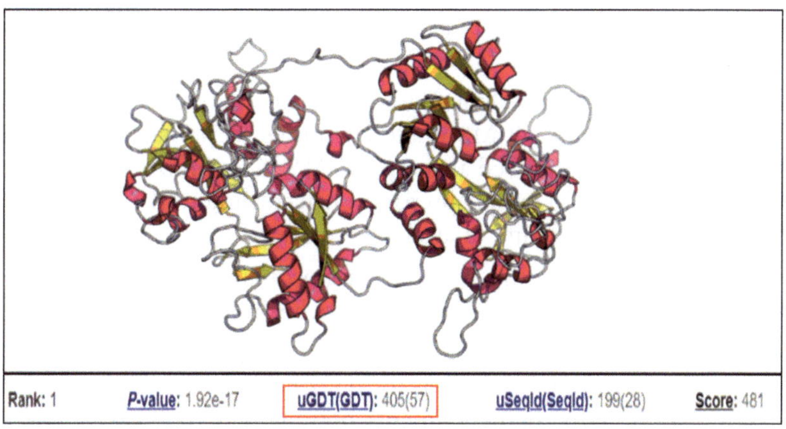

Fig. 5 Structure of transferrin using RaptorX

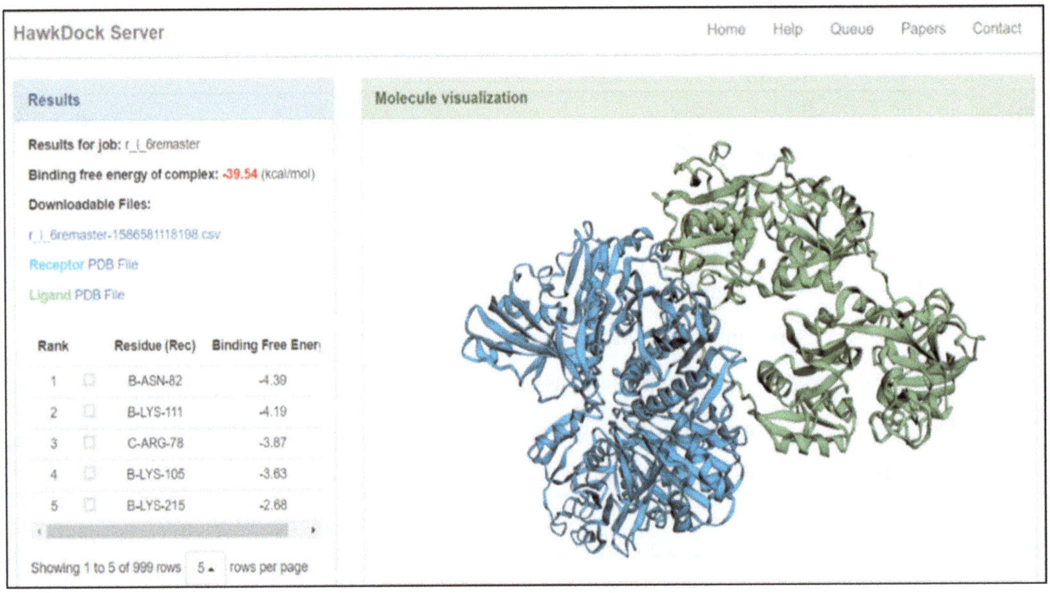

Fig. 6 MM/GBSA analysis of insect transferrin and GAPDH with MRSA

4 Notes

1. Freshly prepared solutions of APS, BAA, tank buffer, resolving gel buffer (Tris–HCl buffer, pH 8.6), protein extraction buffer, and protein elution buffer help to obtain better and faster results. The addition of all reagents in the exact sequence mentioned in the procedure delivers better results.
2. The bacterial culture media should be used within a week. Avoid using if it is older than 1 week.
3. Steam sterilize the 5-mm Whatman's Filter Paper discs before usage. Handle the discs with sterile forceps and transfer under laminar flow.
4. Use analytical-grade chemicals for the preparation of reagents. Use all MS-grade chemicals to prepare reagents and samples employed in nanoLC-MS. Customize the nanoLC flow rate and sample preparation according to the protein of interest and instrument available.
5. Isolating and purifying proteins by PAGE at lower room temperatures (20–23 °C) helps in maintaining gel integrity.
6. While preparing samples for PAGE, adjust the sample/dye ratio according to the concentration of protein in the sample. In the experiments presented here, we kept the ratio on the high side (1:1) to obtain a higher amount of protein in the gel.

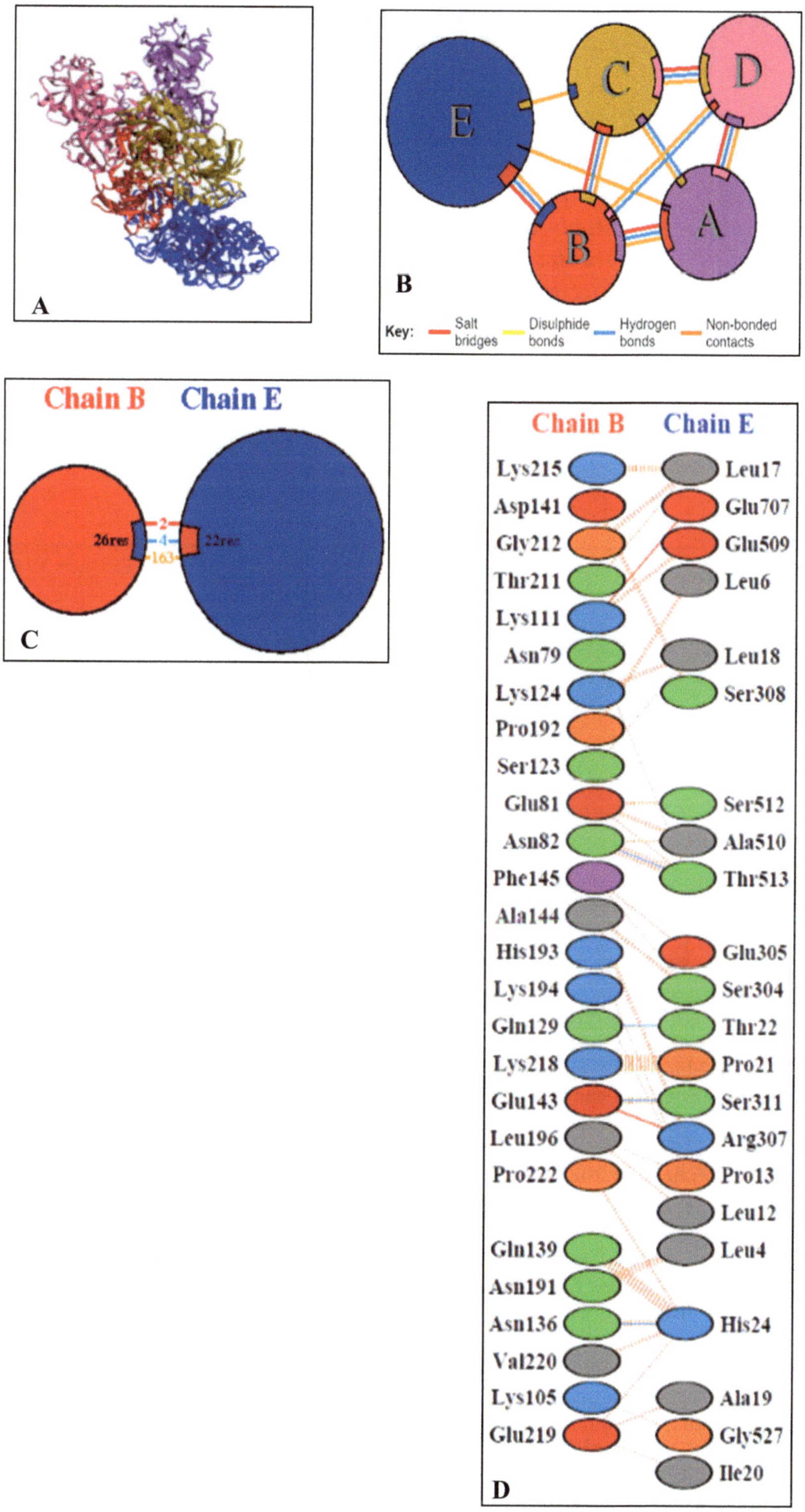

Fig. 7 PDBsum analyses of insect transferrin and GAPDH of MRSA. A. Ribbon model of interaction, B. Picturesque model of interaction, C. Interaction between specific chain B of GAPDH and transferrin, D. Interactions of specific amino acids of chain B of GAPDH and transferrin chain E

7. A native gel has been selected for protein separation. It enables maintenance of protein structure and function, which is required to assess antibacterial activity. A denatured protein (separated by SDS-PAGE) may not show antibacterial activity.
8. Steam sterilization of glassware and media at 15 psi for 15 min helps to prevent contamination during the antimicrobial sensitivity test (AST). The AST is performed under laminar airflow or in a biosafety cabinet to provide a greater level of sterility and accuracy in results.
9. The resin bed produces backpressure, so set the pipette to 10 μL, depress plunger to a dead stop, and slowly release or dispense the plunger throughout the operation.
10. A 5% methanol in 0.1% TFA/water wash can improve desalting efficiency.

Acknowledgments

Our grateful thanks to Ms. Manali Jadhav, Laboratory In-Charge, Department of Mass spectrometry SAIF, IIT-Bombay; Mr. Vyomesh Jawale, Team Lead, Clinical Bioinformatics, 4Base-Care, Bengaluru and "R ward" BMC workers and Mr. Rathod, Zoology Department, Mithibai College.

References

1. Dobermann D, Swift JA, Field LM (2017) Opportunities and hurdles of edible insects for food and feed. Nutr Bull 42(4):293–308. https://doi.org/10.1111/nbu.12291
2. Queiroz LS, Silva NFN, Jessen F, Mohammadifar MA, Stephani R, de Carvalho AF, Casanova F (2023) Edible insect as an alternative protein source: a review on the chemistry and functionalities of proteins under different processing methods. Heliyon 9:e14831. https://doi.org/10.1016/j.heliyon.2023.e14831
3. Kim TK, Yong HI, Kim YB, Kim HW, Choi YS (2019) Edible insects as a protein source: a review of public perception, processing technology, and research trends. Food Sci Anim Resour 39(4):521. https://doi.org/10.5851/kosfa.2019.e53
4. Seabrooks L, Hu L (2017) Insects: an underrepresented resource for the discovery of biologically active natural products. Acta Pharm Sin B 7(4):409–426. https://doi.org/10.1016/j.apsb.2017.05.001
5. Ratcliffe N, Azambuja P, Mello CB (2014) Recent advances in developing insect natural products as potential modern day medicines. Evid Based Complement Alternat Med 2014. https://doi.org/10.1155/2014/904958
6. Lokeshwari RK, Shantibala T (2010) A review on the fascinating world of insect resources: reason for thoughts. Psyche-J Entomol 2010: 1–11. https://doi.org/10.1155/2010/207570
7. Bonnamour A, Blake RE, Liebhold AM, Nahrung HF, Roques A, Turner RM, Yamanaka T, Bertelsmeier C (2023) Historical plant introductions predict current insect invasions. Proc Natl Acad Sci USA 120(24):e2221826120. https://doi.org/10.1073/pnas.2221826120
8. Moges F, Eshetie S, Endris M, Huruy K, Muluye D, Feleke T, G/Silassie F, Ayalew G, Nagappan R (2016) Cockroaches as a source of high bacterial pathogens with multidrug resistant strains in Gondar Town, Ethiopia. BioMed Res Int 2016:2825056. https://doi.org/10.1155/2016/2825056
9. Lavine MD, Strand MR (2002) Insect hemocytes and their role in immunity. Insect Biochem Mol Biol 32(10):1295–1309. https://doi.org/10.1016/S0965-1748(02)00092-9

10. Lee S, Siddiqui R, Ahmed Khan N (2012) Animals living in polluted environments are potential source of antimicrobials against infectious agents. Pathog Glob Health 106(4): 218–223. https://doi.org/10.1179/2047773212Y.0000000033
11. Yi HY, Chowdhury M, Huang YD, Yu XQ (2014) Insect antimicrobial peptides and their applications. Appl Microbiol Biotechnol 98(13):5807–5822. https://doi.org/10.1007/s00253-014-5792-6
12. Boate U, Suotonye B (2020) Cockroach (*Periplaneta americana*): nutritional value as food and feed for man and livestock. Asian Food Sci J 15(2):37–46. https://doi.org/10.9734/afsj/2020/v15i230150
13. Sule SO, Ojetayo TA, Sotolu AO (2020) Cockroach (*Periplanata americana*) meal nutritive composition. Magnesium (Mg) 1180:14–06
14. Khajuie F, Valizadeh R, Naserian A, Dadvar P, Dayani O (2022) Evaluation of cockroach (*Periplaneta americana*) powder as a potential feed ingredient for ruminants: chemical composition, fatty acids profile and ruminal degradability. J Livest Sci Technol 10(1):19–29. https://doi.org/10.22103/JLST.2022.18955.1396
15. Zhao Y, Yang A, Tu P, Hu Z (2017) Antitumor effects of the American cockroach, *Periplaneta americana*. Chin Med 12(1):1–6. https://doi.org/10.1186/s13020-017-0149-6
16. Ali SM, Siddiqui R, Ong SK, Shah MR, Anwar A, Heard PJ, Khan NA (2017) Identification and characterization of antibacterial compound(s) of cockroaches (*Periplaneta americana*). Appl Microbiol Biotechnol 101(1):253–286. https://doi.org/10.1007/s00253-016-7872-2
17. Balasubramanian S, Priya K, Revathi I, Revathi A, Venkatesh P (2017) Screening of antibacterial activity and biochemical assay from haemolymph of cockroach *Blatta orientalis* (Linnaeus, 1758). J Entomol Zool Stud 5(3):753–758
18. Kumar P, Kizhakkedathu JN, Straus SK (2018) Antimicrobial peptides: diversity, mechanism of action and strategies to improve the activity and biocompatibility in vivo. Biomol Ther 8(1):4. https://doi.org/10.3390/biom8010004
19. Wu H, Lu H, Huang J, Li G, Huang Q (2012) EnzyBase: a novel database for enzybiotic studies. BMC Microbiol 12(1):1–5. https://doi.org/10.1186/1471-2180-12-54
20. Hultmark D, Steiner H, Rasmuson T, Boman HG (1980) Insect immunity. Purification and properties of three inducible bactericidal proteins from hemolymph of immunized pupae of *Hyalophora cecropia*. Eur J Biochem 106(1): 7–16
21. Li Z, Mao R, Teng D, Hao Y, Chen H, Wang X, Wang X, Yang N, Wang J (2017) Antibacterial and immunomodulatory activities of insect defensins-DLP2 and DLP4 against multidrug-resistant *Staphylococcus aureus*. Sci Rep 7(1):12124. https://doi.org/10.1038/s41598-017-10839-4
22. Imler JL, Bulet P (2005) Antimicrobial peptides in Drosophila: structures, activities and gene regulation. Mechanisms of epithelial defense. Chem Immunol Allergy 86:1–21. https://doi.org/10.1159/000186648
23. Rabel D, Charlet M, Ehret-Sabatier L, Cavicchioli L, Cudic M, Otvos L, Bulet P (2004) Primary structure and in vitro antibacterial properties of the Drosophila melanogaster attacin C Pro-domain. J Biol Chem 279(15):14853–14859
24. Sagar S, Javle VRK, Chunduri JR (2020) In Silico modelling of antibacterial protein transferrin from *Periplaneta americana* and its interaction analysis with membrane protein of MRSA. Biosci Biotechnol Res Commun 13(2): 642. https://doi.org/10.21786/bbrc/13.2/42
25. Khan NA, Osman K, Goldsworthy GJ (2008) Lysates of *Locusta migratoria* brain exhibit potent broad-spectrum antibacterial activity. J Antimicrob Chemother 62(3):634–635. https://doi.org/10.1093/jac/dkn239
26. Kainulainen V, Loimaranta V, Pekkala A, Edelman S, Antikainen J, Kylväjä R, Korhonen TK (2012) Glutamine synthetase and glucose-6-phosphate isomerase are adhesive moonlighting proteins of Lactobacillus crispatus released by epithelial cathelicidin LL-37. J Bacteriol 194(10):2509–2519
27. Burns A, Olszowy P, Ciborowski P (2016) Biomolecules. In: Proteomic profiling and analytical chemistry, pp 7–24. https://doi.org/10.1016/b978-0-444-63688-1.00002-1
28. Jeffery CJ (2016) Protein species and moonlighting proteins: very small changes in a protein's covalent structure can change its biochemical function. J Proteome 134:19–24. https://doi.org/10.1016/j.jprot.2015.10.003
29. Gallagher SR (2012) SDS-polyacrylamide gel electrophoresis (SDS-PAGE). Curr Protoc Essent Lab Tech 2012(SUPPL.6):1–28. https://doi.org/10.1002/9780470089941.et0703s06

30. Reguera G, Leschine SB (2017) Fast and efficient elution of proteins from polyacrylamide gels using Nanosep® centrifugal devices, vol 5. PALL Laboratory: Department of Microbiology, University of Massachusetts-Amherst, pp 2–6
31. Driscoll AJ, Bhat N, Karron RA, O'Brien KL, Murdoch DR (2012) Disk diffusion bioassays for the detection of antibiotic activity in body fluids: applications for the pneumonia etiology research for child health project. Clin Infect Dis 54(SUPPL. 2):159–164. https://doi.org/10.1093/cid/cir1061
32. Luebker SA, Koepsell SA (2016) Optimization of urea based protein extraction from formalin-fixed paraffin-embedded tissue for shotgun proteomics. Int J Proteom 2016:4324987. https://doi.org/10.1155/2016/4324987
33. Doellinger J, Schneider A, Hoeller M, Lasch P (2020) Sample preparation by easy extraction and digestion (SPEED) – a universal, rapid, and detergent-free protocol for proteomics based on acid extraction. Mol Cell Proteomics 19(1):209–222. https://doi.org/10.1074/mcp.TIR119.001616
34. Amin B, Robinson RA (2020) Dataset of quantitative proteomic analysis to understand aging processes in rabbit liver. Data Brief 31:105701. https://doi.org/10.1016/j.dib.2020.105701
35. Benkert P, Biasini M, Schwede T (2011) Toward the estimation of the absolute quality of individual protein structure models. Bioinformatics 27(3):343–350. https://doi.org/10.1093/bioinformatics/btq662
36. Weng G, Gao J, Wang Z, Wang E, Hu X, Yao X, Cao D, Hou T, Hou T, Cao D, Hou T (2020) Comprehensive evaluation of fourteen docking programs on protein-peptide complexes. J Chem Theory Comput 16(6):3959–3969. https://doi.org/10.1021/acs.jctc.9b01208
37. Weng G, Wang E, Wang Z, Liu H, Zhu F, Li D, Hou T (2019) HawkDock: a web server to predict and analyze the protein-protein complex based on computational docking and MM/GBSA. Nucleic Acids Res 47(W1):W322–W330. https://doi.org/10.1093/nar/gkz397
38. Laskowski RA, Chistyakov VV, Thornton JM (2005) PDBSum more: new summaries and analysis of the known 3D structure of proteins and nucleic acids. Nucleic Acids Res 33:266–268. https://doi.org/10.1093/nar/gki001
39. Moretta A, Salvia R, Scieuzo C, Di Somma A, Vogel H, Pucci P, Sgambato A, Wolff M, Falabella P (2020) A bioinformatic study of antimicrobial peptides identified in the Black Soldier Fly (BSF) Hermetia illucens (Diptera: Stratiomyidae). Sci Rep 10(1):1–14. https://doi.org/10.1038/s41598-020-74017-9
40. Ali F, Kumar H, Alghamdi W, Kateb FA, Alarfaj FK (2023) Recent advances in machine learning-based models for prediction of antiviral peptides. Arch Comput Methods Eng 0123456789:1. https://doi.org/10.1007/s11831-023-09933-w
41. Waghu FH, Idicula-Thomas S (2020) Collection of antimicrobial peptides database and its derivatives: applications and beyond. Protein Sci 29(1):36–42. https://doi.org/10.1002/pro.3714

Chapter 9

Bottom-Up Proteomics Workflow for Studying Multi-organism Systems

Hongxia Bai, Leonard B. Collins, Marcos Rogério André, Edward B. Breitschwerdt, and Taufika Islam Williams

Abstract

In recent years, discovery proteomics has emerged as a pivotal tool in biological research, especially when studying the intricate relationships among multiple organisms. To delve deeper into these interactions, we pioneered a bottom-up proteomics workflow. Using nanoLC-MS/MS and a label-free quantification method, this work specifically examines the differential protein expression in fleas (*Ctenocephalides felis felis*) that have been experimentally infected with *Bartonella henselae*, the causative agent of cat scratch disease (CSD). Our detailed methodology, from protein cleanup to data analysis using the Proteome Discoverer software, is meticulously outlined to aid other researchers in adopting and adapting this workflow for their own multi-organism studies. This versatile protocol serves as a foundational guide for examining multiple proteomes from varied taxonomic lineages, exemplified in our cat-flea-bacterium investigation.

Key words Bottom-up proteomics, nanoLC-MS/MS, Differential protein expression, *Ctenocephalides felis felis* (flea), *Bartonella henselae*

1 Introduction

Proteomics, the comprehensive study of proteins, is an indispens able tool in contemporary biological research. It not only aids in uncovering fundamental biological mechanisms but also paves the way for advances in therapeutic target identification and vaccine development. Particularly intriguing is the realm of multi-organism interactions where proteomics can illuminate intricate interplays, providing insights into phenomena like pathogen transmission. One classic case of such an interaction is observed between fleas and the bacteria they harbor, especially when examining the transmission dynamics of *Bartonella henselae* (*B. henselae*), the causative agent of cat scratch disease [1–3].

Taufika Islam Williams (ed.), *Tissue Proteomics: Methods and Protocols*, Methods in Molecular Biology, vol. 2884, https://doi.org/10.1007/978-1-0716-4298-6_9,

Given the complexity of these interactions, there is a need for robust protocols that can effectively quantify differential protein expressions across different conditions. In response, this work presents a bottom-up proteomics workflow tailored for systems with multiple organisms. This workflow, while initially developed with the flea-cat-*B. henselae* interactions in mind, delves into three scenarios: (a) unfed fleas, (b) fleas feeding on a noninfected cat, and (c) fleas drawing sustenance from a *B. henselae-infected* cat. The goal of the study built on the method described here is not just to understand these interactions but also to identify potential protein targets that might be instrumental in developing a transmission-block vaccine against *B. henselae.* Notably, while the focus of this protocol is specific, its versatility makes it a valuable resource for other label-free proteomics endeavors involving a diversity of distinct proteomes.

2 Materials

Reagents and Solutions

All reagents are recommended to be made fresh using LC/MS-grade water and ACN, unless otherwise noted (*see* **Note 1**).

2.1 Protein Extraction

1. Protein extraction buffer: 0.25 M sucrose, 1 mM $MgCl_2$, 10 mM Tris–HCl, pH 7.4; 4 mL of this buffer is supplemented with 4% SDS and HALT protease inhibitor cocktail (without EDTA) (*see* **Notes 2** and **3**).
2. Protein solubilization buffer: 50 mM ammonium bicarbonate (ABC) with 1% SDC. Weigh out 500 mg of sodium deoxycholate (SDC) and 197.6 mg of ABC and add water to a total volume of 50 mL. Mix the solution continuously until all the constituents have dissolved completely (*see* **Note 4**).
3. Total protein quantification: Bicinchoninic acid (BCA) Protein Concentration Assay [4] (*see* **Note 5**).

2.2 Protein Cleanup

1. Acidified acetone: 100 mM HCl in acetone. Acetone is first acidified by adding 6 N HCl to reach a concentration of 100 mM HCl and then cooled at a −20 °C freezer for 30 min (*see* **Notes 6** and 7).

2.3 Filter-Aided Sample Preparation (FASP) Reagents

The following reagents are appropriate quantities for 20 samples with 200 μg protein per sample and can be scaled as needed.

1. 0.1 M Tris–HCl pH 7.5: Add water to 5 mL of a 1 M Tris–HCl buffer with pH 7.5 to a final volume to 50 mL in a 50-mL Falcon tube; examine the pH with a pH probe (*see* **Note 8**).

2. 0.1 M Tris–HCl pH 8: Add water to 5 mL of a 1 M Tris–HCl buffer with pH 8 to a final volume to 50 mL in a 50-mL Falcon tube.
3. 50 mM DTT in 0.1 M Tris–HCl pH 8: Accurately measure 7.7 mg of dithiothreitol (DTT) and add 1.0 mL of 0.1 M Tris–HCl with pH 8 in a 1.5-mL microcentrifuge tube. Vortex the solution until the DTT is entirely dissolved.
4. 8 M urea in 0.1 M Tris–HCl pH 8: Weigh 12.0 g of urea and transfer it to a 50-mL Falcon tube. Add enough 0.1 M Tris–HCl pH 8 to bring the total volume to 25 mL, ensuring the urea is fully dissolved by mixing vigorously and warming with the palm (*see* **Note 9**).
5. 200 mM IAA (iodoacetamide) solution in 8 M urea in 0.1 M Tris–HCl pH 8: Weigh 55.5 mg of iodoacetamide (IAA) and add it to 1.5 mL of previously prepared 8 M urea in 0.1 M Tris–HCl pH 8 solution in a 2-mL microcentrifuge tube. Ensure that the solution is well mixed until the IAA has completely dissolved (*see* **Note 10**).
6. Preparing 2 M urea with 10 mM $CaCl_2$ in 0.1 M Tris–HCl pH 8: Mix 2.5 mL of 8 M urea with 11.1 mg of calcium chloride ($CaCl_2$) in a 15-mL Falcon tube. Then add 0.1 M Tris–HCl pH 8 to bring the total volume to 10 mL. Mix the solution until all components are dissolved.
7. Trypsin Gold solution, MS grade (100 μg vial): Reconstitute the trypsin by adding 2.5 mL of 0.1 M Tris–HCl pH 7.5 to the vial containing the lyophilized trypsin. Vortex well.
8. Quench buffer preparation: Start by adding 5 mL of Zwittergent 3–6 (10 μg/mL) to a container. Next, add 50 μL of formic acid (HCOOH) and vortex to mix (*see* **Note 11**).
9. Quality control samples: Pierce standard LC/MS-grade BSA Protein Digest and Pierce Standard HeLa Protein Digest [or equivalent standard single protein (BSA) and single proteome (HeLa) digests].

2.4 Other Supplies and Equipment

1. Glass beads for homogenization, 1 mm.
2. Ovens (set at 56 °C and 37 °C).
3. Eppendorf Model 5430R microcentrifuge or equivalent.
4. Benchtop mini centrifuge.
5. SpeedVac vacuum concentrators.
6. Analytical balance.
7. pH meter for FASP solutions preparation.
8. Acclaim PepMap™ C18 trap column or equivalent.
9. PepMap™ C18 analytical column or equivalent.

3 Methods

Standard Precautions

Wear standard laboratory personal protective equipment (PPE) throughout the experiment. This includes safety glasses, a laboratory coat, and disposable gloves. Due to the risk of keratin contamination from laboratory personnel, take additional precautions, such as frequent glove changes, not touching the body with gloved hands, and avoiding leaning over samples. Always use fresh reagents and place samples on ice when not in active use (*see* **Notes 12** and **13**).

3.1 Protein Extraction (2-h Processing Time)

1. Pulverize approximately 325 fleas from each experimental group in liquid nitrogen. Add 4 mL of protein extraction buffer as described in Subheading 2.1.
2. Homogenize each sample in a bead beating instrument (Fisherbrand) with 1-mm disruption glass beads (Research Products International Corp. or equivalent) using the following conditions:
   - 1 cycle at 5 m/s for 40 s.
   - 1 cycle at 5 m/s for 30 s.
   - 1 cycle at 5 m/s for 20 s.
   - 1 cycle at 5 m/s for 10 s.
3. Sonicate samples for 1 min in an ultrasonic cooled bath and vortex for 10 s. Repeat this step for three cycles.
4. Centrifuge the homogenate at 206 *g* for 5 min at room temperature to remove cellular debris.
5. Collect the supernatant and determine protein concentration using the BCA Protein Assay (Thermo Scientific) with BSA as a standard (or another similar total protein quantification workflow).

3.2 Acetone-Based Protein Cleanup (Approximately 2 h; see Notes 6 and 7)

1. Dissolve each sample in 200 μL of protein solubilization buffer, 50 mM ABC with 1% SDC (referred to as ABC/SDC from this point forward).
2. Add 800 μL of cooled, acidified acetone to each sample to precipitate proteins. For varying protein solution volumes, maintain a 4:1 acetone-to-sample volume ratio.
3. Freeze samples at −20 °C for 30 min.
4. Centrifuge samples at 13,000× *g* at 21 °C for 20 min.
5. Remove the supernatant, ensuring not to disturb the protein precipitate.

6. Add 200 μL of acidic acetone to the samples, freeze for 30 min at −20 °C, and then centrifuge at 13,000× *g* at 21 °C for 20 min.
7. Carefully remove the supernatant and air-dry samples in a hood for a maximum of 2 min.
8. Redissolve the samples in 100 μL ABC/SDC. Use vortexing and sonication in an ice-water bath to assist dissolution.
9. Centrifuge samples to remove any undissolved material and collect the supernatant for the following steps.

3.3 Filter-Aided Sample Preparation (FASP) Process [5] (Approximately 1.5 Days Including Overnight Proteolytic Digestion)

Label two sets of Vivacon-500, 10-kDa MWCO filtration units or equivalent (*see* **Note 14**).

1. Total Protein Content.
   Protein concentration of the samples should be measured using the BCA assay (A280 measurements or other total protein quantification workflows can serve as viable alternatives) prior to the FASP process (*see* **Note 15**).
2. Lyophilize and Reconstitute.
   (a) In separate 1.5-mL low-retention microcentrifuge tubes, pipet the volume equivalent of 200 μg of protein for each sample. Evaporate each sample to dryness. Store the dried samples at −20 °C until FASP. *Pause point: Lyophilized samples can be safely stored in the freezer for months.*
   (b) Add 200 μL of 50 mM ABC/SDC to each sample and vortex to solubilize proteins. If necessary, perform alternate sonication (under ice) and vortexing for complete solubilization.
3. Denature.
   (a) Add 15 μL of 50 mM DTT (dithiothreitol) in 0.1 M Tris–HCl pH 8 to each sample. Vortex and spin down.
   (b) Incubate the microcentrifuge tubes at 56 °C for 30 min (*see* **Note 16**).
   (c) Add 200 μL of 8 M urea [in 0.1 M Tris(hydroxymethyl) aminomethane hydrochloride (Tris–HCl), pH 8 buffer] to each sample. Vortex and spin down.
4. Transfer.
   (a) For each sample, passivate a separate Vivacon-500 10-kDa MWCO filtration unit (or equivalent) with 20 μL 0.1 M Tris–HCl pH 8 buffer.
   (b) Transfer each sample onto a separate, passivated, Vivacon-500 10-kDa molecular weight cutoff (MWCO) filtration units.

   (c) Centrifuge the MWCO filters at 12,000× *g* for 15 min at 21 °C. Discard the flow-through (*see* **Note 17**).
   (d) Add an additional 200 μL of 8 M urea to each sample filtration unit. [Do not centrifuge.]

5. Alkylation.
   (a) Add 64 μL of 200 mM IAA (in 8 M urea prepared with 0.1 M Tris–HCl, pH 8) to each filtration unit (on top of the existing urea solution) to yield a final IAA concentration of 50 mM (*see* **Note 10**).
   (b) Incubate the samples at room temperature and in the dark for 60 min.
   (c) Centrifuge the filtration units containing the samples at 12,000× *g* for 15 min at 21 °C to remove all the solution, thus trapping the denatured proteins onto the filters.

6. Washing.
   (a) Add 100 μL of 2 M urea with 10 mM calcium chloride ($CaCl_2$) prepared with 0.1 M Tris-HCl, pH 8 to each filtration unit of sample. Spin the units for 10 min at 12,000× *g* at 21 °C (*see* **Note 18**).
   (b) Perform the above step a total of three times. Discard all the flow-through.

7. Buffer Exchange.
   (a) Add 100 μL of 0.1 M Tris–HCl pH 7.5 to each filtration unit of sample. Spin the units for 10 min at 12,500 × *g* at 21 °C.
   (b) Perform the above step a total of three times. Discard all the flow-through.

8. Remove the filters from the first set of collection tubes and place them on a clean, labeled second set. All eluents should be kept from this point forward.

9. Proteolytic Digestion.
   (a) Add trypsin proteolytic enzyme solution (0.2 μg/μL in 0.1 M Tris–HCl, pH 7.5) at a 1:50 enzyme/protein ratio to each filtration unit.
   (b) Incubate the filtration units containing the samples/control at 37 °C overnight (*see* **Note 19**).

10. Elute and Rinse.
   (a) Add 50 μL of quench buffer [1% HCOOH in 0.001% Zwittergent 3–16 (aqueous solution)] to each sample/control.
   (b) Spin the filtration units for 15 min at 12,500× *g* at 21 °C.

   (c) Add 400 μL of quench buffer to the retained volume of each sample/control.
   (d) Spin the filtration units for 30 min at 13,000× *g* at 21 °C. Repeat centrifugation if needed.

11. Lyophilize.
   (a) Evaporate the samples/control to dryness and store them at −20 °C until further analysis.

Pause point: Lyophilized samples can be safely stored in the freezer for months.

3.4 nanoLC-MS/MS Analyses (see Note 20)

1. Reconstitute lyophilized samples in 200 μL mobile phase A [MPA, 2% ACN in water with 0.1% HCOOH] just prior to analyses. Reconstitute HeLa cells lysis to achieve a final concentration of 0.1 μg/μL using MPA, while employing pure MPA for blanks.
2. In the Roadmap and Sequence Setup page, generate a sample table detailing the sample name, volume of injection, chosen analytical method, and the designated data pathway. It is advisable to intersperse the sample sequence with quality control (QC) samples and blanks between your experimental samples to ensure analytical robustness and integrity. This practice aids in monitoring system performance and detecting potential contamination or carryover (*see* **Note 21**).
3. Inject 2 μL of each sample, QC, or blank into the system for online desalting and reversed-phase nanoLC separation using a "trap and elute" configuration on a Thermo Scientific EASY nanoLC 1200 system (San Jose, CA, USA) coupled to a Thermo Scientific Exploris-480 mass spectrometer (Bremen, Germany) with EASY-Spray ion source (Thermo Scientific, Bremen, Germany). Any nanoLC system interfaced through a nanoelectrospray source to a high-resolution MS instrument with MS2 capabilities can be employed.
4. Perform reversed-phase separation on a 25 cm EASY-Spray™ C18 nanoLC column (2 μm particle size, 100 Å pore size, 75 μm diameter) or equivalent nanoLC column. Use a 2-cm trap column [Acclaim PepMap™ C18 column (or similar pre-column), 3 μm particle size, 100 Å pore size, 75 μm diameter] for sample concentration and desalting.
5. Conduct analytical separations at 300 nL/min on the nanoflow pump, applying a gradient from 5% mobile phase B (MPB, 80% ACN with 0.1% FA) to 95% MPB over 135 min.
6. Collect data using a data-dependent analysis (DDA) instrument method for discovery proteomics, with using a full-MS top speed data-dependent MS/MS (dd-MS/MS) analysis

using the following settings: spray voltage of +1900 V, ion transfer tube temperature of 275 °C, RF lens of 40%, scan range of 375–1600 *m/z*, 120 k MS resolving power, normalized automatic gain control (AGC) target of 300%, MS max injection time (IT) of 120 ms, MS/MS resolving power of 15 k, normalized MS/MS AGC target of 100%, MS/MS max IT of 21 ms, isolation window of 1.5 *m/z*, normalized higher-energy collisional dissociation (HCD) collision energy of 30%, and a dynamic exclusion of 20 s. Exclude unassigned and +1 charges from selection for MS2.

7. Refer to detailed parameters in Thermo Scientific Xcalibur, illustrated in Figs. 1 and 2. While we have used a Thermo Scientific instrument setup, any other instrument with similar data analysis capabilities can be employed.

3.5 Thermo Scientific Proteome Discoverer (PD) Analyses (see Note 22)

1. We describe the use of Thermo Scientific Proteome Discoverer (PD) 2.4.0.305 for proteomics data analysis, though other discovery proteomics software can be used. When using PD, use two separate analysis workflows (AWF) in data interrogation, (a) one for protein identification with each sample and (b) the second to compare the proteome among different groups using label-free quantification (LFQ). For each workflow, generate primary search results from input data files using the processing workflow (PWF) and display the results with the consensus workflow (CWF).

2. For (a) protein identification with each sample: Process and search nanoLC-MS/MS data against target species proteome databases with a 5-ppm precursor mass tolerance and 0.02-Da fragment tolerance with the *Ctenocephalides felis* (412 sequences), *Felis catus* (50,040 sequences), *Bartonella henselae* (4111 sequences), and *Siphonaptera* (2079 sequences) subsets of the UniProtKB/Swiss-Prot databases. Set the database search settings as tryptic cleavage at Arg and Lys residues. Specified modifications include dynamic Met oxidation, as well as static Cys carbamidomethylation. Filter the identifications to a strict protein false discovery rate (FDR) of 1% and relaxed FDR of 5% using the Percolator node in Proteome Discoverer. Refer to Fig. 3 for an example of protein identification from a sample processed in PD.

3. For (b) label-free quantification (LFQ) to compare the proteome among different groups, assign samples with study variables to generate sample groups and generate abundance ratios between the groups to determine the differentially expressed proteins. Use summed protein abundance, after normalizing to the total peptide amount, for ratio calculation. Perform analysis

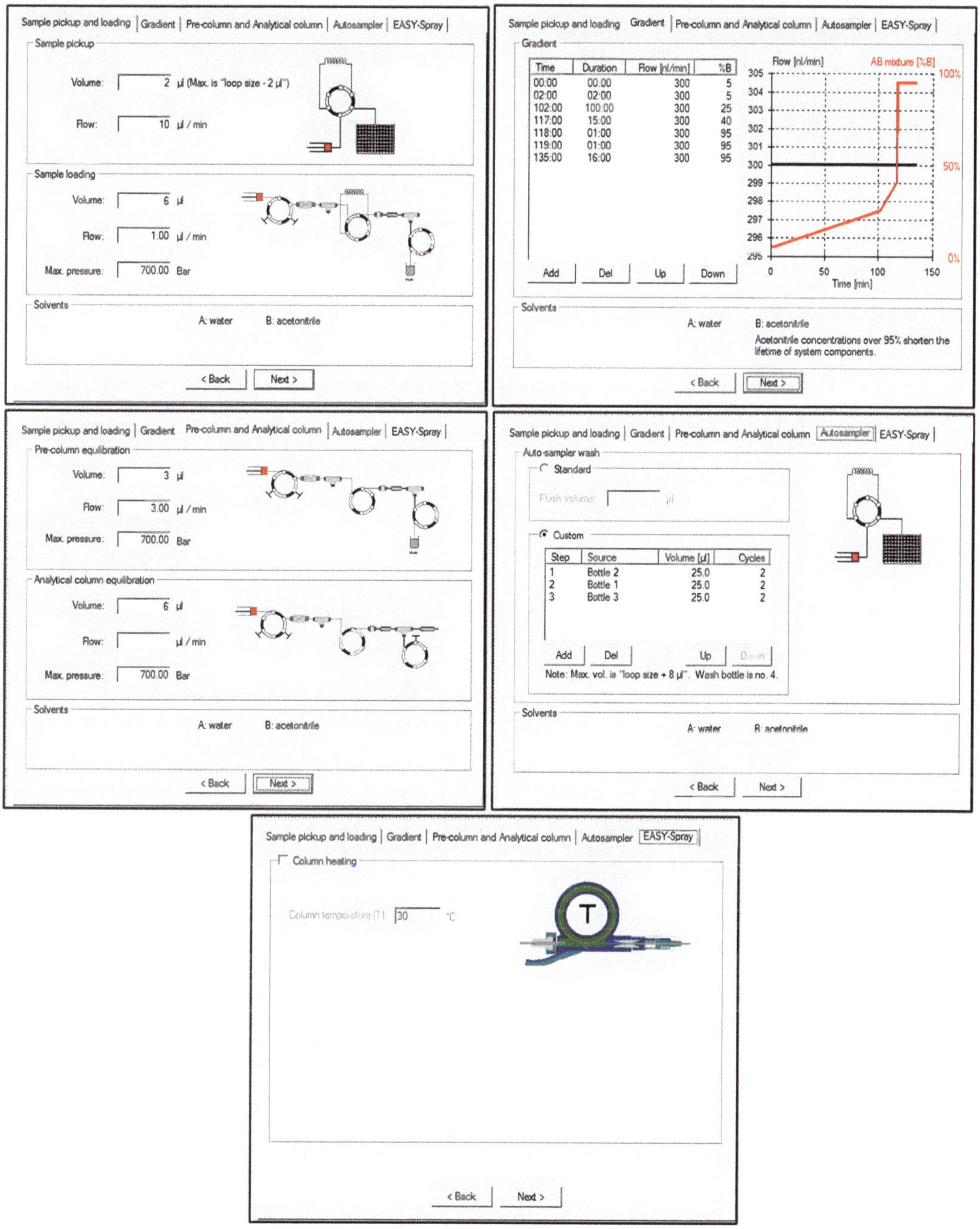

Fig. 1 NanoLC parameters including sample pickup and loading, gradient, pre-column and analytical column, autosampler, and EASY-Spray

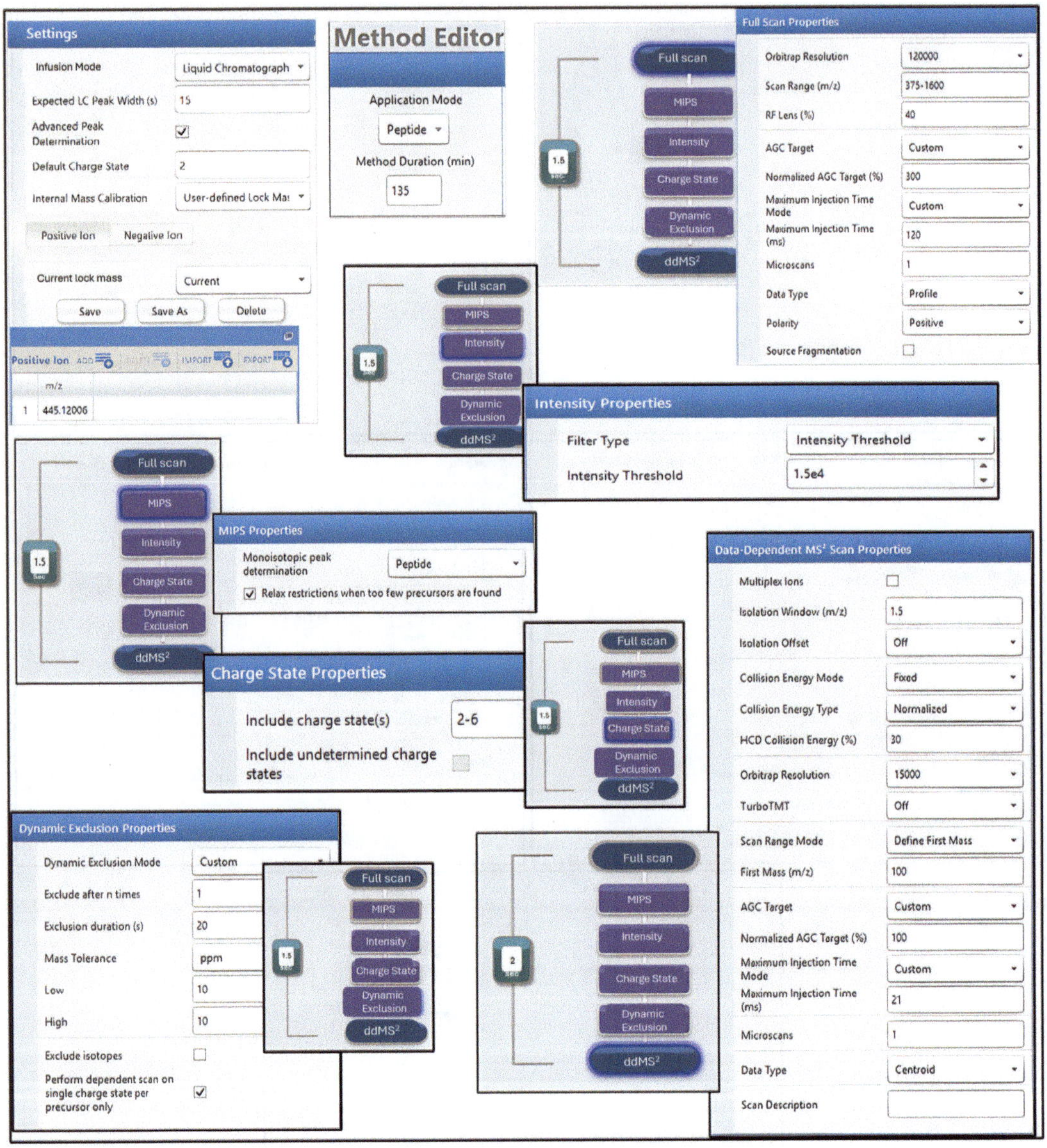

Fig. 2 Mass spectrometric parameters

of variance (ANOVA) test to report the significance of the ratio with the noted *p*-value. The details of AWFs are included in Figs. 4, 5, 6, 7, and 8 and Tables 1, 2, 3, and 4.

4 Notes

1. Buffer Freshness
   To avoid unexpected variability in results, always use fresh buffer solutions and store according to recommended conditions. If you notice inconsistent protein yields or anomalies in results, consider preparing fresh reagents.

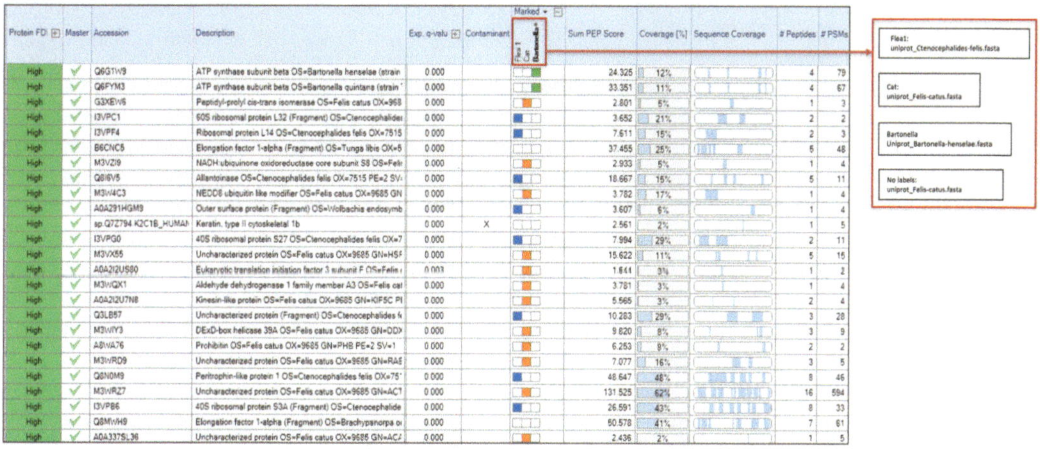

Fig. 3 A representative annotation result

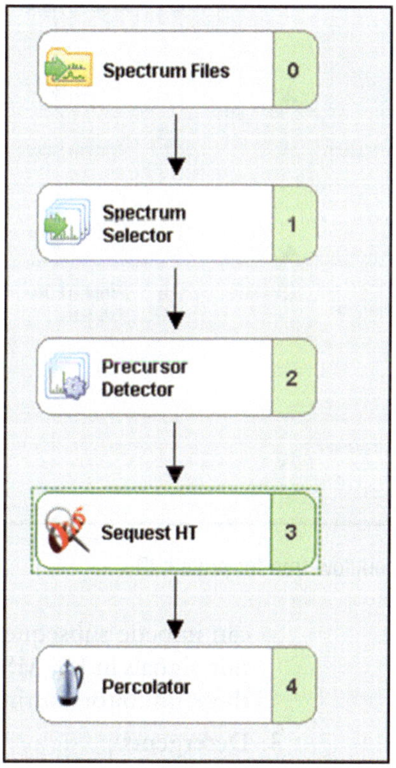

Fig. 4 Processing workflow tree for protein IDs

2. Protease Inhibitor

   Incorporating a protease inhibitor during protein extraction is essential. Proteases in the sample can degrade proteins into peptides, which can confound the proteomic data interpretation. However, if not adequately removed, protease inhibitors

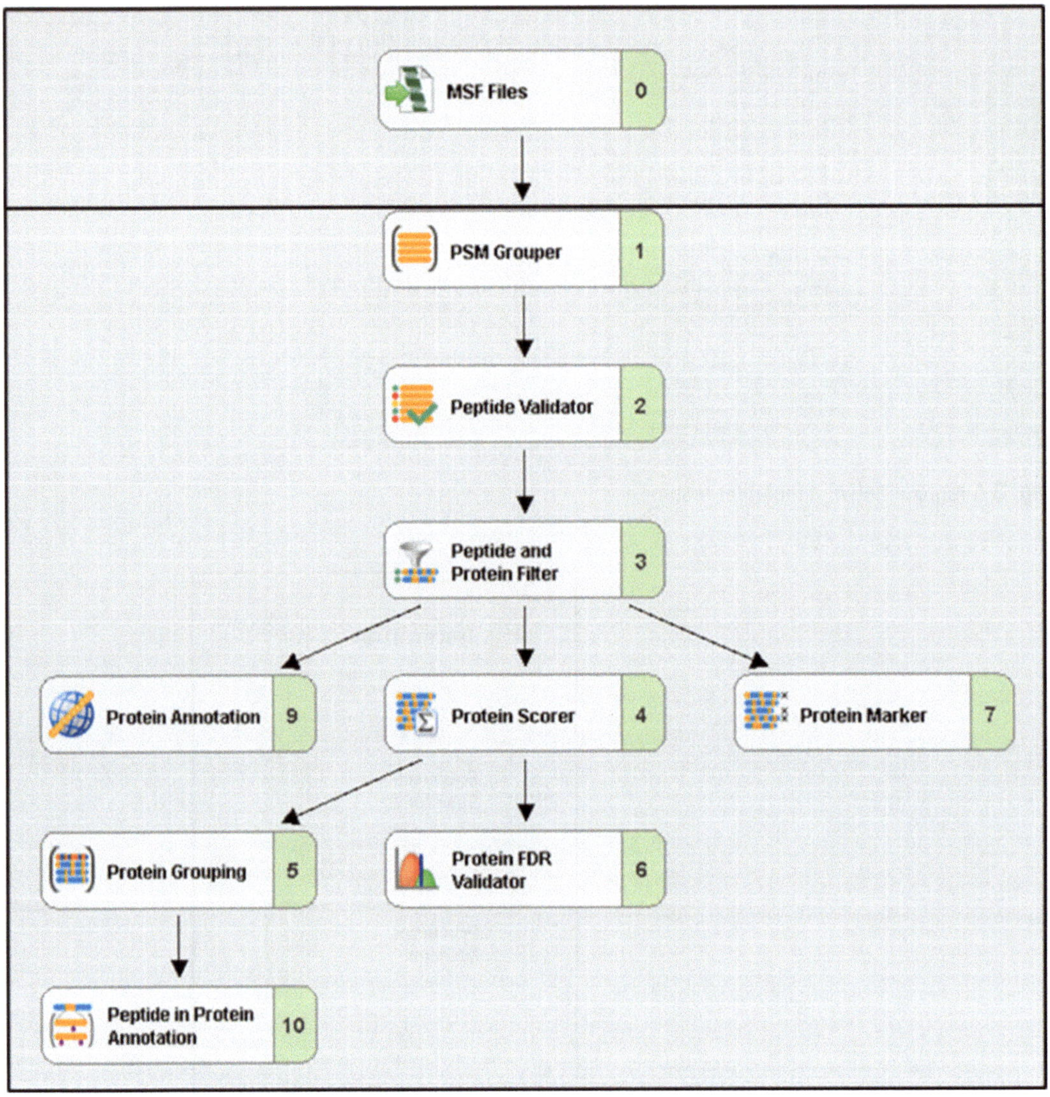

Fig. 5 Consensus workflow tree for protein ID

can impede subsequent proteolytic digestion and reduce peptide signals in LC-MS/MS analysis. Ensure efficient removal of these inhibitors during protein cleanup steps.

3. Detergents
   Ethylenediaminetetraacetic acid (EDTA), as well as other detergents, can be challenging to eliminate. Its presence can interfere with MS readings by suppressing target peptide signals. Avoid using detergents like this.
4. 50 mM ABC with 1% SDC
   This buffer is compatible with the BCA and subsequent proteomics procedures. SDC effectively solubilizes proteins but must be removed prior to nanoLC-MS/MS analysis.

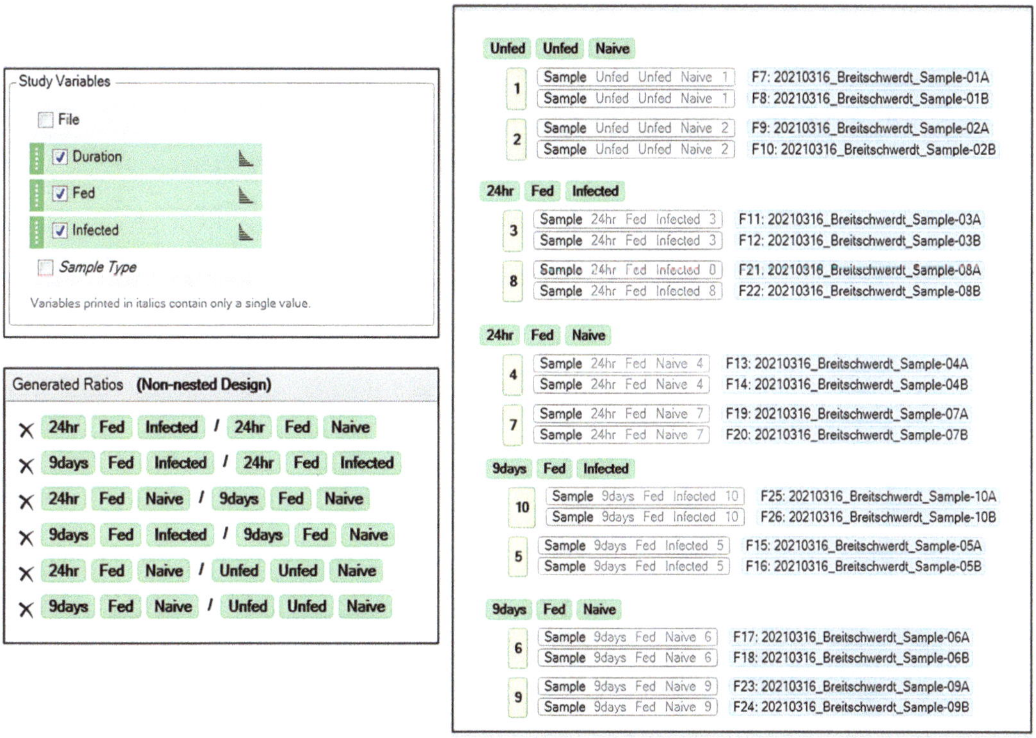

Fig. 6 Label free quantification setup in PD 2.4. Samples are assigned with study variables, and ratios are generated to find the desired differentially expressed proteins based on the hypothesis testing

5. Alternative Protein Quantification Methods
   The BCA and micro-BCA assays are preferred for total protein quantification, but A280 is an acceptable alternative for samples with less than 50 μg. There are other options for total protein quantification.
6. Acetone Precipitation
   Protein cleanup is essential in this protocol. There can be components in the sample from pre-proteomics sample handling, which might hamper tryptic digestion, resulting in diminished peptide signal in the nano-LC-MS/MS analyses. As an example, Fig. 9 shows LC-MS/MS signal before (a) and after (b) acetone precipitation cleanup on study samples. After acetone precipitation, protein solubility can become an issue. Avoid overdrying the protein pellet to ensure it resolubilizes easily. If the sample appears brown, it indicates overdrying. Harsh sonication and vortex might help but can cause nonspecific cleavage of target proteins.
7. Alternative Precipitation Methods
   If acetone does not yield desired results in protein cleanup, consider alternatives like TCA precipitation. The underlying principle remains consistent, but protocol specifics may differ.

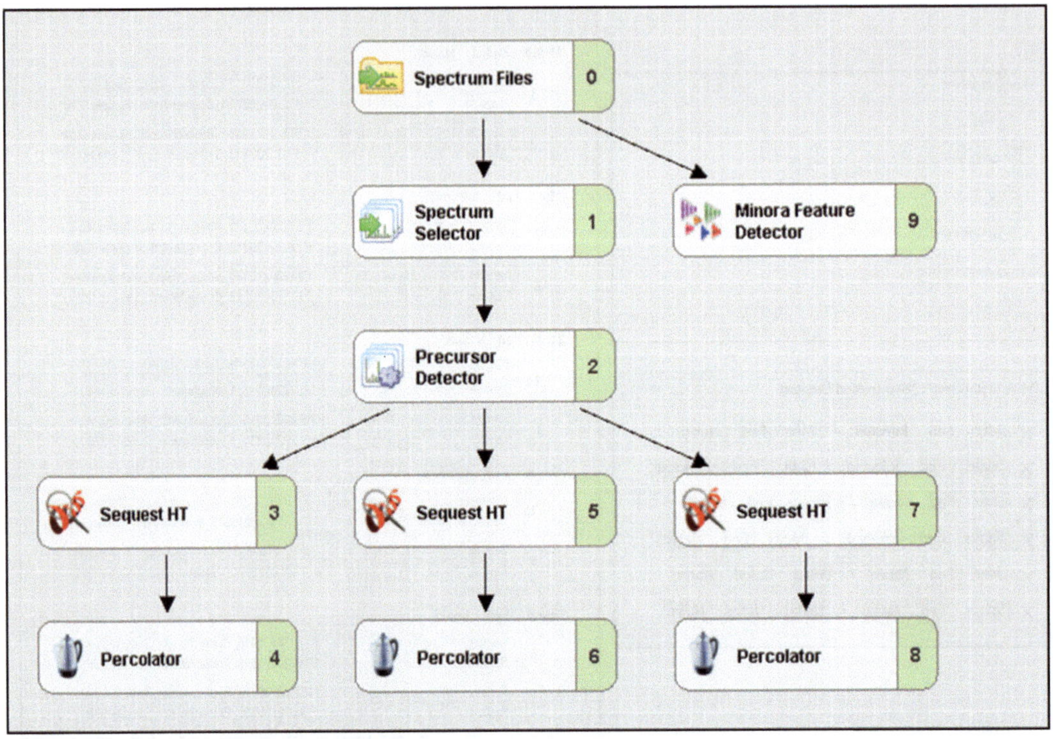

Fig. 7 Processing workflow tree for LFQ

8. pH of the Buffer

   Regularly check the pH of buffer solutions before reagent preparation, as reaction rates and potential side reactions depend on it. A $\pm$ 0.1 pH deviation is tolerable.

9. Dissolution Techniques

   The urea solution used in this method is highly concentrated and might be hard to dissolve. Using palm temperature can facilitate the dissolution process.

10. IAA Sensitivity

    IAA is light sensitive; prepare IAA solution and add it to samples swiftly to prevent degradation. Perform the IAA incubation step in the dark.

11. Quench Buffer

    Zwittergent has been proved to be compatible with MS analysis.

12. Sample Contamination

    The presence of keratin from laboratory personnel is a notorious contaminant in protein samples. Ensure rigorous glove changes, avoid touching any part of the body with gloves on, and avoid leaning over the samples. If a high abundance, unexpected peak that correlates with keratin is observed in the LC-MS/MS chromatogram, consider redoing the sample preparation with stringent contamination avoidance measures.

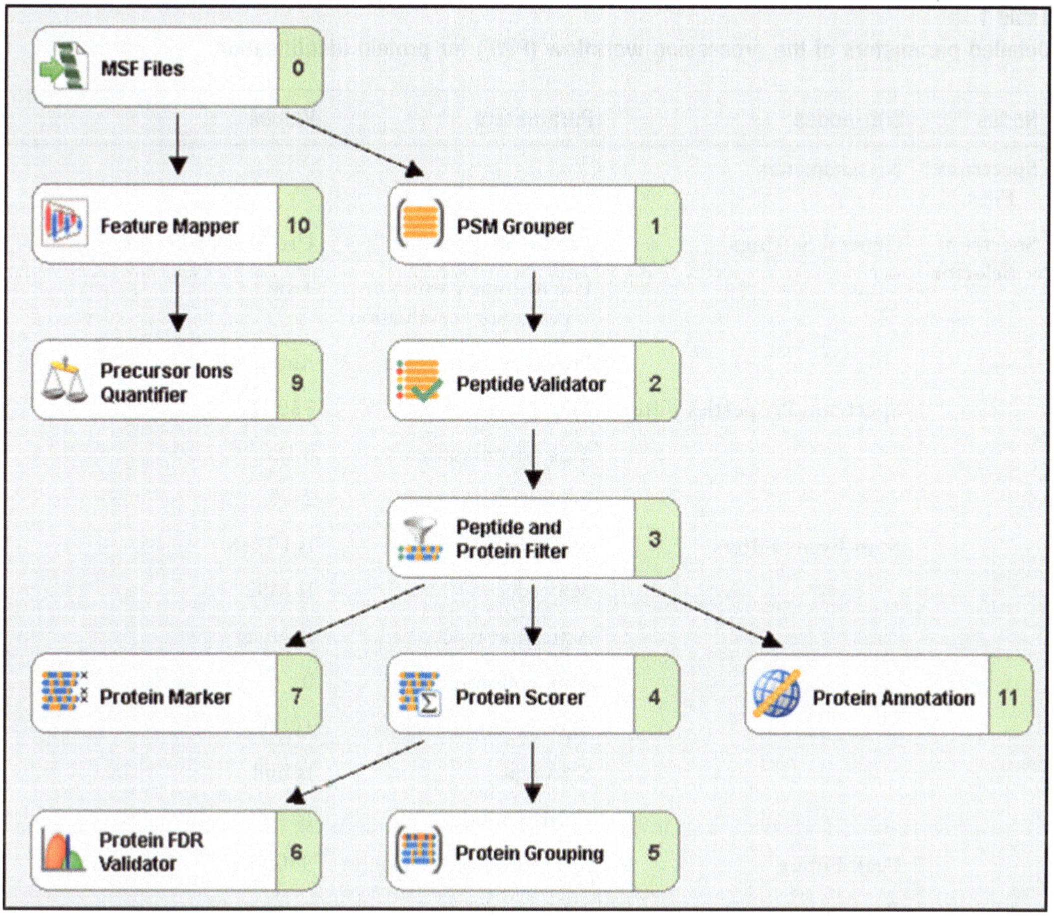

Fig. 8 Consensus workflow tree for LFQ

13. Protein Degradation
Minimize protein exposure to room temperature to avoid degradation. Always handle and store samples on ice unless specified otherwise.

14. Sample Handling
Use low-retention microcentrifuge tubes to reduce protein loss. Properly label tubes and track samples diligently.

15. FASP Protein Quantity
Adequate protein yield is essential for satisfactory protein coverage in FASP. If the total protein amount is less than 50 μg, consider generating more protein before proceeding further downstream on the proteomics workflow.

16. DTT Reactions
Before adding urea, cool centrifuge tubes with DTT reactions to room temperature. Urea is used as a chaotrope in protein denaturation. In the presence of heat and proteins, urea can lead to carbamylation [6].

Table 1
Detailed parameters of the processing workflow (PWF) for protein identification

Nodes	Sub-nodes	Parameters	Values
Spectrum Files	No parameters		
Spectrum Selector	General Settings	Precursor selection	Use MS1 precursor
		Use isotopic pattern in precursor reevaluation	True
		Provide profile spectra	Automatic
	Spectrum Properties Filter	Min. precursor mass	350 Da
		Max. precursor mass	0 Da
		Lowest charge state	2
	Scan Event Filters	Mass analyzer	Is FTMS
		MS order	Is MS2
		Activation type	Is HCD
		Min. collision energy	0
		Max. collision energy	1000
		Scan type	Is Full
		Polarity mode	Is +
	Peak Filters	S/N threshold (FT only)	5000 Da
	Replacements for Unrecognized Properties	Unrecognized properties	Automatic
	Precursor Pattern Extraction	Precursor clipping range before	2.5 Da
		Precursor clipping range after	5.5 Da
Sequest HT	Input Data	Protein database	Contaminants.fasta; *Uniprot_Ctenocephalides_felis. fasta; Uniprot_Felis-catus. Fasta; Uniprot_Bartonella-henselae. fasta; Uniprot_Siphonaptera.fasta*
		Enzyme name	Trypsin (Full)
		Max. missed cleavage sites	3
		Min. peptide length	6
		Max. peptide length	100
	Tolerances	Precursor mass tolerance	5 ppm

(continued)

Table 1 (continued)

Nodes	Sub-nodes	Parameters	Values
		Fragment mass tolerance	0.02 Da
		Use average precursor/ fragment mass	False
	Spectrum Matching	Weight of a ions	0.2
		Weight of b ions	1
		Weight of x ions	0.2
		Weight of y ions	1
	Dynamic Modifications	Max. equal modifications per peptide	3
		Max. dynamic modifications per peptide	4
		1. Dynamic modification	Oxidation/ +15.995 Da (M)
		2. Dynamic modification	Deamidated/ +0.984 Da (N, Q)
	Dynamic Modifications (protein terminus)	1. N terminal modification	Acetyl/ +42.011 Da (N-terminus)
		2. N-terminal modification	Met loss + Acetyl/ -89.030 Da (M)
		3. N-terminal modification	Met loss −131.040 Da (M)
	Static Modifications	1. Static modification	Carbamidomethyl/ +57.021 Da (C)
Percolator	**Target/Decoy Strategy**	Target/decoy selection	Concatenated
		Validation based on	q-value
	Input Data	Maximum delta Cn	0.05
		Maximum rank	0
	FDR Targets	Target FDR (strict)	0.01
		Target FDR (relaxed)	0.05

17. Centrifuge Speed
    Extra centrifugation may be performed according to sample needs. This could be done by increasing centrifugation speed or increasing centrifugation time. It is not recommended that the centrifugation speed exceed 14, 000× *g* as this might cause damage to the filters and/or clogging. This applies to all other

Table 2
Detailed parameters for the consensus workflow for protein identification

Nodes	Sub-nodes	Parameters	Values
MSF Files	Storage Settings	Spectra to store	Identified or quantified
		Feature traces to store	All
	Merging of Identified Peptides and Proteins	Merging mode	Globally by searching engine type
	FASTA Title Line Display	Reported FASTA title line	Best match
	PSM Filter	Maximum delta Cn	0.05
		Maximum rank	0
		Maximum delta mass	10 ppm
PSM Grouper	Peptide Group Modifications	Site probability threshold	75
Peptide Validator	General Validation Settings	Validation mode	Automatic (control peptide level error rate if possible)
		Target FDR (strict) for PSMs	0.01
		Target FDR (strict) for peptides	0.01
Peptide and Protein Filter	Peptide Filters	Peptide confidence at least	High
		Keep lower confident PSMs	False
	Protein Filters	Minimum number of peptide sequences	1
Protein Grouping	Protein Grouping	Apply strict parsimony principle	False
Peptide in Protein Annotation	Flanking Residues	Annotate flanking residues of the peptide	True
	Modifications in Peptide	Protein modifications reported	Only for master proteins
	Modifications in Protein	Modification sites reported	All and specific
		Minimum PSM confidence	High
		Report only PTMs	False
	Positions in Protein	Protein positions for peptides	For all proteins

(continued)

Table 2
(continued)

Nodes	Sub-nodes	Parameters	Values
Protein Scorer	No parameters		
Protein Annotation	Annotation Aspects	1. Aspect	Biological process
		2. Aspect	Cellular component
		3. Aspect	Molecular function
Protein Marker	Contaminant Database	Protein database	Contaminants.fasta
	Marker 1 Database	Column name	Flea 1
		Protein database	Uniprot_Ctenocephalides_felis.fasta;
	Marker 2 Database	Column name	Cat
		Protein database	Uniprot_Felis_catus.fasta;
	Marker 3 Database	Column name	Bartonella
		Protein database	Uniprot_Bartonella-henselae.fasta;
Protein FDR Validator	Confidence Thresholds	Target FDR (strict)	0.01
		Target FDR (relaxed)	0.05

centrifugation steps described in this protocol. Extra centrifugation steps may be taken as needed if extra liquid remains on the filter after centrifugation.

18. Washing and Buffer Exchanges
Washing steps and buffer exchanges are necessary to remove salt, protease inhibitors, and components like urea, which inhibit the trypsin protease. This is also important to prepare the appropriate environment (i.e., pH) for enzymatic digestion.

19. Filtration Precautions
Place some weights on top of filtration units to prevent them from popping open during the digestion period. It is also a good idea to wrap the top of the units with parafilm to minimize sample evaporation. For Vivacon-500 MWCO filtration units, it is advised that digestion solution atop each filter be 200 μL. Sample protein and trypsin protease concentrations should be considered accordingly.

20. nanoLC-MS/MS Parameters
Proper parameter settings in the nanoLC-MS/MS system are crucial for obtaining reliable and accurate data. Ensure that parameters, especially the ion transfer tube temperature, spray voltage, and RF Lens settings, are correctly set. Variations can lead to suboptimal ionization and signal detection.

Table 3
Detailed parameters for the processing workflow for LFQ. Nodes 1, 2, and 4 have the same parameters as the PWF for protein identification

Nodes	Sub-nodes	Parameters	Values
Sequest HT 3	**Input Data**	Protein database	Contaminants.fasta; *Uniprot_Felis-catus.Fasta;*
		Enzyme name	Trypsin (Full)
		Max. missed cleavage sites	3
		Min. peptide length	6
		Max. peptide length	100
	Tolerances	Precursor mass tolerance	5 ppm
		Fragment mass tolerance	0.02 Da
		Use average precursor/ fragment mass	False
	Spectrum Matching	Weight of a ions	0.2
		Weight of b ions	1
		Weight of x ions	0.2
		Weight of y ions	1
	Dynamic Modifications	Max. equal modifications per peptide	3
		Max. dynamic modifications per peptide	4
		1. Dynamic modification	Oxidation/ +15.995 Da (M)
		2. Dynamic modification	Deamidated/ +0.984 Da (N, Q)
	Dynamic Modifications (protein terminus)	1. N-terminal modification	Acetyl/ +42.011 Da (N-terminus)
		2. N-terminal modification	Met loss + Acetyl/ -89.030 Da (M)
		3. N-terminal Modification	Met loss −131.040 Da (M)
	Static Modifications	1. Static modification	Carbamidomethyl/ +57.021 Da (C)
Sequest HT 5	**Input Data**	Protein database	Contaminants.fasta; *Uniprot_Felis-catus.Fasta;*
	Other parameters are the same as the Sequest HT 3		

(continued)

Table 3
(continued)

Nodes	Sub-nodes	Parameters	Values
Sequest HT 7	Input Data	Protein database	Contaminants.fasta; *Uniprot_Felis-catus. Fasta; Uniprot_Siphonaptera. fasta*
	Other parameters are the same as the Sequest HT 3		
Minora Feature Detector	Peak Feature Detection	Min. trace length	5
		Max. delta RT of isotopic pattern multiplets (min)	0.2
	Feature to ID Linking	PSM confidence at least	High

Table 4
Detailed parameters for the consensus workflow for LFQ. Nodes 0, 1, 2, 3, 4, 5, 6, 7, and 11 have the same parameters as the CWF for protein identification

Nodes	Sub-nodes	Parameters	Values
Feature Mapper	Chromatographic Alignment	Perform RT alignment	True
		Maximum RT shift [min]	10
		Mass tolerance	10 ppm
	Feature Linking and Mapping	Minimum S/N threshold	2
Precursor Ions Quantifier	General Quantification Settings	Peptides to use	Unique + razor
	Precursor Quantification	Precursor abundance based on	Intensity
	Normalization and Scaling	Normalization mode	Total peptide amount
	Exclude Peptides from Protein Quantification	For protein rollup	Use all peptides
	Quan Rollup and Hypothesis Testing	Protein abundance calculation	Summed abundances
		N for top N	3
		Protein ratio calculation	Protein abundance based
		Hypothesis test	ANOVA (individual proteins)

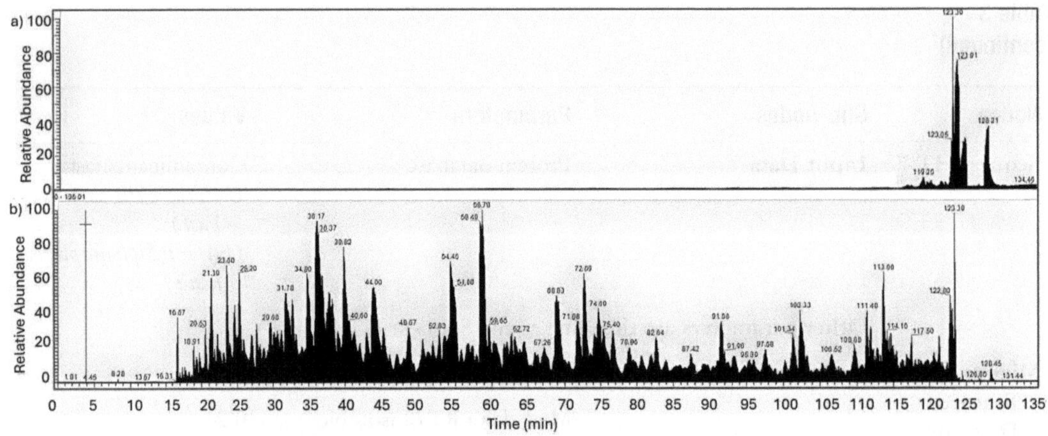

Fig. 9 Mass spectra (**a**) before and (**b**) after acetone-based protein cleanup

21. Quality Controls

    To ensure consistent system performance and data integrity, it is highly recommended to intersperse QC samples, such as standard protein digests from HeLa cells or bovine serum albumin (BSA), among your experimental samples. Utilizing these QC samples facilitates the monitoring of the system's operational efficiency. Important metrics, including retention time and the total count of identified proteins, act as crucial indicators for assessing system stability and performance. These practices contribute to the maintenance of high-quality analytical results.

22. Databases

    Ensure that the correct species databases are selected for protein identification, as incorrect database selection can lead to false identifications or missed identifications altogether. It is crucial to regularly update the database to incorporate newly identified sequences. When choosing between TrEMBL and Swiss-Prot databases, consider that Swiss-Prot offers manually annotated and reviewed entries, ensuring higher accuracy and reliability, while TrEMBL provides automatically annotated entries that may include a broader range of sequences, including isoforms and sequences from sub-taxonomies. Depending on the specificity and sensitivity required for your analysis, you may opt to include isoforms to capture a comprehensive protein diversity, but be aware of the potential for increased complexity in data analysis. Similarly, including sub-taxonomies can enrich the dataset, especially when studying organisms with less well-characterized proteomes, but it necessitates a careful balance between comprehensiveness and data manageability. Always tailor your database choice and inclusion criteria to the goals of your experiment, balancing between the breadth of data captured and the precision of identification desired.

References

1. André MR, Neupane P, Lappin M, Herrin B, Smith V, Williams TI, Collins L, Bai H, Jorge GL, Balbuena TS, Bradley J, Maggi RG, Breitschwerdt EB (2022) Using proteomic approaches to unravel the response of Ctenocephalides felis felis to blood feeding and infection with Bartonella henselae. Front Cell Infect Microbiol 12:828082
2. Breitschwerdt EB, Maggi RG, Chomel BB, Lappin MR (2010) Bartonellosis: an emerging infectious disease of zoonotic importance to animals and human beings. Vet Emerg Crit Care 20:8–30
3. Laukaitis HJ, Macaluso K (2021) Unpacking the intricacies of Rickettsia–vector interactions. Trends Parasitol 37:734–746
4. Contreras M, Villar M, Artigas-Jerónimo S et al (2018) A reverse vaccinology approach to the identification and characterization of Ctenocephalides felis candidate protective antigens for the control of cat flea infestations. Parasit Vectors 11:43
5. Wiśniewski J, Zougman A, Nagaraj N, Mann M (2009) Universal sample preparation method for proteome analysis. Nat Methods 6:359–362
6. Sun S, Zhou JY, Yang W, Zhang H (2014) Inhibition of protein carbamylation in urea solution using ammonium-containing buffers. Anal Biochem 1(446):76–81

References

Chapter 10

Proteomic Analysis of Biological Fluids

Kyle Burk, Kevin Legg, Phillip Danielson, and Glendon Parker

Abstract

Biological fluids are proteinaceous liquids or suspensions released through different body orifices or through penetration of the skin. These fluids are the result of multiple tissues and cell types and contain extensive, highly complex, and dynamic protein populations that reflect both the transcriptional program of the originating cells and a record of the individual's health status. Body fluids are readily accessible to clinicians and researchers, and as such proteomic analyses are an important component of clinical studies, fertility studies, oral health studies, and forensic investigations. Current mass spectrometry (MS) datasets have a dynamic range of up to six orders of magnitude and are as diverse as the originating tissue types. Mass spectrometry has the potential to provide information across a wide range of applications, including basic research into human biology and pathology, biochemical analysis of protein function, biomarker discovery and detection, as well as forensic investigations wherein investigators interpret a protein profile to identify the body site origin of a biological fluid. The method below describes a specimen processing workflow that is flexible in terms of biological fluid type, sample state (e.g., a dried sample extracted from evidence or neat fluid), and level of degradation. The method described here is compatible with both high sensitivity shotgun liquid chromatography-mass spectrometry LC/MS analysis and targeted (qualitative or quantitative) MS-based analysis of biomarker candidates.

Key words Biological fluid, Bodily fluid, Peptide biomarker, Quantitative proteomics, Sexual assault evidence, Forensic proteomics, Semen, Vaginal fluid, Saliva, Serum, Urine

1 Introduction

Biological fluids are an integral component of the vasculature and extracellular matrix. All living cells are in contact with biological fluids. These fluids therefore are an ideal medium for chemical intercellular communication, exchange of nutrients and waste products, and execution of physiological function. As proteins enter the extracellular spaces and are removed through a variety of biological processes, the fluid proteome at any one point will reflect these changes and processes [1].

Biological fluids, while complex, contain a deep information on multiple levels of cellular activity. Since any one cell can express over 10,000 gene products [2], and many different cell types contribute

Taufika Islam Williams (ed.), *Tissue Proteomics: Methods and Protocols*, Methods in Molecular Biology, vol. 2884,
https://doi.org/10.1007/978-1-0716-4298-6_10,

to the protein composition in any biofluid [3], these fluids provide an integrated and current window into any living organism's biological state [4]. This complexity, however, presents researchers with a challenge when attempting to resolve the biological processes reflected in any one biological fluid proteome. Specifically, in addition to the large number of proteins present, the abundance of these proteins spans more than nine orders of magnitude [5], and the dynamic range of peptide abundance in any one mass spectrometry dataset is often more than six orders of magnitude. Moreover, these proteins are often subject to a wide array of posttranslational modifications (PTMs) that further increase the total complexity of peptides in any digested mixture. These include nonenzymatic modifications, such as deamidation, oxidation, and pyroglutamation [6], as well as enzymatic modifications that occur during biosynthesis (e.g., the addition of *N*-linked oligosaccharide and glycosaminoglycan moieties) that increase the number of proteoforms [7] and sterically impede access of proteases to the underlying proteins [8]. This can also introduce novel protein structures in some pathological states [9]. Finally, necrosis and cell damage may introduce intracellular proteins into the extracellular space [10]. An in-depth proteomic analysis can help to integrate these processes and thereby provide a snapshot in time of an individual's cellular physiology at the proteome level [1].

Proteomic analysis of biological fluids can also play an important role in forensic investigations [6]. In fact, the detection of biomarker proteins has been a consistent component of forensic analysis for many decades [11]. Connecting an unknown biological fluid stain to a site of body origin can provide critical context to criminal investigations. This context can provide support for (or refute) investigational hypotheses and ultimately provide a more informed basis for testimony. Traditionally, body fluid characterization has been the starting point for additional forensic analysis such as DNA extraction and typing [12]. The characterization of a biological fluid has historically relied largely on chemical and/or enzymatic-based assays which target abundant protein markers that are characteristic of (but not unique to) the body fluid of interest. More modern approaches often target these same proteins but employ more sensitive immunochromatographic lateral flow assays [13–15]. Despite the advances in detection technology, none of the proteins employed in forensic assays can truly serve as the basis for the confirmatory identification of any specific body fluid. Given the fact that the proteins that are targeted by these assays show a wide distribution of tissue expression (albeit typically at lower levels in nontarget biological fluids) [16], their detection can only provide a presumptive indication of a given body fluid's presence. While biomarkers that are highly specific to a single body fluid have been identified, these are typically expressed at lower levels and thus require more advanced methods such as LC/MS for their reliable detection.

During forensic evidence collection, biological fluids normally present as a dried stain that is rehydrated and transferred either onto a cotton swab at the crime scene or in the laboratory during evidence examination. Alternatively, cuttings of dried stains may be transferred into microcentrifuge vials for rehydration and subsequent analyses. The source material can be any biological fluid. A forensic sample may be degraded, in low abundance, and may consist of a mixture of more than one body fluid. Biological replication, therefore, is not an option, and technical replication may be limited due to the need to preserve the evidentiary sample for subsequent analyses.

The methods described here reflect the development of sample processing pathways that can be broadly applied across a range of biological fluids. This protocol assumes the starting point to be a cotton swab, but the method can also be expanded to incorporate neat biological fluid. The sample is not fractionated, and high concentrations of Tris(2-carboxyethyl) phosphine hydrochloride (TCEP) are used to ensure the cleavage of disulfide bonds in the highly cross-linked spermatozoan acrosome. The result is a protocol for the processing of dried semen, vaginal fluid, and/or saliva on a cotton swab as well as neat whole peripheral blood, blood serum, vaginal and menstrual fluids, semen, saliva, and urine. The protocol can be employed with single source stains, as well as mixtures. The objective of the protocol is to maximize peptide yields either for discovery using label-free quantitation and statistical analysis or quantitative peptide biomarker detection using targeted mass spectrometry. The protocol does not encompass data acquisition, although it is compatible with a wide range of downstream liquid chromatography mass spectrometry platforms [6, 17–22] and analytical frameworks [18, 21, 23–25].

2 Materials

All materials should be of the maximum quality available and, if possible, MS-grade should be used. Water, likewise, should be ultrapure grade, e.g., reverse osmosis-purified water from a Nanopure™ purification system (Thermo Scientific Inc.) or equivalent ultrapure water system. Since only low volumes are needed, small volume formats of liquid reagents should be purchased when possible (*see* **Notes 1** and **2**).

Solutions for Use with Dried Specimens

1. Tris hydrochloride pH 8.0 solution: Prepare a 1 M Tris hydrochloride (Tris–HCl) pH 8.0 stock solution. Dilute to make at least 1.25 mL per sample of 100 mM Tris–HCl, plus 2 mL extra (*see* **Note 3**).

2. Tris(2-carboxyethyl)phosphine hydrochloride (TCEP) stock solution: Prepare a filtered solution at a concentration of 0.5 M in water (*see* **Note 4**).
3. Chloroacetamide (CCA) stock solution: Prepare a filtered solution at a concentration of 0.5 M in water (*see* **Note 5**).
4. Urea in Tris–HCl stock solution: Using the Tris–HCl stock described above, prepare a filtered solution at a concentration of 8 M urea in 100 mM Tris–HCl (*see* **Note 6**).
5. TCEP, urea, Tris–HCl solution: Using the Tris–HCl and TCEP stocks described above, prepare a solution of 5 mM TCEP, 8 M Urea in 100 mM Tris–HCl (*see* **Note** 7).
6. Reconstitution buffer for trypsin and ProteaseMax surfactant: Prepare a filtered solution of 50 mM ammonium bicarbonate (ABC) in water (*see* **Note 8**).
7. ProteaseMax solution: Prepare a 1% w/v ProteaseMax solution by adding 100 μL of filtered 50 mM ABC to a vial of 1 mg of ProteaseMax (*see* **Note 9**).
8. Trypsin solution: Prepare a solution of 0.5 μg/μL of MS-grade trypsin protease solution by adding 40 μL of filtered 50 mM ABC to a vial containing 20 μg trypsin (*see* **Note 9**).
9. Trifluoroacetic acid solutions: Prepare solutions of trifluoroacetic acid (TFA) at concentrations of 10%, 1%, and 0.1% using ultrapure water or equivalent (*see* **Note 10**).
10. Elution buffer: Prepare a 60% acetonitrile (ACN), 0.1% TFA elution buffer (i.e., six parts ACN, one part TFA at 1%, and three parts ultrapure water).

Solutions for Use with Neat Specimens

1. Tris hydrochloride pH 8.0 solution: Prepare a 1 M Tris hydrochloride (Tris–HCl) pH 8.0 stock solution. Dilute to make at least 1 mL per sample of 100 mM Tris–HCl, plus 2 mL extra.
2. Tris(2-carboxyethyl)phosphine hydrochloride (TCEP) stock solution: Prepare a filtered solution at a concentration of 0.5 M in water (*see* **Note 4**).
3. Chloroacetamide (CAA) stock solution: Prepare a filtered solution at a concentration of 0.5 M in water (*see* **Note 5**).
4. Urea in Tris–HCl stock solution: Using the Tris–HCl stock described above, prepare a filtered solution at a concentration of 8 M urea in 100 mM Tris–HCl (*see* **Note 6**).
5. TCEP, urea, Tris–HCl solution: Using the Tris–HCl and TCEP stocks described above, prepare a solution of 5 mM TCEP, 8 M urea in 100 mM Tris–HCl (*see* **Note** 7).

6. Reconstitution buffer for trypsin and ProteaseMax surfactant: Prepare a filtered solution of 50 mM ammonium bicarbonate (ABC) in water (*see* **Note 8**).
7. ProteaseMax solution: Prepare a 0.5% w/v ProteaseMax solution by adding 200 μL of 50 mM ABC to a vial of 1 mg ProteaseMax or by diluting 1% ProteaseMax solution 1:1 with ultrapure water (*see* **Note 9**).
8. Trypsin solution: Prepare a solution of 0.5 μg/μL of MS-grade trypsin protease solution by adding 40 μL of filtered 50 mM ABC to a vial containing 20 μg trypsin (*see* **Note 9**).
9. Trifluoroacetic acid solutions: Prepare solutions of trifluoroacetic acid (TFA) at concentrations of 10%, 1%, and 0.1% using ultrapure water (*see* **Note 10**).
10. Elution buffer: Prepare a 60% acetonitrile (ACN), 0.1% TFA elution buffer (i.e., six parts ACN, one part TFA at 1%, and three parts ultrapure water).

3 Methods

3.1 Solution Preparation, Decontamination, and Quality Control

Extraneous mass spectrometry signals increase the complexity of datasets and reduce confidence in proteomic analysis. These signals have two sources: environmental proteins and chromatographic carryover from previous sample applications. Primary carriers of environmental contamination are the protocol reagents and solutions. Human skin proteins are in the air, on all laboratory surfaces, including the interior surfaces of disposables and instrumentation, and within reagents. Practices that reduce this contamination risk include the use of standard laboratory personal protective equipment (PPE), a protected workspace such as a PCR workstation, and the use of MS-grade reagents for the preparation of stocks and solutions (*see* **Notes 11–14**). Contamination can be further minimized by pre-cleaning all water, aqueous stocks, and solutions by passage through solid-phase extraction (Subheading 3.2). In addition to these assurance measures, quality control can occur through inclusion of reagent blanks in all sample sequences, to monitor environmental contamination, and inclusion of prewash and postwash chromatography conditioning runs to measure sample-to-sample carryover on the instrument.

3.2 Solid-Phase Extraction for Protocol 1 and 2 Solutions

1. Separate the syringe barrel and plunger. Fit the Sep-Pak cartridge to the Luer-Slip fitting of the syringe.
2. Fill the syringe barrel with at least 10 mL of LC/MS-grade methanol (MeOH). Attach the plunger and slowly push the MeOH through the SPE column.

3. Remove the cartridge from the syringe. Separate the syringe barrel from the plunger.
4. Reattach the cartridge onto the Luer-Slip fitting of the syringe.
5. Fill the syringe barrel with ultrapure water.
6. Reinsert the plunger and pass water through the cartridge. The column is now conditioned.
7. To purify the solution, again disassemble the column and syringe (column first). Reattach the cartridge onto the Luer-Slip fitting of the syringe. Fill the barrel with the solution to be purified.
8. Reinsert the plunger and pass the solution through the cartridge. Let the first 2 mL go to waste since this is water or diluted solution. Collect the remaining solution into suitably labeled container(s). If stock solutions are being made, dispense into individually labelled 1.5-mL microcentrifuge vials, followed by freezing at −20°C. Volumes can be approximate.
9. In order to store a cartridge after a solution has been prepared, wash the solution from the cartridge with water. If ultrapure water is used, this can be left on the cartridge for up to 2 weeks at room temperature.

(For additional considerations, *see* **Notes 15** and **16**).

3.3 Protocol 1 for Dried Specimens

3.3.1 Day 1 Laboratory Procedure

For swabs or dried specimens/reference material.

1. To each sample (swatch/1/2 swab/etc.) add 250 μL of 5 mM TCEP in 8 M urea, 100 mM Tris–HCl, and incubate for 30 min, vortexing frequently to facilitate protein recovery.
2. Transfer the sample substrate and any liquid to the basket of a 0.45 μm centrifuge tube filter. If necessary, add additional liquid to ensure the sample is fully submerged. Centrifuge at a maximum benchtop microcentrifuge speed for 10 min. Ensure that all liquid has moved through the filter, and if necessary, centrifuge again.
3. Transfer all filtrate to a 1.5-mL low-protein-binding microcentrifuge tube, discarding the substrate and centrifuge tube filter.
4. Incubate the filtrate in a thermal mixer at room temperature (RT) (~25 °C) for 30 min at 600 rpm.
5. To alkylate the reduced proteins, add 10.5 μL of 0.5 M CAA (final concentration ~20 mM to each 250 μL sample filtrate for a total 260.5 μL total volume).
6. Incubate in a thermal mixer at RT for 30 min at 600 rpm under aluminum foil.
7. Dilute the concentration of urea with 1000 μL of the prepared 100 mM Tris–HCl (*see* **Note 12**).

8. Add 13.5 μL of 1% (w/v) mass spectrometry-compatible trypsin enhancer surfactant, for a final concentration of 0.01% w/v after trypsin addition.
9. Since a protein assay is not being performed, add an excess of trypsin to the samples (20 μL of 0.5 μg/μL MS-grade trypsin for a total of 10 μg of trypsin per tube).
10. Incubate samples in a thermal mixer at 37 °C for 16 h or overnight at 600 rpm.

3.3.2 Day 2 Laboratory Procedure

1. Remove samples from the thermal mixer. Centrifuge at 13,166 rpm for 5 min.
2. Transfer the supernatant to 0.22 μm PVDF filter tubes in 500 μL aliquots and centrifuge for 3 min at 13,166 rpm or equivalent for the first two aliquots of 500 μL. For the final centrifugation, use a 30-min duration. Pool the filtrate for each sample into a low protein binding tube (*see* **Note 13**).
3. Acidify the peptide filtrate to a pH of 1–2 using 10% TFA (in incremental additions of 5–10 μL). After each addition of TFA, vortex the sample for at least 5 s, before taking 2–5 μL to assess the pH using a pH test strip.
4. Centrifuge the microcentrifuge vials at 13,166 rpm or equivalent for 5 min, to ensure no sample clings to the tube lid or walls.
5. Set out one new low-protein-binding tube for each sample containing 30 μL of elution buffer. Prepare one tube containing 500 μL of 0.1% TFA and one tube containing at least 1000 μL of 0.1% TFA for each sample.
6. Perform cleanup, concentration, and elution of peptides using 100 μL C18 tips and cut to an appropriate length using a razor blade. Attach the tip to a 2.5-mL capacity Hamilton gastight syringe by friction fit. Once the cut tip is attached to the gastight syringe, follow the steps below:
   1. First, condition the C18 resin by passing 500 μL of 100% ACN and then 500 μL of 0.1% TFA through the tip. Dispense both solutions into a waste container.
   2. Load peptides onto the resin by passing the entire volume of sample through the resin for 14 cycles.
   3. Wash the resin with 1000 μL of 0.1% TFA. Again, dispense the wash solution into a waste container.
   4. Elute peptides into the low-protein-binding tube containing 30 μL of elution buffer by passing the liquid through the resin for ten cycles.
   5. Store the eluted peptides at −20 °C until sample submission/sample preparation for LC/MS injection and analysis.

3.4 Protocol 2 for Liquid Substrate

3.4.1 Day 1 Laboratory Procedure

For neat specimens/reference material.

1. In a low-protein-binding tube, combine 10 μL of neat specimen with 30.3 μL of 8 M urea 100 mM Tris–HCl.
2. Add 0.5 M TCEP to a final concentration of 5 mM TCEP (i.e., addition of 4.5 μL of 0.5 M TCEP).
3. Incubate in a thermal mixer at RT for 60 min at 600 rpm.
4. To alkylate the reduced proteins, dilute 0.5 M CAA stock to a concentration of 0.25 M with ultrapure water. Add 3.9 μL of 0.25 M CAA (final concentration ~20 mM CAA) to each sample tube.
5. Incubate in a thermal mixer at RT for 30 min at 600 rpm in darkness.
6. Since the tryptic digestion will be inhibited by high concentrations of urea, dilute the urea concentration with 175 μL of the prepared 100 mM Tris–HCl. This decreases the concentration of urea from approximately 6 M to approximately 1 M urea by the trypsin addition step.
7. Add 5 μL of 0.5% (w/v) mass spectrometry-compatible trypsin enhancer surfactant for a final concentration of 0.01% w/v after trypsin addition.
8. Since a protein assay is not being performed, an excess of trypsin is added to the samples (20 μL of 0.5 μg/μL MS-grade trypsin for a total of 10 μg of trypsin per tube). There is no discernable effect of the excess on data generated due to trypsin autolysis.
9. Incubate samples in a thermal mixer at 37 °C for 16 h or overnight at 600 rpm.

3.4.2 Day 2 Laboratory Procedure

1. Remove samples from the thermal mixer. Centrifuge at 13,166 rpm for 5 min.
2. Transfer the supernatant to 0.22 μm polyvinylidene difluoride (PVDF) filter tubes. Centrifuge at 13,166 rpm for 30 min. Discard the basket. Retain the filtrate in its tube. This is now the binding tube.
3. Acidify the peptide filtrate to a pH of 1–2 using 10% TFA (in incremental additions of 5–10 μL). After each addition of TFA, vortex the sample for at least 5 s, before taking 2–5 μL to assess the pH using a pH test strip.
4. Centrifuge the samples again at 13,166 rpm for 5 min, to ensure no sample clings to the tube lid or walls.
5. Set out one new low-protein-binding tube for each sample containing 30 μL of elution buffer. Prepare one tube containing 500 μL of 0.1% TFA for resin conditioning and one tube containing 1000 μL of 0.1% TFA for peptide washing, for each sample.

6. Perform cleanup, concentration, and elution of peptides using 100 μL C18 tips and cut to an appropriate length using a razor blade. Attach the tip to a 2.5-mL capacity Hamilton gastight syringe by friction fit. Once the cut tip is attached to the syringe, the procedure is as follows:
   1. Condition the C18 resin by passing 500 μL of 100% ACN and then 500 μL of 0.1% TFA through the tip.
   2. Load peptides onto the resin by passing the entire volume of sample through the resin for 14 cycles.
   3. Wash the resin with 1000 μL of 0.1% TFA. Again, dispense the wash solution into a waste container. Elute peptides by passing 30 μL of the elution buffer through the resin for ten cycles (*see* **Notes 17–19**).
7. Store the eluted peptides at −20 °C until submission for LCMS injection and analysis (*see* **Notes 20** and **21**).

4 Notes

1. To minimize contamination, organic solvents (e.g., MeOH, ACN) should be dispensed into fresh polypropylene tubes rather than into previously used glassware. Since polypropylene will eventually leach impurities from the container, organic solvents should be stored no longer than approximately 1 week before being discarded. It is recommended, therefore, to decant no more than what will be necessary for the number of samples being processed.
2. Mass spectrometry-compatible trypsin enhancer surfactant comes in vials of 1 mg each (lyophilized). Accordingly, a 1% solution is prepared by the addition of 100 μL of 50 mM ABC to the 1 mg of lyophilized surfactant. This must be prepared before addition to the sample. This produces a 0.01% final concentration during the digestion step.
3. At least 1 mL per sample will be needed to dilute the urea concentration of the sample before tryptic digestion. An additional 250 μL per sample will be needed to make the urea stock solution. The extra 2 mL is to account for volume loss during the SPE cleanup step.
4. Water should be ultrapure grade. Ensure that at least 2.5 μL of 0.5 M TCEP per sample to be processed has been prepared. Remember to prepare the stock solution with an excess of 2 mL, as some will be lost to the Sep-Pak cartridge resin during the cleaning step.
5. Water should be ultrapure grade. Ensure that at least 10.5 μL of 0.5 M CAA per sample to be processed has been prepared.

Other alkylants, such as iodoacetamide or iodoacetic acid, work effectively as well but may increase oxidations in the sample [26]. Remember to make the stock solution with an excess of 2 mL, as some will be lost to the Sep-Pak cartridge resin during the cleaning step.

6. Ensure that at least 247.5 μL of solution per sample to be processed has been prepared. Remember to make the solution with an excess of 2 mL, as some will be lost to the Sep-Pak cartridge resin during the cleaning step.
7. Make enough for at least 250 μL of 5 mM TCEP in 8 M urea, 100 mM Tris–HCl per sample.
8. To make the stock solution, any concentration at or above 50 mM can be made, but it is most convenient to make a higher concentration such as 2 M ABC initially and dilute the filtered stock to the necessary concentration afterward with ultrapure water. Ensure that enough has been prepared so that when diluted to a concentration of 50 mM, there is sufficient solution to account for the number of vials of mass spectrometry-compatible trypsin enhancer surfactant and trypsin to be used. Each vial of surfactant will require 100 μL of ABC solution and will be sufficient for the processing of up to seven samples. Each vial of trypsin will require 40 μL of the buffer and will be sufficient for the processing of up to two samples. For example, processing of a single sample with a reagent blank would require at least 140 μL of 50 mM ABC. Remember to make the 2 M ABC with an excess of 2 mL, as some will be lost to the Sep-Pak cartridge resin during the cleaning step.
9. Prepare the solutions for mass spectrometry-compatible trypsin enhancer surfactant and mass spectrometry-grade trypsin immediately before adding them to the sample.
10. TFA at 10% is used for acidifying the sample on the second day of the protocol. Because TFA at 0.1% is used as a wash step during peptide cleanup, it is simplest to make a serial dilution of 10%, 1%, and 0.1% TFA. Prepare enough solution so that at least 1.5 mL of 0.1% TFA can be made for each sample.
11. All aqueous stocks and solutions should be prepared using high-quality ultrapure water. The process is the same for all aqueous reagents to be prepared and will require, in addition to the reagent, a 10-mL volume syringe and tC18 Sep-Pak cartridge, ultrapure water, and LC/MS-grade MeOH. Once solvent has been added to the syringe, take care not to push air through the column. In the event that this does occur, the air can be removed by washing with water and repeating the low surface tension MeOH step. Only do this step if it is difficult to force the solution through the SPE column. Because this step

employs a C18 solid-phase resin cartridge to remove impurities from aqueous stocks, take care to not purify solutions containing organics or detergents/surfactants.

12. This decreases the concentration of urea from 8 M to around 1.55 M urea when trypsin is added. Tryptic digestion will be inhibited by high concentrations of urea.
13. The maximum speed of most benchtop microcentrifuges is suitable. Because the capacity of the tubes used for filter centrifugation is about 500 μL, three centrifugations are required in this protocol. The filtrate will need to be transferred (and pooled into) to a low-protein-binding tube with 1.5-mL capacity between one centrifugation and the next, so as not to overfill the spin tube. At the end of this step, the filter tubes may be discarded, and a single low-protein-binding tube per sample containing the pooled filtrate should be left.
14. Dispensing reagents from bottles should be done with care; only pour from a bottle or stock solution. If a spatula must be used to dispense reagents from a bottle (e.g., for weighing on a balance), pour the solid reagent from the original bottle into a secondary container so that the original material is untouched. Replace any lid immediately. Disposables, such as 1.5-mL microcentrifuge vials, should likewise only be dispensed from the manufacturer's sealed (ziplocked) container, followed by immediate resealing. These consumables have low retention options available that minimize loss of signal with low level/ trace samples. When solutions are prepared in a new 15-mL polypropylene vial, the graduations on the side of the vial should be used to approximate the correct volume. Cleanliness is the main consideration, as the protocol is sufficiently robust to accommodate slight variations in concentration. Separate volumetric flasks or measuring cylinders should not be used for creating stocks and solutions, since this leads to an increased risk of contamination.
15. This protocol does not include a protein quantitation step, such as a BCA or Bradford protein assay, which would indicate the optimal amount of trypsin (i.e., a 1:50 protease-to-substrate ratio) to be added to the sample. As written, the use of excess trypsin at 10 μg per sample, while expensive, removes the protein assay and potentially a further sample dilution step. This helps to ensure that even samples with high concentrations of protein will be efficiently digested. Smaller amounts of trypsin may work just as well, particularly with dilute samples such as saliva. Peptide data should be routinely analyzed for incomplete digestion using bioinformatic software. A lower-cost sequencing-grade trypsin may be substituted if desired.

16. A protein quantitation step would also help optimize the cleanup efficiency. Particularly for Protocol 1. The step that is most time intensive is the supernatant transfer in Day 2 which takes more time, particularly as more samples are processed. This time could be reduced by either (a) sample centrifugation with a capacity for flow-through volume of 1.5 mL or more or (b) alteration of initial conditions such that dilution to 2 M or less urea in the digestion step on Day 1 results in a total volume of 1 mL or less. Higher concentrations of protein, for example, from blood or semen samples, may overload the C18 pipette tip resin.
17. Concentration and desalting using the C18 should be done carefully and slowly, to provide time for peptides to bind. This reduces the risk of losing peptides by dislodging the C18 resin from the tip.
18. When cutting the C18 tips prior to the cleanup step, it is recommended that the portion of the tip which is to be in contact with solution be covered (e.g., with a fresh laboratory wipe) to ensure that the tip remains clean. Razor blades are used to cut the C18 tips because scissors deform the plastic. The cut should be made so as to allow friction fit of the Hamilton syringe to the tip; do not cut too far down the tip, as it will no longer fit the syringe.
19. Because care must be taken not to introduce air to the C18 resin when using the Hamilton syringe, this step requires fine control when raising and lowering the syringe plunger. This can be quite taxing on the hands.
20. Gastight syringes should be cleaned after each use by passing two full syringe barrels of 60% ACN, 0.1% TFA followed by two full syringe barrels of MeOH and finally two full syringe barrels of ultrapure water through each syringe.
21. This protocol assumes that institutional specific biohazard procedures for biosafety are followed.

Acknowledgments

The authors thank the Fredric Rieders Family Foundation and National Institute of Justice for their support (2020-DQ-BX-0015, PI:Mirna Ghemwari).

References

1. Strimbu K, Tavel JA (2010) What are biomarkers? Curr Opin HIV AIDS 5(6):463–466. https://doi.org/10.1097/COH.0b013e32833ed177
2. Meier F et al (2018) BoxCar acquisition method enables single-shot proteomics at a depth of 10,000 proteins in 100 minutes. Nat

Methods 15(6):440–448. https://doi.org/10.1038/s41592-018-0003-5

3. Sunder M, Leslie SW (2022) Semen analysis. In: StatPearls. StatPearls Publishing. https://www.ncbi.nlm.nih.gov/books/NBK564369/
4. Loo JA, Yan W, Ramachandran P, Wong DT (2010) Comparative human salivary and plasma proteomes. J Dent Res 89(10):1016–1023. https://doi.org/10.1177/0022034510380414
5. Ma C et al (2023) Comprehensive and deep profiling of the plasma proteome with protein corona on zeolite NaY. J Pharm Anal 13(5):503–513. https://doi.org/10.1016/j.jpha.2023.04.002
6. Parker GJ, McKiernan HE, Legg KM, Goecker ZC (2021) Forensic proteomics. Forensic Sci Int Genet 54:102529. https://doi.org/10.1016/j.fsigen.2021.102529
7. Aebersold R et al (2018) How many human proteoforms are there? Nat Chem Biol 14(3):206–214. https://doi.org/10.1038/nchembio.2576
8. Dammen-Brower K et al (2022) Strategies for Glycoengineering therapeutic proteins. Front Chem 10. https://doi.org/10.3389/fchem.2022.863118
9. Reily C, Stewart TJ, Renfrow MB, Novak J (2019) Glycosylation in health and disease. Nat Rev Nephrol 15(6):346–366. https://doi.org/10.1038/s41581-019-0129-4
10. Rock KL, Kono H (2008) The inflammatory response to cell death. Annu Rev Pathol 3(1):99–126. https://doi.org/10.1146/annurev.pathmechdis.3.121806.151456
11. Sensabaugh G (2016) Serology: overview. In: Payne-James J, Byard RW (eds) Encyclopedia of forensic and legal medicine, 2nd edn. Elsevier, Oxford, pp 254–266
12. Sijen T, Harbison S (2021) On the identification of body fluids and tissues: a crucial link in the investigation and solution of crime. Genes 12(11):1728
13. Siegel D (2018) In: Do J (ed) Development of a proteomic assay for menstrual blood, vaginal fluid and species identification. Department of Justice, Washington D.C
14. Danielson P (2011) In: Do J (ed) Isolation of highly specific protein markers for the identification of biological stains: adapting comparative proteomics to forensics. Department of Justice
15. Legg KM et al (2017) Verification of protein biomarker specificity for the identification of biological stains by quadrupole time-of-flight mass spectrometry. Electrophoresis 38(6):833–845. https://doi.org/10.1002/elps.201600352
16. Pilch B, Mann M (2006) Large-scale and high-confidence proteomic analysis of human seminal plasma. Genome Biol 7(5):R40. https://doi.org/10.1186/gb-2006-7-5-r40
17. Leader MST (2023) Molecular serology validation: proteomic body fluid assay validation. In: Examiner NYCOotCM (ed). New York
18. Bader JM, Albrecht V, Mann M (2023) MS-based proteomics of body fluids: the end of the beginning. Mol Cell Proteomics 22(7):100577. https://doi.org/10.1016/j.mcpro.2023.100577
19. Zhang J et al (2024) Mass spectrometry-based proteomic analysis of biological stains identifies body fluids specific markers. Forensic Sci Int 357:112008. https://doi.org/10.1016/j.forsciint.2024.112008
20. Hu S, Loo JA, Wong DT (2006) Human body fluid proteome analysis. Proteomics 6(23):6326–6353. https://doi.org/10.1002/pmic.200600284
21. McKiernan HE et al (2019) NextGen serology: leveraging mass spectrometry for protein-based human body fluid identification. In: Applications in forensic proteomics: protein identification and profiling. ACS Publications, pp 47–80
22. Zhao M et al (2018) A comparative proteomics analysis of five body fluids: plasma, urine, cerebrospinal fluid, amniotic fluid, and saliva. Proteomics Clin Appl 12(6):1800008. https://doi.org/10.1002/prca.201800008
23. Legg KM (2016) Validation of a mass spectrometry based serological assay. In Report No.: 67484-LS-RIF, U.S. Army Research Office. https://apps.dtic.mil/sti/citations/trecms/AD1195287
24. Liu W et al (2019) Informatics approaches to forensic body fluid identification by proteomic mass spectrometry. In: Applications in forensic proteomics: protein identification and profiling. ACS Publications, pp 81–90
25. Legg KM et al (2014) Discovery of highly specific protein markers for the identification of biological stains. Electrophoresis 35(21–22):3069–3078. https://doi.org/10.1002/elps.201400125
26. Muller T, Winter D (2017) Systematic evaluation of protein reduction and alkylation reveals massive unspecific side effects by iodine-containing reagents. Mol Cell Proteomics 16(7):1173–1187. https://doi.org/10.1074/mcp.M116.064048

Chapter 11

Analyzing Various Biological Fluids with a Single Automated Proteomic Workflow for Biomarker Discovery

Ane Laura Pedersen, Marion Ernest, Michael Affolter, and Loïc Dayon

Abstract

Protein biomarker discovery in human biological fluids has greatly developed over the past two decades thanks to technological advances allowing deeper proteome coverage and higher sample throughput, among others. While blood samples are most commonly investigated due to their moderate ease of collection and high information content, other biological fluids such as cerebrospinal fluid (CSF) and urine are highly relevant for specific pathologies, such as brain and urologic diseases, respectively. Independently of the biofluid of interest, platforms that can robustly handle a large number of samples are essential in the discovery phase of a clinical study.

We have previously described a scalable automated proteomic pipeline (ASAP^2) for the sample preparation of hundreds to thousands of blood plasma, serum, and CSF samples before liquid chromatography (LC)-mass spectrometry (MS) analysis. Here, we describe how the workflow was further adapted to milk and urine samples, with small modifications at the beginning of the workflow. For blood and CSF samples, an optional immuno-affinity depletion step for abundant-protein removal constitutes the first step of the workflow. In the analysis of milk, a defatting step is incorporated before the samples are further processed with ASAP^2, while acetone precipitation is used for the analysis of urine samples. The main sample preparation steps then remain identical for all sample types, are automated on a liquid handling workstation, and include reduction of disulfide bridges, alkylation of free thiols, protein digestion, isobaric labeling of peptides, sample pooling, and purifications. The workflow is completed by LC-MS analysis of the samples and subsequent data processing.

Key words Automation, Biofluid, Clinical research, CSF, Human, Isobaric tagging, Mass spectrometry, Milk, Plasma, Proteomics, Serum, Urine

1 Introduction

Proteomic analysis of biological fluids has emerged as a powerful tool for the discovery of biomarkers intended for disease diagnosis, prognosis, monitoring, and treatment response prediction. Reversed-phase (RP)-liquid chromatography (LC)-mass spectrometry (MS) is the most common, versatile technique used for hypothesis-free proteomic measurements in biofluids [1]. In biomarker discovery clinical studies, a large number of samples should

Taufika Islam Williams (ed.), *Tissue Proteomics: Methods and Protocols*, Methods in Molecular Biology, vol. 2884, https://doi.org/10.1007/978-1-0716-4298-6_11,

be analyzed to increase the chances of positive outcomes of the study [2]. High sample throughput is necessary and relies heavily on automation of sample preparation and analysis.

Blood plasma (or serum) is by far the most studied biofluid, but others such as cerebrospinal fluid (CSF), saliva, tears, milk, and urine have also emerged to be valuable fluids as sources of biomarkers [3]. Particularly, CSF presents an interest when studying brain disorders, with ~20% of total CSF proteins being derived from the central nervous system [4]. Urine is easy and non-invasive to collect and contains specific information about urogenital diseases, as well as connections to blood-derived proteins [5]. Milk can give insights into mammary health, in addition to providing information about optimal infant nutrition [6, 7].

For bottom-up proteomics, disulfide bridges in proteins are reduced, and thiols are alkylated before protein digestion and purification of the generated peptides [8]. However, depending on the sample analyzed, additional steps are necessary to optimize the protein extraction, while others might be beneficial to increase the proteome coverage. For instance, biological fluids generally have a high dynamic range of protein concentrations. Consequently, the detection of low-abundant proteins is often hindered by high-abundant proteins. For blood plasma and serum analysis, immuno-affinity depletion of the most abundant proteins results in a reduction of the dynamic range of the sample and thus a deeper proteome coverage. This technique is also efficient for reducing the dynamic range in CSF [9]. For milk, the high lipid content can interfere with protein extraction and protein detection. It is therefore necessary to incorporate a defatting step into the workflow when analyzing milk. Urine has a low and highly variable protein concentration. Additionally, several other components such as salts and small molecules are present and can interfere with protein analysis [10].

We developed "a scalable automated proteomic pipeline" ($ASAP^2$) for the analysis of blood plasma samples from large cohorts [11]. We have adapted the workflow for the analysis of serum, CSF, milk, and most recently urine samples, which will be described here within [12–14]. $ASAP^2$ is a multi-modular approach which allows for the introduction or not of optional steps in the analytical workflow. The core workflow (Fig. 1) consists of six mandatory steps: (1) spiking of a quality control internal standard; (2) reduction, alkylation, and digestion of proteins into peptides using trypsin/Lys-C; (3) labeling with tandem mass tag (TMT) reagents [15]; (4) sample pooling; and (5), (6) two purification steps [RP solid-phase extraction (SPE) and strong-cation exchange (SCX) SPE]. For milk samples, defatting is included as a step prior to the core workflow, while acetone precipitation is done for urine samples. These steps are currently carried out manually but could be incorporated into $ASAP^2$ with only slight modifications to the

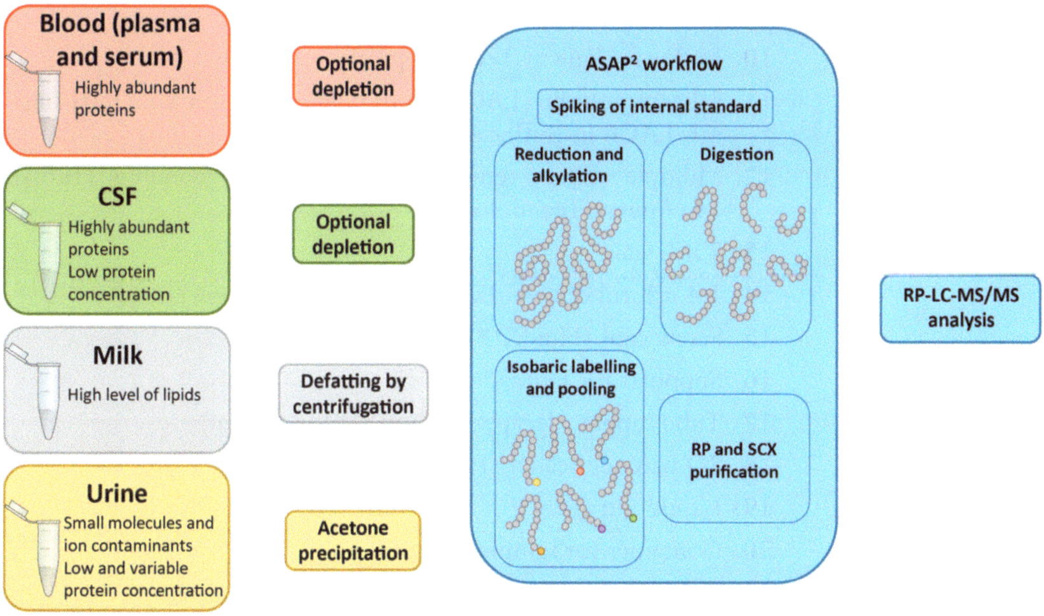

Fig. 1 Overview of the biofluids integrated into ASAP^2. The specificity of each biofluid and the strategy used to integrate the sample into the workflow is shown by color-coded boxes

liquid handling workstation. Additionally, the optional step of immuno affinity depletion of abundant proteins for blood plasma, serum, and CSF samples can be added upstream of the main workflow. A laboratory information management system (LIMS) complements the experimental set-up for easy sample handling and tracking.

2 Materials

2.1 Depletion of Abundant Proteins

1. 12-channel 10–100 μL multi-pipette.
2. Sterile, clear 96-well filter plate with 0.22 μm pore size polyvinylidene difluoride membrane.
3. Vibrating platform.
4. *Vacuum* manifold.
5. Polypropylene 96-well plate (V-bottom).
6. Adhesive seal.
7. Speed-vac centrifuge system equipped with an ultra-low temperature refrigerated vapor trap.
8. High-performance LC (HPLC) system composed of a 1260 Infinity Bio-inert Quaternary LC, a 1290 Infinity Flexible Cube, a two-position/ten-port valve, and a fraction collector module (Agilent Technologies) or an equivalent system, allowing a two-column configuration as described previously [16].

9. Buffer A (Agilent Technologies).
10. Buffer B (Agilent Technologies).
11. β-Lactoglobulin (LACB) from bovine milk solution: 0.01 mg/mL LACB in Buffer A (*see* **Note 1**).
12. Multiple affinity removal system (MARS) columns Human 14 (Agilent Technologies).
13. 2 mL vials.
14. 1.4 mL tubes (FluidX™, Brooks Life Sciences).
15. Screw caps (FluidX™).
16. Supports for FluidX™ tubes.
17. Tube capper/decapper (FluidX™) (optional).
18. 2D barcode reader (FluidX™).
19. Liquid nitrogen.
20. Container for sample snap freezing.
21. Low-temperature freezer (−80 °C).
22. LIMS software.

2.2 Buffer Exchange

1. 2D barcode reader (FluidX™).
2. Vibrating platform.
3. Support for FluidX™ tubes.
4. Tube capper/decapper (FluidX™) (optional).
5. Speed-vac centrifuge system equipped with an ultra-low temperature refrigerated vapor trap.
6. Strata-X 33u Polymeric RP (30 mg/1 mL) cartridges (Phenomenex).
7. *Vacuum* manifold and SPE cartridge holder.
8. Conditioning buffer: 0.08% trifluoroacetic acid (TFA) in acetonitrile (CH_3CN).
9. Washing buffer: 0.1% TFA in water (18.2 MiΩ · cm at 25 °C obtained with a Milli-Q apparatus).
10. Elution buffer: 70% CH_3CN, 0.08% TFA in water.
11. 12-channel 50–1200 μL multi-pipette.
12. Pipette tips 50–1250 μL, with two different lengths (76 and 103 mm).
13. 2 mL 96-DeepWell plate.
14. 96-well sealing mats.
15. Low-temperature freezer (−80 °C).
16. LIMS software.

2.3 Defatting

1. Refrigerated centrifuge.
2. 0.5-mL polypropylene tubes.
3. Sodium dodecyl sulfate (SDS) solution: 2% (*w/v*) SDS in water.

2.4 Acetone Precipitation

1. Ice-cold acetone (stored at −20 °C).
2. Plate shaker.
3. Freezer.
4. 12-channel 15–300 μL multi-pipette.
5. Speed-vac centrifuge system equipped with an ultra-low temperature refrigerated vapor trap.
6. Pierce™ bicinchoninic acid (BCA) protein assay kit or equivalent.

2.5 Reduction, Alkylation, and Proteolytic Digestion

1. 1.5-mL polypropylene tubes.
2. Triethylammonium bicarbonate (TEAB) buffer: 100 mM TEAB in water obtained from 1 M TEAB stock solution, pH 8.5.
3. SDS solution: 2% (*w/v*) SDS in water (for depleted blood plasma/serum and depleted CSF sample preparation).
4. SDS solution with LACB: 2% (*w/v*) SDS in water supplemented with 0.12 μg/μL LACB (for urine, nondepleted blood plasma/serum and nondepleted CSF sample preparation).
5. TEAB buffer with LACB: 100 mM TEAB, supplemented with 0.12 μg/μL LACB (for milk sample preparation).
6. Tris(2-carboxyethyl) phosphine hydrochloride (TCEP) solution: 20 mM TCEP in water.
7. Iodoacetamide (IAA) solution: 150 mM IAA in CH_3CN.
8. Trypsin solution: 0.25 μg/μL sequencing grade modified trypsin/Lys-C (Promega) in 100 mM TEAB.
9. Four-channel Microlab STAR liquid handling system (Hamilton) or equivalent system.
10. Tube racks, temperature-controlled mixer, temperature-controlled dark chamber, and reservoirs mounted on the deck of the liquid handling system.
11. 96-well sealing mats.
12. Centrifuge.
13. LIMS software.

2.6 Isobaric Labeling

1. TMT reagents (Thermo Scientific) (*see* **Note 2**).
2. Hydroxylamine solution: 5% (*w/v*) hydroxylamine in water obtained from a 50 wt.% hydroxylamine solution in water (99.999%).

3. RP loading buffer: 5% CH_3CN, 0.1% TFA in water.
4. 5.0-mL, 1.5-mL, and 0.5-mL polypropylene tubes.
5. Four-channel Microlab STAR liquid handling system (Hamilton) or equivalent system.
6. Tube racks and temperature-controlled mixer mounted on the deck of the liquid handling system.
7. Vortex mixer and centrifuge.
8. 96-well sealing mats.
9. LIMS software.

2.7 Sample Purifications

1. Oasis hydrophilic-lipophilic-balanced (HLB) 1 cc (30 mg) extraction cartridges (Waters).
2. Strata-X-C 33u polymeric strong cation (30 mg/1 mL) cartridges (Phenomenex).
3. Adhesive seal and SPE cartridge holder for plate formats.
4. 2.0-mL 96-DeepWeel plate.
5. RP conditioning buffer: 95% CH_3CN, 0.1% TFA in water.
6. RP loading buffer: 5% CH_3CN, 0.1% TFA in water.
7. RP elution buffer: 50% CH_3CN, 0.1% TFA in water.
8. SCX loading buffer: 25% CH_3CN, 0.1% TFA in water.
9. SCX elution buffer: 25% CH_3CN, 400 mM ammonium acetate (NH_4OAc) in water.
10. Four-channel Microlab STAR liquid handling system (Hamilton) or equivalent system.
11. *Vacuum* manifold mounted on the deck of the handling system.
12. Speed-vac centrifuge system equipped with an ultra-low temperature refrigerated vapor trap.
13. LIMS software.

2.8 RP-LC-MS/MS Analysis

1. Vanquish NEO UHPLC (Thermo Scientific) or equivalent system.
2. Acclaim™ PepMap™ 300 μm × 5 mm (C18, 5 μm, 100 Å) pre-column and Acclaim™ PepMap™ 75 μm × 50 cm (C18, 2 μm, 100 Å) analytical column (Thermo Scientific) or equivalent columns.
3. Stainless steel nanobore emitter (40 mm, OD 1/32″) mounted on a Nanospray Flex™ Ion Source (Thermo Scientific) or equivalent system.
4. PRSO-V1 column oven (Sonation) or equivalent system.

5. LC eluant A and weak wash solution: 2% CH_3CN, 0.1% formic acid (FA) in water.
6. LC eluant B and strong wash solution: 80% CH_3CN, 0.1% FA in water.
7. Seal wash solution: 75% isopropanol, 0.1% FA in water.
8. LC vials (Thermo Scientific) (*see* **Note 3**).
9. Orbitrap Fusion Lumos Tribrid (Thermo Scientific) or equivalent mass spectrometer.
10. LIMS software.

2.9 Data Analysis

1. Mascot (version 2.8.2 or above) with Mascot Daemon (version 2.8.0 or above, *see* **Note 4**) (Matrix Sciences) [17] or an equivalent search engine.
2. ProteoWizard msConvert 3.0 [18] or an equivalent MS raw file converter if needed.
3. Human UniProtKB/Swiss-Prot database (*see* **Note 1**).
4. Scaffold Q+ (version 5.0.0 or above) (Proteome Software) or equivalent proteomics software allowing quantitative analysis of TMT data.
5. Excel (Microsoft) or equivalent.

3 Methods

3.1 General Practice

ASAP^2 was first developed for the discovery of protein biomarkers in plasma by analyzing cohorts of 100 s to 1000s of samples [19]. It was further expanded to study other biological fluids, such as blood serum and CSF, with little to no modifications. We recently adapted the workflow to even more diverse biofluids, such as milk and urine, by incorporating sample-specific steps to make these samples amenable to the core ASAP^2 (*see* Fig. 1). Preparation of blood plasma/serum and CSF can incorporate an optional first step of immunoaffinity depletion, performed using an antibody-based column mounted on a LC system. This step is linked to, and followed by, mandatory buffer exchange before the continuation of the workflow. The preparation of milk contains an initial defatting step, while for the preparation of urine, acetone precipitation is incorporated at the beginning of the workflow. Additionally, a BCA assay is used to estimate protein amount after precipitation of the proteins contained in urine. The rest of the workflow is identical for all sample types and includes reduction, alkylation and digestion, TMT labeling/pooling, RP, and SCX SPEs. These steps are fully automated and carried out with a liquid handling platform which dispenses pre-defined volumes of reagents with a four-channel pipetting arm, also acting as a gripper to move plates

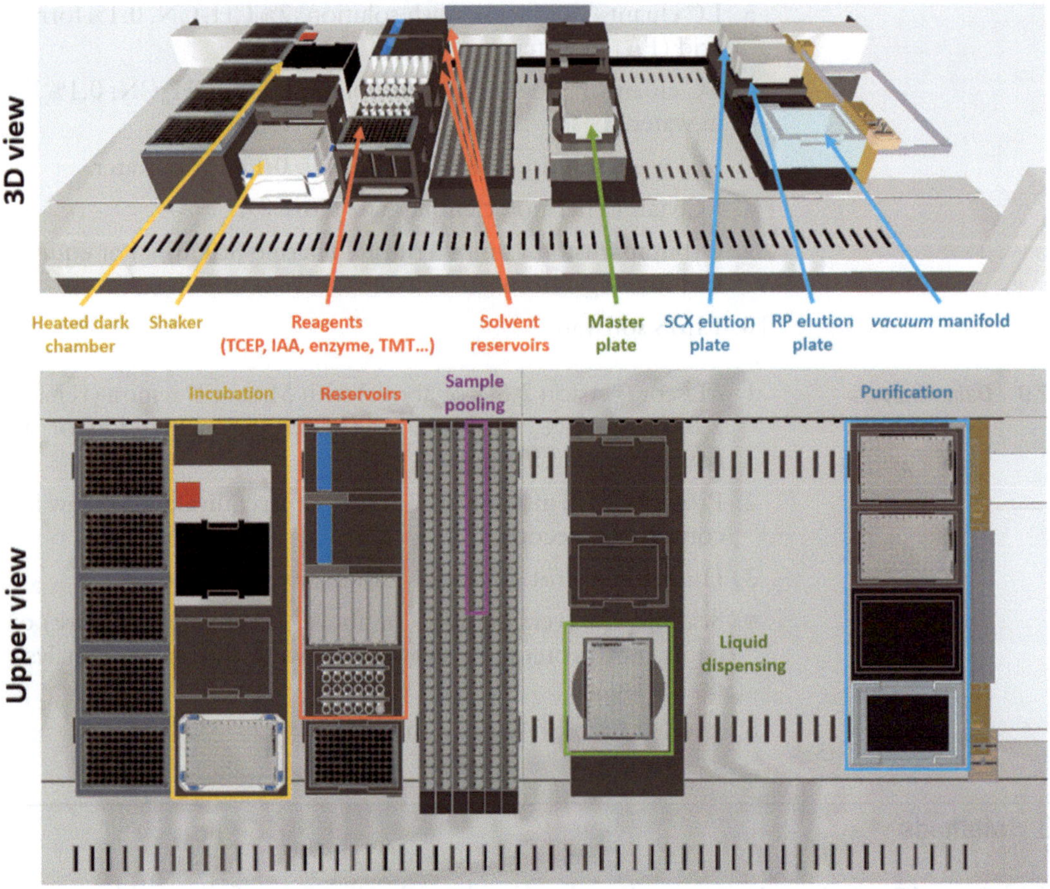

Fig. 2 Overview of the liquid handling station

across the robotic deck. The platform also includes a 96-well plate shaker, a heater, a dark chamber, a *vacuum* manifold, and reservoirs for stocking solvents (*see* Fig. 2). The procedures presented here within can equally be performed manually without the use of a liquid handling platform. However, this generally implies a lower throughput and is more prone to handling errors.

3.2 Depletion of Abundant Proteins

The depletion of abundant proteins is an optional step that can be performed for blood plasma, serum, and CSF according to the following method (*see* **Note 5**). The procedure depletes serum albumin, IgG, alpha-1-antitrypsin, IgA, serotransferrin, haptoglobin, fibrinogen, alpha-2-macroglobulin, alpha-1-acid glycoproteins, IgM, apolipoprotein A-I, apolipoprotein A-II, complement C3, and transthyretin from the sample matrices. It is recommended to perform this step when less abundant proteins are of relevance to the research and/or an increased proteome coverage of the fluids is desired.

Table 1
Recommended sample preparation details for each biological fluid typically prepared in our laboratory

Sample	Details	Typical protein concentration	Starting volume	LC-MS analysis: volume of LC eluant A for dissolving final sample
Blood plasma	Depletion		15 μL	250 μL
Blood plasma	No depletion	60–80 μg/μL	1.5 μL	1000 μL
Blood serum	Depletion		15 μL	250 μL
Blood serum	No depletion	60–80 μg/μL	1.5 μL	1000 μL
CSF	Depletion		400 μL for drying	200 μL
CSF	No depletion	0.15–0.6 μg/μL	100 μL for drying	375 μL
Urine	–	0.1–0.2 μg/μL	100 μg of protein calculated from BCA after precipitation	750 μL
Milk	–	10–15 μg/μL	40 μL[a]	900 μL

[a]Only 1/8 of this volume is used after the defatting step

1. Track the identity of the tubes containing samples to be depleted and check with a LIMS (*see* **Note 6**).
2. Transfer 105 μL of LACB from bovine milk solution (*see* **Note 1**) in wells of a 0.22 μm filter plate mounted on a polypropylene 96-well plate (*see* **Note 7**).
3. Add the samples to the previously filled wells and shake for 30 s. For detailed information about sample volumes, *see* Table 1 (*see* **Note 8**).
4. Filter the samples using a *vacuum* manifold for 1 min (*see* **Note 9**).
5. Spin down the diluted/filtered samples by performing a pulse centrifugation of the reception plate and apply an adhesive mat to seal the plate.
6. Keep diluted/filtered samples refrigerated until use.
7. Check that the HPLC system is functioning properly per the manufacturer's recommendations. Refill Buffers A and B. Check volumes in waste bottles and empty them if needed. Mount two MARS depletion columns (*see* **Notes 10** and **11**). Equilibrate the system with 100% Buffer A at a flow rate of

Table 2
LC conditions for immuno-affinity depletion of abundant proteins

Gradient			
Time (min)	**Buffer A (%)**	**Buffer B (%)**	**Flow rate (μL/min)**
0.00	100	0	125
9.50	100	0	125
9.51	100	0	1000
11.50	100	0	1000
11.51	0	100	1000
16.00	0	100	1000
16.01	100	0	1000
17.00	Stop time		
Flexible cube			
Time (min)	**Function**	**Parameter**	
0.00	Pump volume	Pump 4.5 mL, 1 mL/min	
12.00	Pump volume	Pump 4.9 mL, 1 mL/min	
Injection volume	100 μL		
Thermostat auto-sampler	4 °C		
Thermostat fraction collector	4 °C		
DAD	280 nm		
Peak width	>0.1 min (2 s resp. time) (5 Hz)		
Fraction collector			
4.00 min	Time-based. Collecting time slices 300.0 s		

125 μL/min. Ensure LC pressures are appropriate and stable. Monitor for leaks.

8. Place the 1.4 mL collection tubes in the fraction collector. Scan the barcodes of the collection tubes and register them in a LIMS (*see* **Note 6**).
9. Place a vial filled with Buffer A for blank injections (see **Note 12**).
10. Prepare the sample injection sequence.
11. Place the polypropylene 96-well plate containing the diluted samples in the auto-sampler of the HPLC system, maintained at 4 °C.
12. Deplete samples using the HPLC method indicated in Table 2 by systematically collecting unbound fractions that elute first

from the MARS column. Bound fractions containing abundant proteins are eluted with Buffer B and discarded (*see* **Note 13**).

13. After sample depletion, collect the tubes containing the depleted CSF samples from the fraction collector. Close the tubes with screw caps.
14. Snap-freeze the samples with liquid nitrogen in an appropriate container.
15. Store the tubes at −80 °C.
16. Record the analytical step and register any problems encountered in a LIMS after completing the process (*see* **Note 6**).

3.3 Buffer Exchange

The buffer exchange is mandatory after the depletion of abundant proteins (*see* **Note 5**) as at this stage, the proteins are in a buffer not compatible with the following reduction, alkylation, and proteolytic digestion.

1. Take depleted samples from the −80 °C freezer and place them on a 96-well rack.
2. Let the samples thaw at room temperature for 90 min. If necessary, place the samples in a container filled with cold water and agitate on a vibrating platform to accelerate the process.
3. In a LIMS, record the sample barcodes and confirm that their identities match the previously registered collection tubes used during the depletion process. Record the step in a LIMS (*see* **Note 6**).
4. Rapidly centrifuge the tubes/96-well rack to spin down the liquid. Uncap the tubes with the capper/decapper.
5. Use strata-X 33u Polymeric RP cartridges mounted on a cartridge holder and a *vacuum* manifold.
6. Using a 12 channel automatic pipette, condition the cartridges with 1 mL conditioning buffer (*see* **Note 14**).
7. Equilibrate the cartridges using the same 12-channel automatic pipette with 1 mL washing buffer.
8. Load the samples, introducing tips slowly into the cartridges using the multi-pipette fitted with longer tips.
9. Rinse the emptied sample tubes with 300 μL washing buffer and slowly load the washing solutions onto the cartridges.
10. Place a new clean 2.0 mL 96-well plate in the *vacuum* manifold and elute samples with 1 mL elution buffer.
11. At the end of the elution step, slowly increase the *vacuum* and wait for 1 min.
12. Recover the 96-DeepWell plate containing the buffer-exchanged samples and lyophilize the samples overnight in

the speed-vac centrifuge (we usually do not heat the centrifuge).

13. Recover the 96-DeepWell plate containing dried samples and seal it with a mat.
14. Store the bar-coded 96-DeepWell plate at −80 °C. Confirm completion of the step and register any problems encountered in a LIMS (*see* **Note 6**).

3.4 Defatting

Defatting is a mandatory step in the preparation of milk samples (it is not performed for other sample types) (*see* **Note 5**) to remove a high proportion of lipids and extract the protein fraction for proteomic analysis.

1. Start recording the step in a LIMS (*see* **Note 6**).
2. Place 40 μL of milk sample in a tube.
3. Add 40 μL of SDS 2% to the sample. Vortex.
4. Centrifuge for 1 h at 1000× *g*, 4 °C.
5. Pipette 10 μL of the skimmed milk, protein-containing phase (*see* **Note 15**) into a bar-coded 96-DeepWell plate.

3.5 Acetone Precipitation

The acetone precipitation is a mandatory step in the preparation of urine samples (it is not performed for other sample types) (*see* **Note 5**) in order to get rid of the salts and metabolites as well as extract the proteins that are at rather low concentrations in urine.

1. Start recording the step in a LIMS (*see* **Note 6**).
2. Centrifuge urine samples at 1000× *g*, 4 °C for 10 min.
3. Pipette 200 μL of the supernatant in a bar-coded 96-DeepWell plate (*see* **Note 16**).
4. Add 600 μL of cold acetone and shake the plate for 5 min.
5. Keep the plate at −20 °C overnight.
6. Centrifuge for 15 min at 2200× *g*, 4 °C.
7. Using the multichannel pipette, remove the supernatant without disturbing the pellet (*see* **Note 17**).
8. Dry the precipitated protein pellets for 10 min in the speed-vac centrifuge (we usually do not heat the centrifuge).
9. Redissolve the pellets in 100 μL of 100 mM TEAB. Perform a BCA to determine the appropriate volume to use for $ASAP^2$ (*see* **Note 16**).

3.6 Reduction, Alkylation, and Proteolytic Digestion

From this point, all the sample types follow the same preparation. Yet, the volume and nature of solvents used at the beginning of this step will vary depending on how the sample was treated previously.

1. Choose the correct **step 4** below (A–C) and refer to Table 1 for sample-specific starting volumes.

2. Start recording the step in a LIMS (*see* **Note 6**).
3. Place the 96-DeepWell plate containing the samples on the liquid handling workstation deck (Fig. 2, "Master Plate").
4. A. For *blood plasma, serum, CSF (nondepleted), and urine* samples, fill the TEAB reagent reservoir with the TEAB buffer and the SDS reagent tube with the SDS solution with LACB (Fig. 2, "Reagents") (*see* **Notes 18** and **19**).

   B. For *blood plasma, serum, and CSF (depleted)* samples, fill the TEAB reagent reservoir with the TEAB buffer and the SDS reagent tube with the SDS solution (Fig. 2, "Reagents") (*see* **Notes 18** and **19**).

   C. For *milk* samples, fill the TEAB reagent reservoir with the TEAB buffer and the SDS reagent tube with the TEAB buffer with LACB (Fig. 2, "Reagents") (*see* **Notes 18** and **19**).
5. Fill the TCEP tube with the TCEP solution (Fig. 2, "Reagents") (*see* **Notes 18** and **19**).
6. Start the automated run. The robot adds 5 μL solution from the SDS reagent tube and 95 μL solution from the TEAB reagent reservoir to each well. After shaking the plate for 30 s, the robot adds 5.3 μL TCEP solution to each well. Once all the solutions have been dispensed, the robot shakes the plate for 30 s and incubates it for 1 h at 55 °C (Fig. 2, "Heated dark chamber") (*see* **Note 20**).
7. Fill the IAA tube with the IAA solution (Fig. 2, "Reagents") shortly before the end of the previous incubation (*see* **Notes 18** and **19**).
8. After waiting for samples to cool down to room temperature for 2 min, the robot adds 5.5 μL IAA solution to each well, shakes the plate for 30 s, and incubates the samples at room temperature for 1 h in the dark (Fig. 2, "Heated dark chamber") (*see* **Note 20**).
9. Fill the trypsin/Lys-C reactant tube (Fig. 2, "Reagents") (*see* **Notes 18** and **19**). The robot adds 10 μL trypsin solution to each well, places a plastic cover on the 96-DeepWell plate, and incubates overnight at 37 °C with gentle shaking (*see* **Note 20**).
10. The next morning, stop the automated program and incubation process.
11. Discard used and/or empty laboratory ware.
12. Quickly spin down the samples in the 96-DeepWell plate by performing a pulse centrifugation.
13. Record the step completion and any problems encountered in a LIMS (*see* **Note 6**).

3.7 Isobaric Labeling

1. Start recording the step in a LIMS (*see* **Note 6**).
2. Place the 96-DeepWell sample plate on the liquid handling workstation (Fig. 2, "Master Plate").
3. Spin down the tubes containing the thawed, dried TMT reagents (*see* **Note 2**). Dissolve lyophilized TMT reagents in an adequate volume of CH_3CN. Typically, 0.8 mg of TMT reagent is needed for each sample to perform labeling and should be dissolved in 41 μL of CH_3CN (*see* **Note 21**). Vortex for 30 s and spin down the liquid.
4. Transfer the dissolved TMT reagents to labeled [e.g., 126, 127, 128, 129, 130, and 131 in the case of TMT 6-plex (*see* **Note 2**)] 0.5-mL or 1.5-mL tubes (*see* **Note 22**). Place the open tubes on the deck of the liquid handling system (Fig. 2, "Reagents").
5. Start the automated run. The robot adds 41 μL of TMT solution to each sample in the 96-DeepWell plate in line with the design of the TMT experiments. Samples are incubated for 1 h at room temperature while shaking (*see* **Note 20**).
6. Fill the hydroxylamine reagent tube with 1 mL of hydroxylamine solution (Fig. 2, "Reagents").
7. To quench the TMT reaction and reverse occasional labeling of tyrosine, serine, and threonine residues, the robot adds 8 μL of the hydroxylamine solution to each sample and incubates at room temperature for 15 min while shaking (*see* **Note 20**).
8. Next, the robot mixes/pools samples [6 by 6 in the case of TMT 6-plex (*see* **Note 2**), *see* Fig. 3, [20]] in a clean 5-mL tube (Fig. 2, "Sample Pooling"). In the case of 96 initial samples and using TMT 6-plex, 16 TMT experiments will be generated at this stage.
9. Fill the dedicated reservoir with RP loading buffer (Fig. 2, "Solvent reservoirs").
10. The robot washes each well of the 96-DeepWell plate with 120 μL of RP loading buffer and adds the washing liquid to the corresponding TMT experiment.
11. Discard used and/or empty laboratory ware.
12. Record the isobaric labeling step completion and any problems encountered in a LIMS after the process ends (*see* **Note 6**).

3.8 Sample Purification

3.8.1 RP Purification

1. Start recording the step in a LIMS (*see* **Note 6**).
2. Fill the stock bottles containing the RP loading buffer and the RP elution buffer. Fill the RP conditioning buffer reservoir (Fig. 2, "Solvent reservoirs") (*see* **Note 23**).

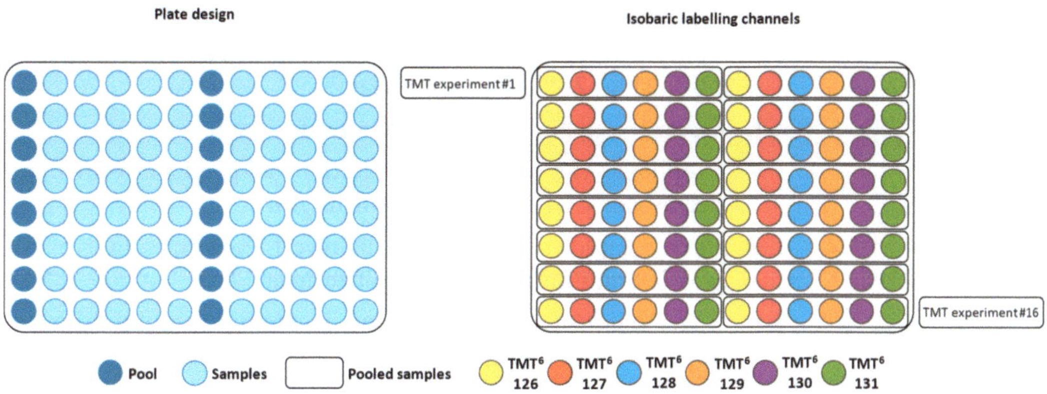

Fig. 3 Plate design used in ASAP^2. Left: Initial plate layout with pools used as internal references. Right: Representation of the labeling layout. Pooled samples making up one TMT experiment are shown by dashed boxes

3. Place the Oasis HLB cartridges on the holder of the *vacuum* manifold (Fig. 2, "*Vacuum* manifold") positioned on the deck of the liquid handling workstation for purification of each TMT experiment. Place a clean, bar-coded 96-DeepWell plate to collect purified samples (Fig. 2, "RP elution plate").
4. Start the automated run. The liquid handler dilutes the peptide samples with 2.7 mL RP loading buffer. Next, the robot prepares the cartridges with 2 × 0.95 mL RP conditioning buffer and 4 × 0.95 mL RP loading buffer. Samples are loaded [aspirating 5 × 0.86 mL from the 5-mL tubes containing the TMT experiments (Fig. 2, "Sample Pooling")] and samples are washed with 4 × 0.9 mL RP loading buffer.
5. Peptides are eluted with 2 × 0.8 mL RP elution buffer in the clean 96-DeepWell plate, which is bar-coded and registered in the LIMS (Fig. 2, "RP elution plate").
6. Discard used and/or empty laboratory ware.
7. Record the RP purification completion and any problems encountered during the process in a LIMS (*see* **Note 6**).

3.8.2 SCX Purification

1. Start recording the step in a LIMS (*see* **Note 6**).
2. Fill the stock bottles containing the SCX loading buffer and the SCX elution buffer (*see* **Note 23**).
3. Place the SCX cartridges on the holder of the *vacuum* manifold (Fig. 2, "*Vacuum* manifold") together with a clean, bar-coded 96-DeepWell plate (Fig. 2, "SCX elution plate").
4. Start the automated run. The robot equilibrates the cartridge with 4 × 0.93 mL SCX loading buffer. The robot loads the samples (aspirating 2 × 0.95 mL from the 96-DeepWell plate containing the Oasis HLB-purified samples) and washes the samples with 4 × 0.9 mL SCX loading buffer.

5. Peptides are eluted with 2 × 0.8 mL SCX elution buffer into the clean 96-DeepWell plate, which is bar-coded and registered in a LIMS (Fig. 2, "SCX elution plate").
6. Discard used and/or empty laboratory ware.
7. Lyophilize the samples overnight in a speed-vac centrifuge (we usually do not heat the centrifuge).
8. After complete evaporation, dissolve the samples in 1.6 mL SCX loading buffer, and lyophilize again.
9. Cover the 96-DeepWell plate with a sealing mat and keep the dried peptide samples at −80 °C before RP-LC-MS/MS analysis.
10. Record the SCX purification completion and any problems encountered during the process in a LIMS (*see* **Note 6**).

3.9 RP-LC-MS/MS Analysis

1. Start recording the step in a LIMS (*see* **Note 6**).
2. Prepare the sample sequence. Two blank (i.e., LC eluant A) samples are inserted every five sample injections (*see* **Note 24** and **25**).
3. Dissolve the prepared samples in an appropriate volume of LC eluant A (*see* Table 1 for details). Shake the 96-DeepWell plate for 5 min. Take 20 μL from each well into pre-labeled LC vials. Place the vials at 5 °C in the LC auto-sampler. Store the remaining sample volumes at −80 °C for further use.
4. Inject 2.5 μL of sample per analysis.
5. Run the RP-LC for 180 min using a gradient of LC eluant A and LC eluant B (as indicated in Table 3) at a flow rate of 300 nL/min.
6. Perform RP-LC-MS/MS analysis (*see* **Note 25**). Using an Orbitrap Fusion Lumos Tribrid mass spectrometer (*see* **Note 26**), data are acquired using a data-dependent method. A positive ion spray voltage of 1900 V and a transfer tube temperature of 275 °C are set up. For MS survey scans in profile mode, the Orbitrap resolution is 120,000 at $m/z = 200$ [automatic gain control (AGC) target of 2×10^5] with a m/z scan range from 300 to 1500, RF lens set at 30%, and maximum injection time of 100 ms. For MS/MS with higher-energy collisional dissociation (HCD) at 35% of the normalized collision energy, the AGC target is set to 1×10^5 (isolation width of 0.7 in the quadrupole), with a resolution of 30,000 at $m/z = 200$, first mass at $m/z = 100$, and a maximum injection time of 105 ms with Orbitrap acquiring in profile mode. A duty cycle time of 3 s (top speed mode) is used to determine the number of precursor ions to be selected for HCD-based MS/MS. Ions are injected for all available parallelizable time. Dynamic exclusion is set to 60 s within a ± 10 ppm window. A lock mass of $m/z = 445.1200$ is used.

Table 3
LC gradient for RP-LC-MS/MS analysis

Time (min)	LC eluant A (%)	LC eluant B (%)	Flow rate (μL/min)
0.00	*Run*		
0.00	93.7	6.3	0.300
2.00	93.7	6.3	0.300
87.00	82.5	17.5	0.300
129.00	74.5	25.5	0.300
157.00	60.0	40.0	0.300
157.00	*Column wash*		
157.10	2.0	98.0	0.300
165.00	2.0	98.0	0.300
165.00	*Column equilibration*		
	Fast equilibration in pressure control, equilibration factor 2.0		

7. Record completion of the analytical step and indicate any problems encountered during the process in a LIMS (*see* **Note 6**).

3.10 Data Analysis

1. Start the search with Mascot Daemon using Mascot as the search engine and ProteoWizard msConvert for the conversion of MS raw files. The search is performed against the Human UniProtKB/Swiss-Prot database (*see* **Note 1**). Set variable amino acid modifications to oxidized methionine, deamidated asparagine/glutamine, and TMT-labelled peptide amino terminus. TMT-labelled lysine is set as a fixed modification together with carbamidomethylation of cysteine. Trypsin is selected as the proteolytic enzyme, with a maximum of two potential missed cleavages. Peptide and fragment ion tolerance should be set to 10 ppm and 0.02 Da, respectively.
2. Load Mascot result files into Scaffold Q+ by queuing files from the Mascot server. Further, search the files with X!Tandem using Percolator for scoring. Set both peptide and protein false discovery rates (FDRs) to 1% with a requirement for two identified peptides to report protein identification.
3. Relative quantitative protein values are exported into an Excel-compatible format from Scaffold Q+ as Log_2 of the protein fold-change ratios with respect to the reference TMT channel, i.e., mean Log_2 values after isotopic purity correction (*see* **Note 27**). Normalization between samples and experiments is applied or not depending on the experiment.

4. Some quality checks are applied to the data, such as checking that all TMT experiments and replicates have a similar number of identified and quantified proteins and determining the standard deviation and errors in the values measured for the internal standard, LACB (*see* **Note 1**).
5. Proceed to further data processing and analysis depending on the nature of the study (*see* **Note 28**).

4 Notes

1. LACB from bovine milk is used as an internal protein standard and spiked in a fixed amount into all human samples at the earliest possible point of sample preparations. For blood plasma, serum, and CSF, this is done during the depletion step, while for urine and milk, this is done before the reduction step. We recommend including the LACB sequence in the most recent release of the Human UniProtKB/Swiss-Prot database. In the case where other organisms are studied, the database should be adapted accordingly. In this case, the internal standard might need to be adapted. It should be absent from the samples under investigation and should not generate tryptic peptides with sequences similar to the proteome studied. This can be checked by performing a BLAST sequence similarity search for the tryptic peptides generated by the internal standard in the proteome under study.
2. For the TMT reagents, different multiplexing exists. The reader is free to choose the one that best fits his or her experimental set-up, for instance, the TMTpro 16-plex [21] or 18-plex. We initially developed the ASAP^2 workflow using TMT 6-plex [11].
3. It is recommended to use auto-sampler-approved LC vials when working with the Vanquish NEO system to avoid clogging the needle of the HPLC system.
4. The Mascot Daemon interface allows for batch launching of database searches and integration of ProteoWizard msConvert for raw file conversion.
5. Herein, we describe how our ASAP^2 workflow was adapted to handle the sample preparation of different biofluids. This was achieved by performing specific optional or mandatory steps at the beginning of the sample preparation. More precisely, the depletion of abundant proteins is an optional step that can be included in the sample preparation of blood plasma, serum, and CSF. In the case where depletion of abundant proteins is performed, it must be followed by buffer exchange. Defatting is a step that is mandatory for milk samples and should be

omitted for all other biofluids considered herein. On the other hand, acetone precipitation is mandatory for urine samples while unnecessary for all other biofluids.

6. To minimize the risk of errors in both wet and dry laboratory steps in large-scale research studies, a customized LIMS software application was implemented in our laboratory. With ASAP^2, bar-coded laboratory ware is used at every step, and the LIMS is used for sample tracking and to record each experimental procedure.
7. Considering the gradient needed (*see* Table 2) for each sample and taking into account blank samples (typically injected every four samples to control for the absence of sample carry-over), it is recommended to process 48 samples per day.
8. For the preparation of CSF, a rather large volume of fluid is needed (*see* Table 1). Therefore, we recommend drying the sample first in order to have a more convenient volume for injection into the depletion HPLC system.
9. When samples are very viscous making the filtration difficult, it is recommended to repeat the filtration step and to pool the filtrated samples for depletion.
10. We evaluated the impact of using commercial MARS immunoaffinity LC columns for human plasma samples in terms of proteome coverage and investigated the specificity of the depletion of 14 highly abundant human plasma proteins. Typically, around 93% of the protein content is removed after depletion of the 14 specified proteins. The MARS columns should be changed every 300 injections.
11. The use of two columns in parallel allows to speed up the process. Typically, using two columns of 50 mm instead of one of 100 mm in our older set-up allows us to double the number of depleted samples per day [19].
12. It is recommended to inject two blank samples containing only Buffer A every eight plasma, serum, or CSF samples, i.e., one blank sample every four samples on each column to allow washing the lines and columns and checking for eventual contaminants. For this, a chromatogram using UV detection at 254 nm is acquired, and the fraction collection is disabled in the method.
13. Check that the collection window is well defined for the elution of the unbound fraction.
14. During buffer exchange, it is important to progressively apply the vacuum and verify that there is no remaining liquid in the cartridges before proceeding to the next steps.
15. When milk is centrifuged at 1000× *g* for 1 h, the fat is separated. One observes a pellet at the bottom of the tube, a liquid

protein-containing phase in the middle, and a fat layer at the top. When pipetting, the pipette should be introduced slowly through the fat layer to avoid disturbing it, before the protein-containing phase can be aspirated.

16. The protein concentration in urine is usually low and highly variable. Therefore, we recommend performing a BCA assay, according to the manufacturer's instructions, to determine the protein concentration after acetone precipitation and adapt the sample volume used for the ASAP^2 to correspond to 100 μg of proteins. Depending on the samples, 200 μL might not be sufficient to obtain this protein amount after precipitation. When working with new, unfamiliar samples, we recommend performing the precipitation on a pool to estimate the correct starting volume of urine before analyzing the full set of individual samples.
17. It can be difficult to aspirate the maximum volume of acetone without disrupting the pellet. Alternatively, one can aspirate a lower volume, i.e., 500 μL of acetone, before repeating subheading 3.5, **steps 4**–7.
18. We recommend overfilling the reagent tubes and reservoirs with an excess volume to avoid the lack of solutions.
19. We recommend preparing and placing solvents/reagents only a few minutes before the beginning of each step to avoid moisture condensation.
20. The liquid handling workstation includes a shaker/heating unit (Fig. 2, "Shaker"). The shaker speed is static and defined a priori for all processes. The shaker/heating unit can be switched on or off and is temperature-controlled. The dark chamber (Fig. 2, "Heated dark chamber") is temperature-controlled and can also be switched on or off.
21. TMT reagents are water-sensitive. One should therefore take care to let the reagents reach room temperature before opening the tubes.
22. For the TMT labeling process, reagents should be prepared as quickly as possible to prevent evaporation of the CH_3CN used to dissolve the reagents. One should take care to use different laboratory ware for each reagent. The size of the tubes should be adapted to the volume of dissolved reagents, i.e., 0.5-mL tubes when working with a few TMT experiments and 1.5-mL tubes when the volume of dissolved reagents is above 400 μL.
23. We recommend preparing fresh solvents every 3 weeks and purging the solvent lines before the RP and the SCX steps. This step is implemented in both the RP purification and the SCX purification on the liquid handling workstation.

24. It is a good practice to randomize the injection sequence for each replicate to avoid introducing a potential batch effect to samples analyzed early or late in the sequence.
25. The mass spectrometer is calibrated every week, following standard operating procedures and well-defined quality checks. We recommend using a complex standard peptide mix (e.g., obtained from the digestion of a complex protein sample such as HeLa cells) to control the performance of the RP-LC-MS/MS instrumentation and fine-tune it. The number of MS/MS scans, number of identified proteins and peptides, reference peptide retention time, peak width, intensity and ion injection time, mass accuracy, and multiplier voltage values are examples of readouts we record and track on a weekly basis.
26. $ASAP^2$ was originally developed using a hybrid linear ion-trap-Orbitrap (LTQ-OT) Elite (Thermo Scientific) mass spectrometer. It is currently performed using an Orbitrap Fusion Lumos Tribrid mass spectrometer. Other mass spectrometers can be used such as tandem time-of-flight (TOF-TOF) and quadrupole (Q)-TOF allowing MS/MS and detection of reporter ions in the low-mass range for isobaric labeling. Other MS acquisition parameters than the ones described in this chapter will be required.
27. A pool of samples (used as a biological reference) is mandatory to link TMT experiments in clinical studies where a large number of samples are compared. This biological reference is used for the calculation of the relative protein difference between all samples (Fig. 3).
28. The data analysis will depend heavily on the nature of the study, the samples, and the questions asked. We have previously described the data analysis for biomarker discovery in blood plasma/serum and CSF using $ASAP^2$ [22, 23].

References

1. Noor Z, Ahn SB, Baker S, Ranganathan S, Mohamedali A (2021) Mass spectrometry-based protein identification in proteomics—a review. Brief Bioinformatics 22:1620–1638
2. Hernández B, Parnell A, Pennington SR (2014) Why have so few proteomic biomarkers "survived" validation? (sample size and independent validation considerations). Proteomics 14:1587–1592
3. Dayon L, Cominetti O, Affolter M (2022) Proteomics of human biological fluids for biomarker discoveries: technical advances and recent applications. Expert Rev Proteomics 19:131–151
4. Begcevic I, Brinc D, Drabovich AP, Batruch I, Diamandis EP (2016) Identification of brain-enriched proteins in the cerebrospinal fluid proteome by LC-MS/MS profiling and mining of the human protein atlas. Clin Proteomics 13:1–13
5. Zhao M, Li Y, Yang Y, Guo Z, Sun Y, Shao C, Li M, Sun W, Gao Y (2017) A comprehensive analysis and annotation of human normal urinary proteome. Sci Rep 7:3024
6. Dingess KA, Li C, Zhu J (2021) Human milk proteome: what's new? Curr Opin Clin Nutr Metab Care 24:252–258

7. Schneider SS, Aslebagh R, Ngounou Wetie G, Sturgeo R, Darie CC, Arcaro KF (2014) Using breast milk to assess breast cancer risk: the role of mass spectrometry-based proteomics. In: Woods A, Darie C (eds) Advancements of mass spectrometry in biomedical research. Advances in experimental medicine and biology, vol 806. Springer, Cham
8. Shuken SR (2023) An introduction to mass spectrometry-based proteomics. J Proteome Res 22:2151–2171
9. Guldbrandsen A, Vethe H, Farag Y, Oveland E, Garberg H, Berle M, Myhr K-M, Opsahl JA, Barnes H, Berven FS (2014) In-depth characterization of the cerebrospinal fluid (CSF) proteome displayed through the CSF proteome resource (CSF-PR). Mol Cell Proteomics 13: 3152–3163
10. Thomas S, Hao L, Ricke WA, Li L (2016) Biomarker discovery in mass spectrometry-based urinary proteomics. Proteom Clin Appl 11:358–370
11. Dayon L, Núñez Galindo A, Corthésy J, Cominetti O, Kussmann M (2014) Comprehensive and scalable highly automated MS-based proteomic workflow for clinical biomarker discovery in human plasma. J Proteome Res 13:3837–3845
12. Lan J, Núñez Galindo A, Doecke J, Fowler C, Martins RN, Rainey-Smith SR, Cominetti O, Dayon L (2018) Systematic evaluation of the use of human plasma and serum for mass-spectrometry-based shotgun proteomics. J Proteome Res 17:1426–1435
13. Macron C, Lane L, Núñez Galindo A, Dayon L (2018) Deep dive on the proteome of human cerebrospinal fluid: a valuable data resource for biomarker discovery and missing protein identification. J Proteome Res 17:4113–4126
14. Dayon L, Macron C, Lahrichi S, Núñez Galindo A, Affolter M (2021) Proteomics of human milk: definition of a discovery workflow for clinical research studies. J Proteome Res 20: 2283–2290
15. Thompson A, Schäfer J, Kuhn K, Kienle S, Schwarz J, Schmidt G, Neumann T, Hamon C (2003) Tandem mass tags: a novel quantification strategy for comparative analysis of complex protein mixtures by MS/MS. Anal Chem 75:1895–1904
16. Schneider S, Schuhun B (2017) Reducing cycle time for affinity removal of high-abundant proteins in human plasma. Agilent Technologies Application Note, publication number 5991-4721EN
17. Perkins DN, Pappin DJC, Creasy DM, Cottrell JS (1999) Probability-based protein identification by searching sequence databases using mass spectrometry data. Electrophoresis 20: 3551–3567
18. Chambers MC, Maclean B, Burke R, Amodei D, Ruderman DL, Neumann S, Gatto L, Fischer B, Pratt B, Egertson J, Hoff K, Kessner D, Tasman N, Shulman N, Frewen B, Baker TA, Brusniak M-Y, Paulse C, Creasy D, Flashner L, Kani K, Moulding C, Seymour SL, Nuwaysir LM, Lefebvre B, Kuhlmann F, Roark J, Rainer P, Detlev S, Hemenway T, Huhmer A, Langridge J, Connolly B, Chadick T, Holly K, Eckels J, Deutsch EW, Moritz RL, Katz JE, Agus DB, MacCoss M, Tabb DL, Mallick P (2012) A cross-platform toolkit for mass spectrometry and proteomics. Nat Biotechnol 30:918–920
19. Cominetti O, Núñez Galindo A, Corthésy J, Valsesia A, Irincheeva I, Kussmann M, Saris W, Astrup A, McPherson R, Harper ME, Dent R, Hager J, Dayon L (2018) Obesity shows preserved plasma proteome in large independent clinical cohorts. Sci Rep 8:16981
20. Dayon L, Hainard A, Licker V, Turck N, Kuhn K, Hochstrasser DF, Burkhard PR, Sanchez JC (2008) Relative quantification of proteins in human cerebrospinal fluids by MS/MS using 6-plex isobaric tags. Anal Chem 80: 2921–2931
21. Thompson A, Wölmer N, Koncarevic S, Selzer S, Böhm G, Legner H, Schmid P, Kienle S, Penning P, Höhle C, Berfelde A, Martinez-Pinna R, Farztdinov V, Jung S, Kuhn K, Pike I (2019) TMTpro: design, synthesis, and initial evaluation of a proline-based isobaric 16-plex tandem mass tag reagent set. Anal Chem 91:15941–15950
22. Dayon L, Núñez Galindo A, Cominetti O, Corthésy J, Kussmann M (2017) A highly automated shotgun proteomic workflow: clinical scale and robustness for biomarker discovery in blood. In: Greening D, Simpson R (eds) Serum/plasma proteomics. Methods in molecular biology, vol 1619. Humana Press, New York
23. Núñez Galindo A, Macron C, Cominetti O, Dayon L (2019) Analyzing cerebrospinal fluid proteomes to characterize central nervous system disorders: a highly automated mass spectrometry-based pipeline for biomarker discovery. In: Brun V, Couté Y (eds) Proteomics for biomarker discovery. Methods in molecular biology, vol 1959. Humana Press, New York

Chapter 12

A Multi-enzyme Protocol Improves Total Proteome Coverage in Extracellular Matrix

Ying Xi, Leonard B. Collins, Hongxia Bai, Andreea Biehl, Camilo Mora-Navarro, Donald Freytes, and Taufika Islam Williams

Abstract

Extracellular matrix (ECM) from decellularized mammalian tissues has been used in many therapeutic applications. The tissue-specific composition of the ECM is critically associated with therapeutic performance. However, ECM translation needs to be improved because of the complex composition and limited understanding of ECM repairing mechanisms due partly to incomplete proteomic interrogation of ECM samples. In this chapter, we describe a multi-enzyme, bottom-up proteomics workflow employing trypsin, Lys-C, collagenase, and elastase to enhance the digestion of ECM and increase total protein coverage. The outcomes from the reported approach, in a standardized manner, enable users to pinpoint changes in the ECM composition, thereby facilitating the establishment of mechanistic correlations between ECM composition and its effects.

Key words Decellularized extracellular matrix (ECM), Bottom-up proteomics, Collagenase, Trypsin/Lys-C, Elastase, Hydrogel, Mass spectrometry

1 Introduction

Extracellular matrix (ECM), derived from decellularized mammalian tissues such as cardiac, tendons, and skin, represents a complex and naturally derived biomaterial. ECM possesses tissue regenerative properties associated with their composition. Decellularized ECM serves as a reflection of the tissue homeostatic environment and biomolecular niche, offering versatile potential for transformation into various scaffolds, including therapeutic hydrogels. ECM as a hydrogel or powder has been applied in different ways to repair, regenerate, and reconstruct damaged tissues following injury. A few successful clinical uses have been to enhance the healing of the myocardial, burn wounds, repair dura mater, and replace heart valves [1–3]. Despite its versatile applications [4, 5], the mechanism of tissue healing has been difficult to elucidate given the complex

Taufika Islam Williams (ed.), *Tissue Proteomics: Methods and Protocols*, Methods in Molecular Biology, vol. 2884, https://doi.org/10.1007/978-1-0716-4298-6_12,

composition and architecture of ECM [6, 7]. Much of the ECM is composed of collagen and, depending on the tissue, a substantial amount of elastin and other tissue-specific biomolecules. There are 28 members of the collagen protein family in mammals, and several types of collagens exist in numerous possible isoforms. Collagen is ubiquitous, making up about 30% of the total protein content of mammalian tissues [8]. Elastin content can vary considerably across tissues from about 4% dry weight in the tendons to 57% in the aorta. Though elastin is formed from a single protein, tropoelastin, the extensively cross-linked supramolecular assemblies are much larger than the 60–70 kD precursor [9].

The complexity and insolubility of ECM pose a primary obstacle to comprehensively describing its proteome by mass spectrometry–based analytical methods [10]. Some types of collagens, and elastin in particular, may be insoluble in typical proteomic buffer solutions. Although trypsin or trypsin/Lys-C combinations are commonly used for protein digestion, collagenase and elastase have been successfully employed to more thoroughly explore the ECM proteomes [10–12]. Our laboratory observations suggest that introducing each of these enzymes to the proteomics workflow will enhance digestion and overall proteome coverage of ECM [13]. Herein, we demonstrate that a carefully selected multi-protease method can increase collagen and non-collagen protein coverage by using porcine-derived ECM as a model for their relevance to humans [14]. Collagenase is used to more effectively degrade collagen and release non-collagen proteins that might be enclosed in the superstructure. Elastase, trypsin, and Lys-C are added to degrade elastin and finally create tryptic peptides from all the proteins. Mass spectrometry data acquisition and Proteome Discoverer data analysis are then performed in a standard fashion.

2 Materials

Prepare all buffers and requisite solutions using deionized, LC/MS-grade water and LC/MS-grade reagents at room temperature unless indicated otherwise. Protocols to generate ECM powders can be found in Kuljanin and Biehl [11, 12].

2.1 Filter-Aided Sample Preparation (FASP) with Multiple Proteases

1. Decellularized ECM: 0.2 mg of freeze-dried powder.
2. 50 mM ammonium bicarbonate (ABC) with 1% sodium deoxycholate (SDC).
3. 0.1 M Tris–HCl, pH 7.5.
4. 0.1 M Tris–HCl, pH 8.0.
5. 8 M urea, 0.1 M Tris–HCl, pH 8.0.
6. 2 M urea, 10 mM $CaCl_2$, 0.1 M Tris–HCl, pH 8.0.

7. Purified Class I and Class II collagenase mixture from *Clostridium histolyticum*: The collagenase should be reported as having minimal contaminating protease activity. Our preferred collagenase source is collagenase HA (Vitacyte, Indianapolis, IN). Add 5 mL cold water to the contents of one vial, and keep it on ice. Gently invert the vial occasionally, and allow 30 min to completely dissolve. The final concentration in collagen degradation activity units (CDA) is 0.5 to 0.72 million CDA units/mL. The stock protease solution can be divided into aliquots and stored frozen at −20 °C until use.
8. Elastase, Type I from porcine pancreas: Our protocol was developed using crude elastase solution (Sigma-Aldrich, Inc., Saint Louis, MO), but a similar reagent (elastase, Type I from porcine pancreas) can be used from a different source. Centrifuge an aliquot of the crude solution at 5000× *g* for up to 10 min to gain a clear solution.
9. Trypsin/Lys-C mixed protease.
10. Dithiothreitol (DTT): 50 mM in 0.1 M Tris–HCl, pH 8 for the reduction of protein disulfide bonds.
11. Iodoacetamide (IAA): 200 mM in 8 M urea, 0.1 M Tris–HCl, pH 8.0 for alkylation of free sulfides.
12. Centrifugal 10 kDa molecular weight cut-off (MWCO) filters: Pall Omega Nanosep are preferred (Cytiva, Marlborough, MA). Centrifugation speeds and times may be higher when using MWCO filters from a different manufacturer.
13. Centrifugal vacuum concentrator or lyophilizer.

2.2 NanoLC-MS/MS

1. Mobile phase A (MPA): 98% water, 2% acetonitrile, 0.1% formic acid.
2. Mobile phase B (MPB): 80% acetonitrile, 20% water, 0.1% formic acid.
3. Analytical column: Thermo Scientific EASY-Spray™ C_{18} nanoLC column, 2 μm particle size, 75 μm inner diameter, 500 mm length, 100 Å particle size. Other, similar C_{18} nanoLC analytical columns can be used.
4. Trap column: Thermo Scientific Acclaim PepMap™ 100 C_{18} nanoLC column, 3 μm particle size, 75 μm inner diameter, 20 mm length, 100 Å particle size. Other, similar C_{18} nanoLC trap columns can be used.
5. Mass spectrometer: Thermo Scientific Exploris 480 (Thermo Scientific, Bremen, GE) or other high resolution/accurate mass (HR/AM) mass spectrometer with MS^2 capability that is suitable for a proteomics workflow.

2.3 Data Analysis

1. Proteome Discoverer or any other proteomics sequencing data analysis software may be used for data analysis.

3 Methods

This protocol is appropriate for decellularized ECM and is intended for Discovery Proteomics applications. Procedures are carried out at room temperature unless otherwise specified.

3.1 Sample Preparation

1. Measure 0.2 mg ± 0.1 mg of sample on top of a 10 kDa MWCO filter. If weighing solid material is difficult, first, weigh 1 mg powder into a 2-mL microcentrifuge tube. Add sufficient 50 mM ABC, 1% SDC to the tube to reach 1 mg/mL concentration. Vortex and sonicate to dissolve as much material as possible. Solids will remain but attempt to make the suspension homogenous. If necessary, allow the material to continue soaking overnight under refrigeration. Use a pipettor fitted with a 1-mL pipet tip to transfer 200 μL of the suspension to the MWCO filter. Proceed to **step 3**.
2. Add 200 μL 50 mM ABC, 1% SDC to the filter containing the weighed, dry ECM. Use the pipet tip to push any unsubmerged solid material into the solution.
3. Add 15 μL of 50 mM DTT in 0.1 M Tris–HCl pH, 8.0.
4. Incubate the sample in an oven or heating block at 56 °C for 30 min.
5. Remove the sample from heat and let it cool down to room temperature.
6. Centrifuge at 10,000× *g* for 10 min at 21 °C to remove DTT solution. Add additional centrifugation time or increase the *g* force as needed.
7. Add 200 μL of 8 M urea in 0.1 M Tris–HCl, pH 8.0.
8. Centrifuge at 10,000× *g* for 10 min at 21 °C. Add additional centrifugation time or increase force as needed. Discard flow-through solutions.
9. Add an additional 200 μL of 8 M urea in 0.1 M Tris–HCl, pH 8.0 to the 10 kDa MWCO filter.
10. Add 64 μL of 200 mM IAA to the MWCO filter. Gently mix with the pipettor.
11. Incubate the sample in the dark at room temperature for 1 h.
12. Centrifuge at 10,000× *g* for 10 min at 21 °C. Add additional centrifugation time or increase the *g* force as needed. Discard flow-through solutions.

13. Wash the MWCO filter with 100 μL of 2 M urea with 10 mM $CaCl_2$.
14. Centrifuge at 10,000× *g* for 10 min at 21 °C. Add additional centrifugation time or increase the *g* force as needed.
15. Repeat **steps 17** and **18** two more times. Discard the flow-through solutions.
16. Wash each MWCO filter with 100 μL of 0.1 M Tris–HCl pH 7.5.
17. Centrifuge at 10,000× *g* for 10 min at 21 °C. Add additional centrifugation time or increase the *g* force as needed.
18. Repeat **steps 21** and **22** two more times. Discard the flow-through solutions.
19. Prepare 0.02 μg/μL trypsin/Lys-C protease solution in 0.1 M Tris–HCl, pH 7.5.
20. Add 200 μL trypsin/Lys-C protease solution to the sample. This gives a protease-to-protein ratio of 1:50 when starting with a 200 mg sample.
21. Incubate the sample overnight in a 37 °C oven.
22. Remove the sample from the oven and let it cool down to room temperature.
23. Centrifuge at 10,000× *g* for 20 min at 21 °C to collect the eluent as Fraction 1. Add additional centrifugation time or increase the *g* force as needed.
24. Lyophilize Fraction 1 until it is dried and store it at −20 °C until nanoLC-MS/MS.
25. Prepare a 0.01 μg/μL solution of Trypsin/Lys-C in 0.1 M Tris–HCl, pH 7.5. Add 200 μL to the undigested material that has remained on the MWCO filter.
26. Add 2 μL of clarified elastase (~0.08 units) and 2 μL of collagenase (~1200 CDA units).
27. Incubate the samples at 37 °C in an oven or thermal mixer for at least 2 h and not more than 4 h.
28. Remove from heat and allow to cool to room temperature.
29. Centrifuge at 10,000× *g* for 20 min at 21 °C to collect the eluent as Fraction 2. Add additional centrifugation time or increase the *g* force as needed.
30. Lyophilize Fraction 2 until it is dried and store it at −20 °C until nanoLC-MS/MS.

3.2 NanoLC-MS/MS

1. On the day of nanoLC-MS/MS analysis, dissolve Fractions 1 and 2 individually with 200 μL MPA. Transfer at least 20 μL of Fractions 1 and 2 into separate autosampler vials

with low-volume inserts and place them in the nanoLC instrument.

2. nanoLC program: The nano-liquid chromatography method is set up in a trap-and-elute configuration. Peptides are first loaded onto a trapping column and then eluted through the analytical column for the separation of peptides in a time-dependent manner.
   (1) Trapping program: Inject a 2 μL volume of sample onto the trap column using constant pressure at 700 bar and a total of 6 μL MPA. The injection volume should be suitable for the instrument and allow from 200 ng to 2 μg total material to be loaded.
   (2) Analytical separation program: Set the flow rate to 300 nL/min and the column temperature to 45 °C. Program the mobile phase gradient as follows: begin with 0% MPB and hold for 2 min, increase MPB linearly to 25% over 75 min, increase to 40% over 10 min, increase to 95% over 1 min, and hold at 95% for 16 min to wash the columns. Verify the Thermo Scientific nanoLC-1200 is programmed to equilibrate the trap and analytical columns before injection of the next sample.
3. MS method: Use the following acquisition parameters for the Thermo Scientific Exploris 480 Orbitrap mass spectrometer: 1.8 kV positive ion mode spray voltage, ion transfer tube temperature of 275 °C, master scan cycle time of 2.5 s, *m/z* scan range of 300 to 1500 at 120 K resolution, 300% normalized AGC Target, 120 ms maximum MS^1 injection time, RF lens of 40%, 15 K mass resolving power for data-dependent MS^2 scans, 1.5 *m/z* isolation window, 30% normalized HCD collision energy, 100% normalized AGC Target, 21 ms maximum injection time, and dynamic exclusion applied for 20 s periods.

3.3 Data Analysis Method

1. Process raw nanoLC-MS/MS files with Proteome Discoverer 2.5 software (PD, Thermo Scientific, San Jose, CA) or equivalent. Note that freeware such as MaxQuant can be used.
2. Install a proper protein database that contains sequences for all proteins of interest. PD uses databases of FASTA file format. UniProt is a good source for protein databases for various organisms.
3. Create a custom cleavage reagent that includes cleavage sites from trypsin and elastase. Identify A, I, K, L, R, and V as cleavage sites with P as an inhibitor. Cleavage occurs at the C-terminus, denoted as 1 in PD.

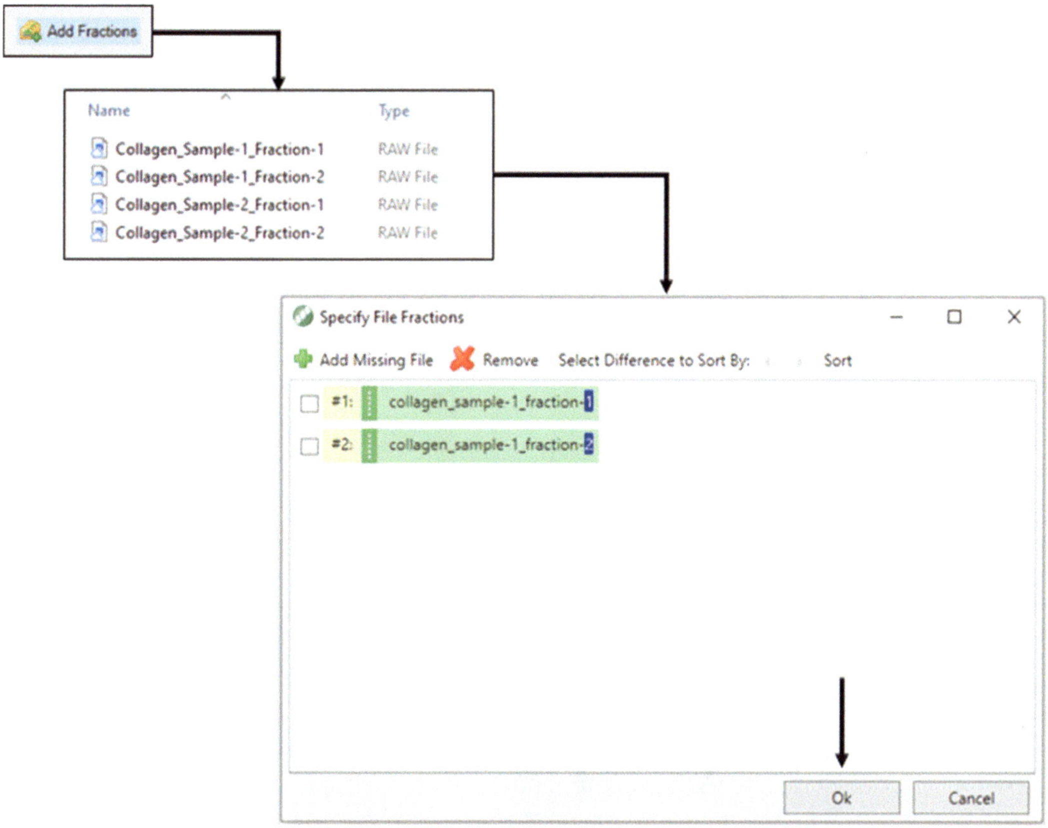

Fig. 1 Use the "Add Fractions" feature of PD to treat two raw files as a single sample

4. Create a custom cleavage reagent for collagenase by setting cleavage at G with no inhibitors. Collagenase cleaves at the N-terminus, denoted as 0 in PD.
5. Sample Fractions 1 and 2 can be analyzed separately or combined into a single file for total protein identification, as shown in Fig. 1. To analyze the two Fractions as one sample using PD:
   (1) Click the icon "Add Fractions" within the Study tab.
   (2) Select the two raw files corresponding to Fraction 1 and Fraction 2 of a single sample. Having a fraction number as part of the file name assists with their pairing. PD will create a new file name and treat the Fractions as one unique sample.
   (3) Click "OK" in the dialog box.
6. After setting up an analysis, drag and drop the new sample name from the Samples tab into the Processing Step input field.
7. Create a Processing Workflow according to Fig. 2. Please *see* **Note 7** for details on creating a label-free quantification workflow.

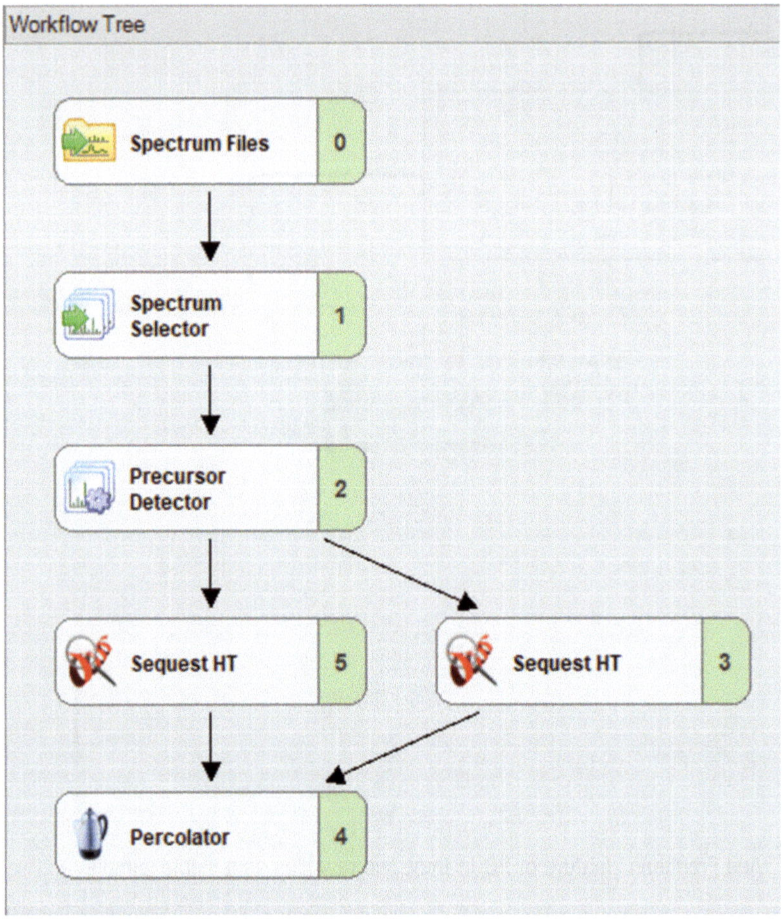

Fig. 2 A basic Processing Workflow tree is shown using Proteome Discoverer. The respective Sequest HT search nodes use collagenase and trypsin/elastase as cleavage reagents. Separate the enzymes by search node to allow peptide spectral matches from each to be displayed in the final PSM results

8. Set up the database search engine using the following parameters: collagenase (full) for the first search node, trypsin/elastase (full) for the second search node, maximum of six missed cleavage sites, minimum peptide length of 60 amino acids, 5 ppm precursor mass tolerance, 0.02 Da fragment mass tolerance, maximum of eight equal modifications, and maximum of eight dynamic modifications, which are oxidation of lysine, methionine, and proline, deamidation of asparagine and glutamine, the addition of galactosyl or glucosylgalactosyl to lysine, conversion of lysine to allysine, and static carbamidomethylation of cysteine. Details of each Sequest HT search node in the Processing Workflow are given in Fig. 3.
9. Include a contaminant database that contains sequences for human keratins, trypsin, Lys-C, collagenase, and elastase in the Sequest node using trypsin.

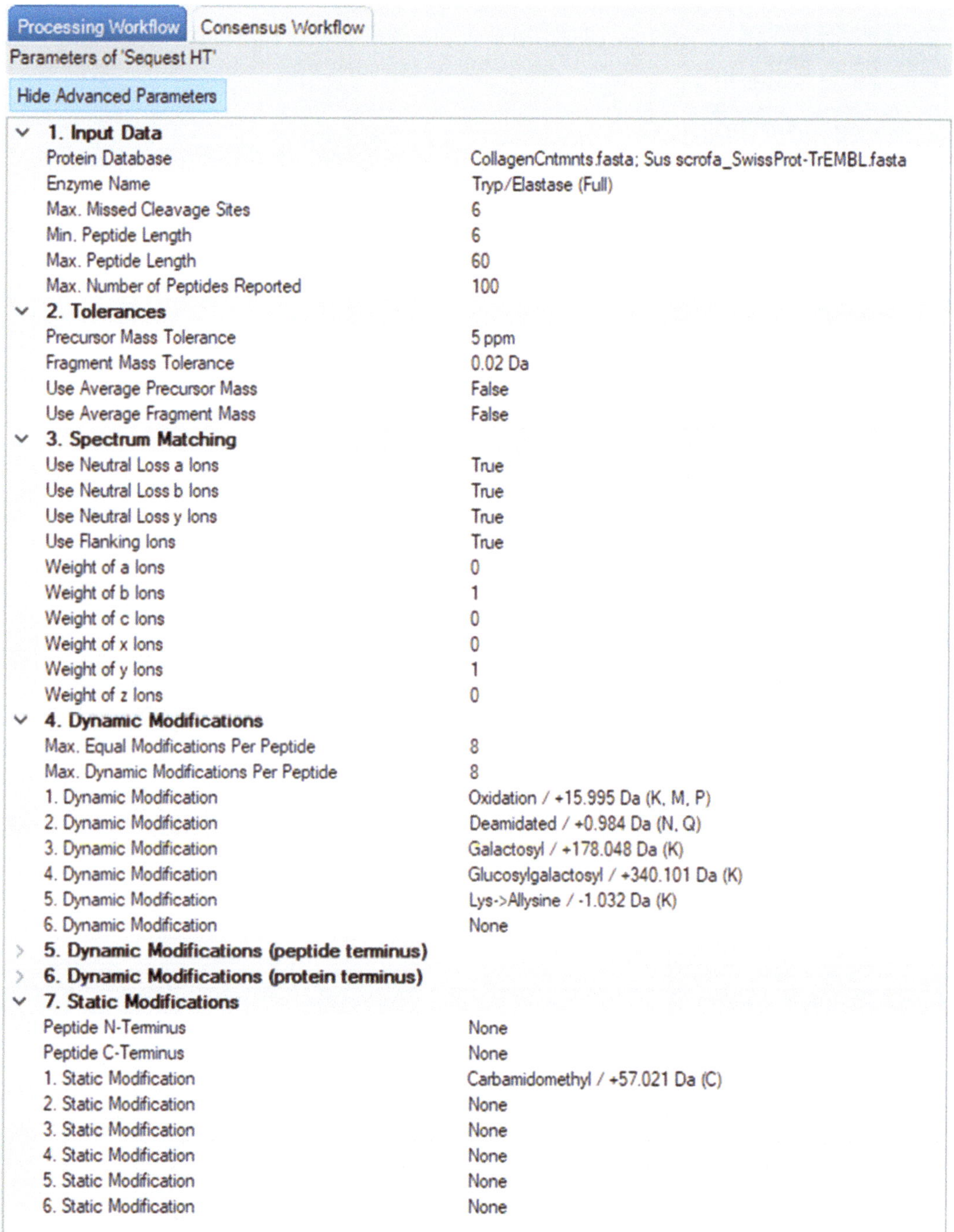

Fig. 3 The details of each Sequest HT search node are shown here. Collagen peptides can be extensively modified. To account for these possibilities, use a higher number for missed cleavages and equal modifications than would normally be used for non-collagen proteins

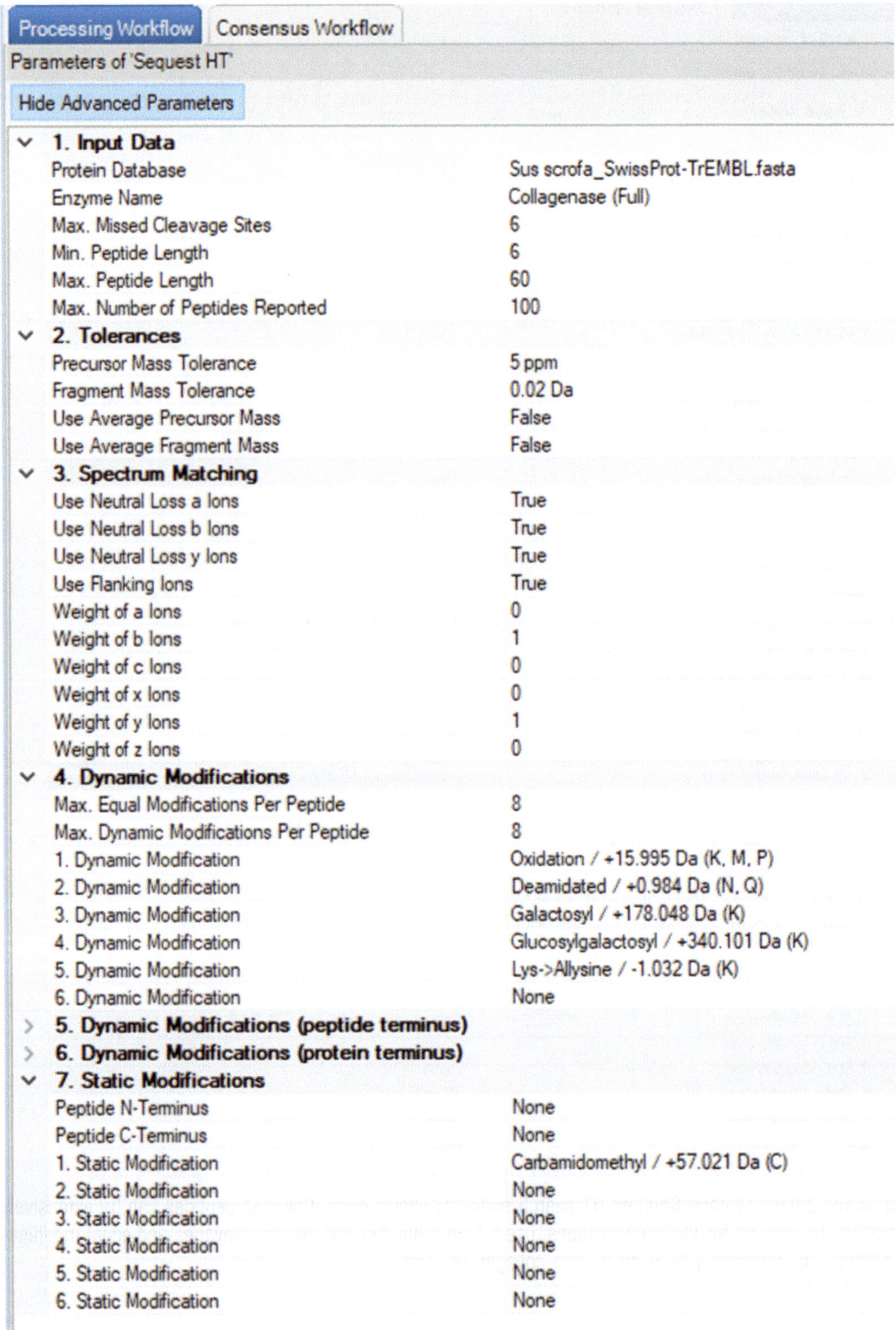

Fig. 3 (continued)

10. Include appropriate statistical tests to ensure high confidence in protein identifications. For PD, peptides are validated by the Percolator node with a *q*-value set to 0.05 and a strict false discovery rate (FDR) set to 0.01. Connect each node to the same percolator.

4 Notes

1. Powdered native collagen and ECM can carry static electricity, which causes the material to adhere to plastic microcentrifuge tubes and MWCO filters. Using a glass scoop helps reduce the tendency of the material to "fly away" during transfer.
2. If using ABC/SDC to hydrate and resuspend material, complete the transfer of the sample to a MWCO filter with a pipet tip that has an opening large enough for suspended solids to pass through. Clip the end of a 1-mL pipet tip with scissors about 3–4 mm up from the bottom to create a larger space if necessary. The proteases will digest undissolved material.
3. MWCO filters may elute fluid at different rates. Remove filters from the centrifuge that have allowed all solvent to pass through before spinning slower-eluting filters again. We increase centrifuge force up to 12,500× *g* when necessary and increase spin time in 10–15 min increments.
4. Trypsin and Lys-C will degrade collagenase during digestion. A large excess of collagenase is used, which will lead to many tryptic peptides from collagenase being present in the final digested sample.
5. Our nanoLC and mass spectrometer instrument parameters are optimized for adequate peptide separation and data-dependent acquisition for discovery proteomics. Any nanoLC/MS methodology suitable for comprehensive peptide detection and mapping can be used.
6. Our aim is to capture as much information about collagen, elastin, and non-collagen proteins as possible from a single sample using a bottom-up proteomics assay. To that end, we use protein databases that contain a comprehensive list of protein sequences. The reviewed databases for some animals may not contain (all) collagen protein sequences, but a database that includes gene-translation entries might. For example, *Sus scrofa* in Swiss-Prot does not contain collagen proteins, but TrEMBL includes them as of this writing [15].
7. To achieve label-free quantification, include the Minora Feature Detector (*see* Fig. 4) in the processing workflow and the Feature Mapper and Precursor Ions Quantifier nodes in the consensus step. The Minora Feature Detector links

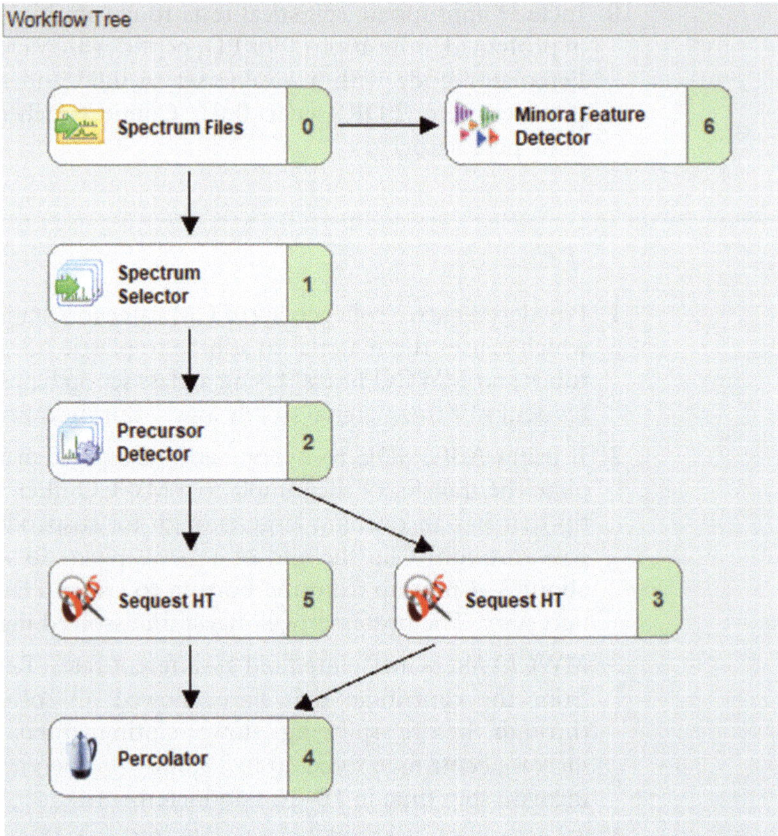

Fig. 4 This workflow tree shows two Sequest HT search nodes, each containing one of collagenase or elastase/trypsin as the cleavage reagent. Set all other parameters for these nodes the same as described in Subheading 3.3 **step 5**. This workflow may allow more accurate LFQ results by including peptides generated by collagenase and elastase as well as trypsin. Expect processing times to be much longer when performing chromatographic feature mapping with the Minora Feature Detector

chromatographic peaks to their respective PSMs and is required for precursor ion quantification. The Feature Mapper identifies chromatographic peaks across samples and aligns retention times. The Precursor Ions Quantifier defines the peptides that are used for quantification, the method of calculating protein abundances, and the statistical method to test for differences across samples.

References

1. Traverse JH, Henry TD, Dib N et al (2019) First-in-man study of a cardiac extracellular matrix hydrogel in early and late myocardial infarction patients. JACC Basic Transl Sci 4(6):659–669. https://doi.org/10.1016/j.jacbts.2019.07.012
2. Spang MT, Christman KL (2018) Extracellular matrix hydrogel therapies: in vivo applications and development. Acta Biomater 68:1–14
3. Yao Q, Zheng YW, Lan QH et al (2019) Recent development and biomedical applications of decellularized extracellular matrix biomaterials. Mater Sci Eng C 104:109942

4. Kalluri R (2016) The biology and function of fibroblasts in cancer. Nat Rev Cancer 16:582–598. https://doi.org/10.1038/nrc.2016.73
5. Poltavets V, Kochetkova M, Pitson SM et al (2018) The role of the extracellular matrix and its molecular and cellular regulators in cancer cell plasticity. Front Oncol 8. https://doi.org/10.3389/fonc.2018.00431
6. Doan N-D, DiChiara AS, Del Rosario AM et al (2019) Mass spectrometry-based proteomics to define intracellular collagen interactomes. In: Sagi I, Afratis N (eds) Collagen. Methods in molecular biology, vol 1944. Humana Press, New York, pp 95–114
7. Ragelle H, Naba A, Larson BL et al (2017) Comprehensive proteomic characterization of stem cell-derived extracellular matrices. Biomaterials 128:147–159. https://doi.org/10.1016/j.biomaterials.2017.03.008
8. Ricard-Blum S (2011) The collagen family. Cold Spring Harb Perspect Biol 3(1): a004978. https://doi.org/10.1101/cshperspect.a004978
9. Mithieux S, Weiss A (2005) Elastin. Adv Protein Chem 70:437–461. https://doi.org/10.1016/s0065-3233(04)70013-3
10. Sato N, Taniguchi T, Goda Y et al (2016) Proteomic analysis of human tendon and ligament: solubilization and analysis of insoluble extracellular matrix in connective tissues. J Proteome Res 15:4709–4721. https://doi.org/10.1021/acs.jproteome.6b00806
11. Ali K, Huang Y, Amanah A et al (2022) In vitro biocompatibility and degradation analysis of mass-produced collagen fibers. Polymers (Basel) 14. https://doi.org/10.3390/polym14102100
12. Kuljanin M, Brown C, Raleigh M et al (2017) Collagenase treatment enhances proteomic coverage of low-abundance proteins in decellularized matrix bioscaffolds. Biomaterials 144: 130–143. https://doi.org/10.1016/j.biomaterials.2017.08.012
13. Biehl A, Gracioso Martins A, Davis Z et al (2023) Towards a standardized multi-tissue decellularization protocol for the derivation of extracellular matrix materials. Biomater Sci 11: 641–654. https://doi.org/10.1039/d2bm01012g
14. Carvalho T (2023) Two US surgical teams transplant functional pig kidneys into humans in xenotransplantation success. Nat Med 29: 2671–2672. https://doi.org/10.1038/d41591-023-00078-8
15. Mora-Navarro C, Garcia M, Sarker P et al (2022) Monitoring decellularization via absorbance spectroscopy during the derivation of extracellular matrix scaffolds. Biomed Mater 17:015008. https://doi.org/10.1088/1748605X/ac361f

Chapter 13

A Protocol to Disclose the Protein Fingerprint of Commercial White Wines Based on Proteomic Tools

Antonio Samper-Herrero, Susana Sellés-Marchart, and Roque Bru-Martínez

Abstract

Proteins remaining in commercial wines are responsible for the protein haze in white wine unless they are effectively removed before bottling. To avoid this undesirable phenomenon, techniques of precipitation and filtration are applied in the white wine making process to eliminate a large part of them (fining processes) (Ribéreau-Gayon et al., Handbook of enology, vol 2, 3rd edn. Wiley, Hoboken, pp 369–406, 2021). After finding treatments, a minute amount of these genetically informative biomolecules remains in the commercialized product, which makes its extraction and identification extremely difficult. The protocol described here allows for the extraction and identification of proteins present in commercial white wines using a proteomic workflow of dialysis-lyophilization-precipitation-nanoLC-MS/MS analysis. Its application to three monovarietal commercial white wines allows us a varietal differentiation of the analyzed samples and the ability to detect and identify potentially allergenic proteins. To have this information before the consumption of wine is of great importance because (1) a varietal differentiation could be used for the authentication of the wine and (2) the identification of potentially allergenic proteins responsible for adverse reactions (from the grapes or the fining agents used) will allow for the protection of allergic consumers.

Key words Commercial white wine, Wine authentication, Allergenic proteins, Proteomic tools

1 Introduction

Wine is a beverage resulting exclusively from the partial or complete alcoholic fermentation of fresh grapes [2]. Winemakers have learned to reproduce these biochemical processes that take place in this transformation and through the application of technological workflows (winemaking processes) they are able to obtain wines with very particular organoleptic characteristics.

These biochemical transformations, based on the fermentation of the sugars and release of pigments and flavors present in the grape berry, allow us to obtain a drink in which many biomolecules coexist. The balance between these biomolecules is essential to

Taufika Islam Williams (ed.), *Tissue Proteomics: Methods and Protocols*, Methods in Molecular Biology, vol. 2884,
https://doi.org/10.1007/978-1-0716-4298-6_13,

render one of the most appreciated and consumed drinks throughout the world sensory pleasant and healthy [3]. Proteins are one of these classes of biomolecules present coming from both the grape berry and the yeasts used in fermentation. Its presence in wine plays a crucial role in the formation of haze [4–6], a common defect of commercial white wines that makes them unacceptable to consumers and therefore reduces the value of the final product. Fining agents are substances used during the winemaking process to enhance clarity and stability in the wines, removing (a) haze-forming proteins, which are typically chitinases and thaumatin-like proteins (TLPs) [7–9], and (b) other components of wine that can also contribute to haze formation, including polyphenols, polysaccharides, and organic acids [10].

For a long time, products containing milk (caseinates or milk powders) or egg (ovalbumin or egg-white powders) proteins have been used as fining agents, but it is well known that its presence in the final product increases the risk of allergic reactions in sensitized people. To protect allergic consumers, several countries (Australia and New Zealand since 2002 [11], the European Union since 2012 [12], and Canada since 2012 [13]) have required that wines carry a declaration on their labels regarding the presence of recognized protein allergens, such as fining agents derived from milk and eggs. In this situation, many winemakers opted for other fining agents (plant proteins, bentonite, and magnetic nanoparticles) that do not (or are not known to) cause these allergic reactions [10, 14, 15]. After the fining process, the concentration of proteins present in the commercialized product is minute, which makes its extraction and identification extremely difficult.

In scientific literature, we can find many protocols to achieve these two objectives [16]. The protocol that we explain in this chapter was developed for the need to extract and identify proteins in commercial white wines where other protocols tested did not provide optimal results. There is a great variety of winemaking processes that are used, and therefore, there is a great variety of commercial white wines with very different amounts and profiles of proteins. Our purpose in this chapter is to share a protocol that has been used in recent years to extract and identify proteins in commercial white wines. The results shown in this chapter demonstrate that using a proteomic workflow involving dialysis-lyophilization-precipitation-nanoLC-MS/MS analysis and following the novel protocol described, the extraction and identification of proteins present in commercial white wines are possible.

Knowing which proteins are present when we drink a glass of wine is of great importance because (1) it could allow us a varietal differentiation that could be used for the authentication of monovarietal commercial wines and (2) it would allow us to identify potentially allergenic proteins responsible for adverse reactions (coming from the grape or the fining agents used).

2 Materials

2.1 Wine Samples

Three commercial monovarietal Spanish white wines (Chardonnay, Verdejo, and Viura) were studied. All wines were purchased from local stores.

2.2 Equipment

1. Refrigerated centrifuge.
2. Ultrasonic processor with a 1.0-mm diameter sonotrode.
3. SpeedVac centrifuge.
4. Lyophilizer.

2.3 Protein Extraction

If not specified, reagents are of analytical grade.

1. Dialysis membrane with a 12 kDa molecular weight cut-off (MWCO).
2. Resuspension buffer: phosphate-buffered saline (PBS): 0.8% (w/v) NaCl, 0.02% (w/v) KCl, 0.144% (w/v) Na_2HPO_4, 0.024% (w/v) KH_2PO_4, pH 7.4 (*see* **Note 1**).
3. Precipitation solutions: solution 1: 2% (w/v) sodium deoxycholate (SDC) in distilled water. Solution 2: 24% (w/v) trichloroacetic acid (TCA) in distilled water (*see* **Notes 2** and **3**).
4. *RC DC*™ Protein Assay (Bio-Rad Laboratories, Hercules, USA) or equivalent protein quantification assay.

2.4 Sodium Dodecyl Sulfate (SDS) Polyacrylamide Gel Electrophoresis (PAGE)

1. Running buffer: 0.3% (w/v) Tris base, 1.44% (w/v) glycine, 0.01% (w/v) SDS.
2. Resolving gel: amount required for 14% acrylamide gel: mix 4.67 mL of 30% (w/v) acrylamide-bisacrylamide (37, 5:1), 2.5 mL 1.5 M Tris–HCl, pH 8.8, 100 μL of 10% (w/v) SDS, and 2.65 mL of distilled water. Add 50 μL of 10% (w/v) ammonium persulfate (APS) and 10 μL of tetramethylethylenediamine (TEMED) to start the polymerization. Add the mix to the cassette and let the gel polymerize for 1 h (*see* **Notes 4–6**).
3. Stacking gel: amount required for 4% acrylamide gel: mix 650 μL of 30% (w/v) acrylamide-bisacrylamide (37, 5:1), 1.25 mL 0.5 M Tris–HCl, pH 6.8, 50 μL of 10% (w/v) SDS, and 3 mL of distilled water. Add 50 μL of 10% (w/v) APS and 10 μL of TEMED for polymerization (*see* **Note 7**).
4. SDS-PAGE sample buffer: 1 mL 0.5 M Tris–HCl, pH 6.8, 1.6 mL of 10% (w/v) SDS, 0.8 mL glycerol, 0.4 mL of 2-mercaptoethanol, 0.2 mL of 1% (w/v) bromophenol blue, and 4 mL distilled water.
5. Fixing solution: 40% (v/v) methanol, 10% (v/v) acetic acid in distilled water.

6. Colloidal Coomassie stain solution: dissolve 10 g of $(NH_4)_3PO_4$ in 90 mL of distilled water, add 1.2 mL H_3PO_4 85% (w/w), and add distilled water up to 100 mL. Separately dissolve 0.1 g Coomassie G-250 in 2 mL of distilled water. When Coomassie is dissolved, add it to the 100 mL of previously prepared phosphate solution and finally add 25 mL methanol. Keep stirring until it is added to the gel [17].

2.5 Protein Digestion

1. Resuspension solution: 6 M urea.
2. Reduction solution: 0.2 M dithiothreitol (DTT) in 25 mM ammonium bicarbonate (*see* **Note 8**).
3. Alkylation solution: 0.2 M iodoacetamide (IAM) in 25 mM ammonium bicarbonate (*see* **Note 9**).
4. Sequencing Grade Modified Trypsin (*see* **Note 10**).

2.6 Peptide Desalting

Water and reagents used are LC-MS grade.

1. Pierce™ C18 Spin Columns.
2. Activation solution: 50% (v/v) acetonitrile (ACN). 400 μL per sample.
3. Equilibration solution: 0.5% (v/v) trifluoroacetic acid (TFA), 5% (v/v) ACN. 400 μL per sample.
4. Binding solution: 5% (v/v) ACN, 0.5% (v/v) TFA (*see* **Note 11**).
5. Washing solution: 0.5% (v/v) TFA, 5% (v/v) ACN. 400–800 μL per sample.
6. Elution solution: 70% (v/v) ACN, 0.1% (v/v) of formic acid (FA). 40 μL per sample.

2.7 nanoLC-MS/MS

Water and reagents used are LC-MS grade.

1. Mobile phase A: 0.1% (v/v) FA.
2. Mobile phase B: 0.1% (v/v) FA, 90% ACN.
3. Reversed-phase analytical column (recommended): Agilent AdvanceBio Peptide Mapping column, 2.1 mm × 250 mm, 2.7 μm particle size.
4. UHPLC system.
5. Q-TOF mass spectrometer.

2.8 Protein Identification

1. Search engine for peptide ion fragmentation spectrum-based protein identification (*see* **Note 12**).
2. Two subsets of the UniProtKB database: a *Vitis vinifera* subset and an Allergen subset (*see* **Notes 13** and **14**).

2.9 Data Analysis

1. Software for label-free quantitative proteomics analysis of nanoLC-MS/MS runs (*see* **Note 15**).

3 Methods

3.1 Protein Extraction

1. Dialyze 100 mL of commercial white wine against 20 volumes of distilled water for 24 h at room temperature under continuous stirring. Three water changes must be made at least (*see* **Notes 16** and **17**).
2. Completely lyophilize the dialysis retentate (*see* **Note 18**).
3. Resuspend the lyophilized sample in the smallest amount of resuspension buffer (*see* **Note 19**).
4. Quantify proteins (*see* **Note 20**).
5. To precipitate the proteins, add 42.5 μL of 2% (w/v) SDC solution per mL of resuspended sample, vortex, and keep for 15 min in an ice bath.
6. Add 1.67 mL of 24% (w/v) TCA solution per mL of resuspended sample, vortex, and keep for 30 min in an ice bath.
7. Centrifuge at 21,000× *g* for 15 min (4 °C) and discard the supernatant.
8. Add 1 mL of chilled acetone to the pellet.
9. Use an ultrasonic processor to sonicate five times for 2 s. Vortex (*see* **Note 21**).
10. Centrifuge at 21,000× *g* for 15 min (4 °C) and discard the supernatant.
11. Repeat **steps 8** and **9** and store overnight at 4 °C (*see* **Note 22**).
12. Centrifuge at 21,000× *g* for 15 min (4 °C) and discard the supernatant.
13. Repeat **steps 8** and **9** and centrifuge at 21,000× *g* for 15 min (4 °C) and discard the supernatant.
14. Air-dry the pellet at room temperature, and if not used immediately, store it at −20 °C.

3.2 SDS-PAGE

SDS-PAGE is recommended to confirm the presence of proteins before performing nanoLC-MS/MS analysis.

1. Perform SDS-PAGE in a 14% acrylamide gel. For each sample, mix 30 μg of precipitated proteins with SDS-PAGE sample buffer and heat to 95 °C for 5 min.
2. Load the samples onto the gel and run the electrophoresis at 200 V, constantly until the bromophenol blue as a marker reaches the glass limit (*see* **Note 23**).

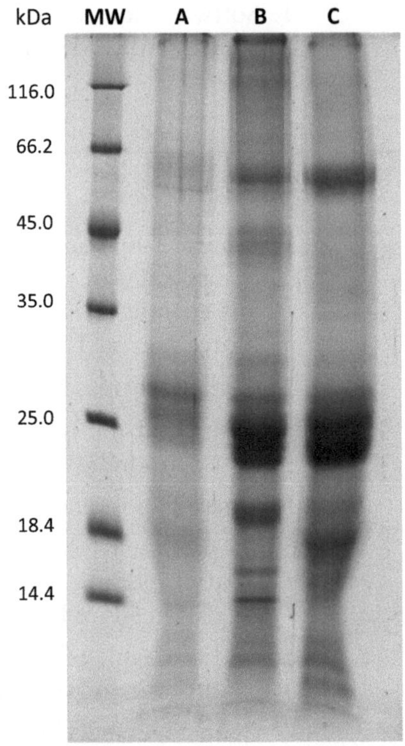

Fig. 1 SDS-PAGE 14% gel of monovarietal white wine proteins (30 μg per lane) stained with colloidal Coomassie Brilliant Blue. *MW* molecular mass marker; A, Chardonnay; B, Verdejo; C, Viura

3. Stop the electrophoresis, transfer the gel into a bowl with fixing solution, and incubate it in an orbital shaker at room temperature for at least 1 h (*see* **Note 24**).
4. Incubate with distilled water in an orbital shaker for at least 10 min.
5. Repeat **steps 3** and **4** twice.
6. Transfer the gel into a bowl with Coomassie stain solution. Incubate overnight in an orbital shaker at room temperature.
7. Remove the excess stain solution with hot distilled water (Fig. 1).

3.3 Sample Preparation for MS Analysis

3.3.1 Protein Digestion

1. Resuspend 50 μg of precipitated protein in 25 μL of 6 M urea (*see* **Note 25**).
2. Reduce disulfide bridges by adding 5 μL of 0.2 M DTT. Vortex and incubate at 37 °C for 1 h.
3. Alkylate free cysteine thiol groups by adding 20 μL of 0.2 M IAM. Vortex and incubate it *in the dark* at room temperature for 1 h.

4. Add 20 μL of DTT to consume excess IAM.
5. Add 25 mM ammonium bicarbonate to reduce the concentration of urea to 0.6 M (*see* **Note 26**).
6. Before adding trypsin, adjust the pH to 7–9 (*see* **Note 27**).
7. Digest samples to obtain peptides by adding trypsin in a 1:30 (w/w) trypsin-protein ratio (i.e., 1 μg trypsin per 30 μg protein). Vortex and incubate at 37 °C overnight.
8. Complete digestion by adding trypsin in a 1:60 (w/w) trypsin-protein ratio (i.e., 1 μg trypsin per 60 μg protein). Vortex and incubate at 37 °C for 3–5 h.
9. Stop digestion, acidifying the sample with 2–3 μL of pure formic acid, to a pH between 1 and 3 (*see* **Note 28**).
10. Dry down samples in a SpeedVac centrifuge.

3.3.2 Peptide Desalting

1. For desalting samples with Pierce™ C18 Spin Tubes spin columns, follow the manufacturer's instructions (*see* **Note 29**).
2. Eventually, dry down the eluted peptides in a SpeedVac centrifuge and resuspend them in 11 μL of LC mobile phase A (3% ACN and 0.1% FA) for MS analysis.

3.4 nanoLC-MS/MS Analysis

The protocol described here is used with an Agilent 1290 Infinity UHPLC equipped with an AdvanceBio Peptide Mapping column (2.1 mm × 250 mm, 2.7 μm particle size) coupled to an Agilent 6550 hybrid spectrometer Q-TOF equipped with a Jet Stream® source. To carry out the data analysis by components (PCA), three analyses were carried out on each sample (analytical replicates).

1. 11 μL injections are programmed to ensure reproducible sample injection in the autosampler.
2. Peptides are separated in the formerly indicated analytical column using a 3–40% gradient of mobile phase B for 80 min operating at 50 °C and a flow rate of 0.4 mL/min.
3. Peptides are introduced to the mass spectrometer from the nanoLC by using a Jet Stream source (Agilent Technologies) operating in positive-ion mode (3.5 kV) and in high-sensitivity mode. Source parameters employed: gas temperature 250 °C, drying gas 14 L/min, nebulizer 35 psig, sheath gas temperature 250 °C, sheath gas flow 11 L/min, capillary voltage 3.5 kV, fragmentor voltage 360 V, and OCT 1 RF Vpp 750 V.
4. The data are acquired with Agilent MassHunter Workstation Software, LC/MS Data Acquisition B.08.00 (Build 8.00.8058.0) operating in Auto MS/MS mode whereby the 20 most intense ions (charge states, 2–5) within 300–1700 *m/z* mass range above a threshold of 1000 counts are selected for MS/MS analysis. MS/MS spectra (50–1700 *m/z*) are collected

with the quadrupole set to "narrow" resolution and are acquired until 25,000 total counts are collected or for a maximum accumulation time of 333 ms. To assure the desired mass accuracy of recorded ions, continuous internal calibration are performed during analyses with the use of signals *m/z* 322.0481 (detected *m/z* $[C_6H_{18}N_3O_6P_3 - H]^+$) and 1221.9906 (detected *m/z* $[C_{24}H_{18}O_6N_3P_3F_{36}\text{-}H]^+$).

3.5 Protein Identification

1. Each MS/MS spectral data set is processed to determine monoisotopic masses and charge states, to merge MS/MS spectra with the same precursor ($\Delta m/z < 1.4$ Da and chromatographic $\Delta t < 15$ s) and to select high-quality spectra with the Extraction tool of the SpectrumMill Proteomics Workbench (Agilent). The reduced data set is searched against the UniProtKB *Vitis vinifera* subset and the UniProtKB Allergen subset (*see* **Note 30**) in the identity mode with the MS/MS search tool of the SpectrumMill Proteomics Workbench and with the following settings: trypsin, up to two missed cleavages, carbamidomethylation of cysteines as fixed modifications, oxidation of methionine as variable modifications, and mass tolerance of ±20 ppm for precursor and ± 50 ppm for product ions. The precursor mass shifts are set between −18 Da and 177 Da to take into consideration variable modifications such as the presence of sodium and potassium adducts. Peptide hits are validated in the peptide mode to achieve a false discovery rate (FDR) of <1.2% and then in the protein mode according to the score settings recommended by the manufacturer. Positive identifications were considered only when two or more peptides were matched, and their summed score was >20.

3.6 Data Analysis

1. Perform the principal component analysis (PCA) (*see* **Note 31**).

4 Notes

1. Resuspension buffer can be prepared and stored at 4 °C.
2. Solution 1 can be prepared and stored at −20 °C.
3. Solution 2 can be prepared and stored at 4 °C.
4. For the first gel, 10% (w/v) APS solution can be prepared fresh and what is not used can be stored at −20 °C in 120 μL aliquots.
5. Add TEMED just before pouring the gel.
6. Avoid bubbles during casting. Once the gel has finished loading into the cassette, add 2-propanol so that the front of the gel is straight.

7. Before loading the stacking gel, remove 2-propanol (you can use blotting paper).
8. The solution must be made fresh.
9. The solution must be made fresh and kept in a dark place.
10. Prepare trypsin solution (according to the manufacturer's recommendations) to reach the final concentration of 0.1 μg/μL enzyme.
11. This is the composition of the Binding Solution when the sample is a pellet.
12. The results shown in Tables 1 and 2 were obtained using the SpectrumMill Proteomics Workbench. Similar software can be used.
13. Link to *Vitis vinifera* subset: https://www.uniprot.org/uniprotkb?query=Vitis+Vinifera.
14. Link to Allergen subset: https://www.uniprot.org/uniprotkb?query=Allergen.
15. The results shown in Fig. 2 were obtained using Progenesis QI for Proteomics v2.0. Similar software can be used.
16. The volume of 100 mL is indicative. It is possible that for some wines where the protein concentration is very small, it may be necessary to work with a larger volume to be able to extract and precipitate the amount of protein necessary to perform the nanoLC-MS/MS proteomic analysis.
17. It may be that the volume after dialyzing is greater than 100 mL.
18. It is very important that the sample is completely lyophilized before it is added to the resuspension buffer.
19. For every 100 mL of dialyzed and lyophilized wine, 2 mL of resuspension buffer is usually used. As all the proteins present in the lyophilized sample must be precipitated, it is very important to know the final volume in which it has been resuspended.
20. The best quantification results were obtained using the *RC DC*™ Protein Assay Kit. Bovine serum albumin was used as a protein standard.
21. We used an ultrasonic processor with a sonotrode with a diameter of 1.0 mm adjusted at 1 cycle and 90% amplitude. During sonication, keep samples on ice, since sonication generates heat which can degrade proteins.
22. Keeping it overnight in absolute acetone allows for better elimination of the remains of the reagents used in protein precipitation.
23. Other default parameters: amperage 500 mA, power 150 W.
24. It is a modification of the method used in [17].

Table 1
Proteins identified in the monovarietal varietal white wines analyzed using the subset *Vitis vinifera* from the UniProtKB database

Name	Accession number	Theoretical mass (Da)	Peptide number	Score	Grape variety
Beta-fructofuranosidase	F6HAU0	60,223.5	20	394.92	Viura
			19	378.37	Verdejo
			4	70.01	Chardonnay
Beta-fructofuranosidase	A0A438DX78	63,888.9	8	140.41	Viura
			6	109.97	Verdejo
GH18 domain-containing protein	D7SSN5	39,738.7	12	220.80	Viura
Chitotriosidase-1	A0A438C8K0	52,221.8	11	207.19	Viura
Thaumatin-like protein	A0A438JJ53	24,877.1	9	169.67	Viura
			6	117.84	Verdejo
Thaumatin-like protein	A0A438BZP1	24,828.2	6	138.13	Viura
			6	123.84	Verdejo
Thaumatin-like protein	Q7XAU7	24,851.2	6	127.99	Viura
Thaumatin-like protein	Q9M4G6	24,963.4	5	89.24	Viura
			11	198.35	Verdejo
Thaumatin-like protein	A5C9F1	24,850.1	9	173.46	Viura
			6	122.21	Verdejo
Thaumatin-like protein	F6HUH1	24,915.4	5	84.53	Viura
			11	204.15	Verdejo
Thaumatin-like protein	A0A438BZN0	26,333.9	2	36.74	Chardonnay
Chitotriosidase-1	A0A438C8K0	52,221.8	11	207.19	Viura
Class IV chitinase	O7XAU6	28,383.7	4	79.57	Viura
			5	100.12	Verdejo
Oryzain alpha chain	F6HMA2	62,195.5	2	32.91	Viura
			3	60.79	Verdejo
			2	35.17	Chardonnay
Peroxidase	A0A438DZJ5	27,483.9	4	58.70	Verdejo
Peroxidase	A0A438JVD2	42,581.8	3	46.56	Verdejo
ATP synthase subunit beta	A0A438K995	67,253.7	3	45.44	Verdejo
			2	34.33	Chardonnay

Table 2
Proteins identified in the monovarietal varietal white wines analyzed using the subset Allergen from the UniProtKB database

Name	Accession number	Theoretical mass ((Da)	Peptide number	Score	Species
Major allergen beta-lactoglobulin	B5B0D4	20,368.4	5	88.98	Bovin
Alfa-S1-casein	P02662	24,585.5	4	65.85	Bovin
Thaumatin-like protein	P83958	25,236.9	2	33.91	*Actinidia chinensis*
			3	39.94	*Actinidia chinensis*

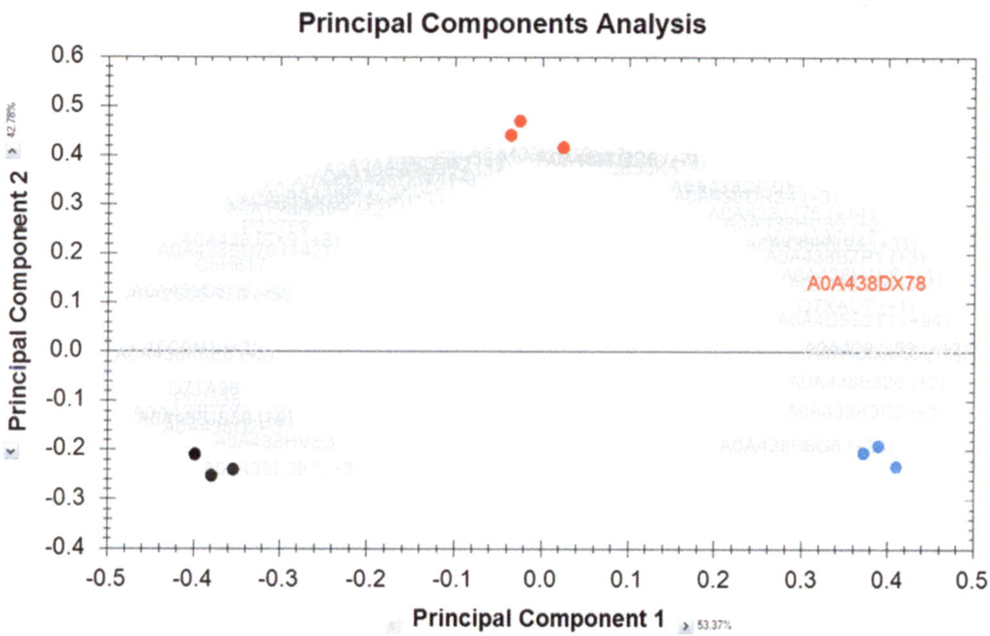

Fig. 2 Results of principal component analysis (PCA). PCA biplot of samples (dots) and variables (accession numbers) showing clustering of three analytical replicates by a variety of commercial white wines to which they belong. The PCA was plotted using those protein spots with a score > 15 that differed significantly in abundance levels (ANOVA, $p < 0.01$) between isolates in the classified groups. Viura (blue dots); Verdejo (red dots); Chardonnay (black dots)

25. Incubate at 4 °C with vigorous shaking overnight until the pellet is completely dissolved.
26. This step is necessary to ensure that trypsin acts at a urea concentration that is not greater than 0.6 M.
27. If the pH is not between 7 and 9, 1 M ammonium bicarbonate should be added until pH is in that optimal range for trypsin action, avoiding excessive dilution of the sample.
28. At this pH, the enzyme is inactivated, and proteolytic digestion will cease.
29. Prior to nanoLC-MS/MS analysis, peptide desalting is important to eliminate interfering salts.
30. Link to UniProt databases used to obtain the results shown in Tables 1 and 2: https://ftp.uniprot.org/pub/databases/uniprot/previous_releases/release-2023_03/knowledgebase/
31. The results shown in Fig. 2 were obtained using the software package Progenesis QI for Proteomics (Nonlinear Dynamics, Waters). For detailed handling of the software, refer to user guides and online help (http://www.nonlinear.com/progenesis/qi-for-proteomics/v3.0/user-guide).

Acknowledgements

This work was supported by Valencian Conselleria d'Innovació, Universitats, Ciencia y Societat Digital grant CIAICO/2021/167; European Funds for Regional Development (FEDER).

References

1. Ribéreau-Gayon P, Glories Y, Maujean A, Dubourdieu D (2021) The chemistry of wine stabilization and treatments. In: Handbook of enology, vol 2, 3rd edn. Wiley, Hoboken, pp 369–406
2. OIV International Code of Oenological Practices. Basic Definition. Issue 2021/01, I.1.3-1. Available online: https://www.oiv.int/standards/international-code-of-oenological-practices/part-i-definitions/wines/basic-definition
3. Rajdl D, Racek J, Trefil L, Siala K (2007) Effect of white wine consumption on oxidative stress markers and homocysteine levels. Physiol Res 56(2):203–212. https://doi.org/10.33549/physiolres.930936
4. Berg HW, Akiyoshi M (1961) Determination of protein stability in wine. Am J Enol Vitic 12:107–110. https://doi.org/10.5344/ajev.1961.12.3.107
5. Hsu JC, Heatherbell DA (1987) Heat-unstable proteins in wine I. Characterization and removal by bentonite fining and heat treatment. Am J Enol Vitic 38:11–16
6. Waters EJ, Wallace W, Williams PJ (1991) Heat haze characteristics of fractionated wine proteins. Am J Enol Vitic 42:123–127. https://doi.org/10.5344/ajev.1991.42.2.123
7. Waters EJ, Shirley NJ, Williams PJ (1996) Nuisance proteins of wine are grape pathogenesis-related proteins. J Agric Food Chem 44:3–5. https://doi.org/10.1021/jf9505584
8. Vincenzi S, Marangon M, Tolin S, Curioni A (2011) Protein evolution during the early stages of white winemaking and its relations with wine stability. Aust J of Grape Wine Res

17:20–27. https://doi.org/10.1111/j.1755-0238.2010.00113.x

9. Pocock KF, Hayasaka Y, McCarthy MG, Waters EJ (2000) Thaumatin-like proteins and chitinases, the haze-forming proteins of wine, accumulate during ripening of grape (*Vitis vinifera*) berries and drought stress does not affect the final levels per berry at maturity. J Agric Food Chem 48:1637–1643. https://doi.org/10.1021/jf9905626
10. Van Sluyter SC, McRae JM, Falconer RJ, Smith PA, Bacic A, Waters EJ, Marangon M (2015) Wine protein haze: mechanisms of formation and advances in prevention. J Agric Food Chem 63:4020–4030. https://doi.org/10.1021/acs.jafc.5b00047
11. Australia-New Zealand Food Standards Code (2002) Standard 1.2.-Mandatory Warning and Advisory Statements and Declaration; http://www.comlaw.gov.au/Details/F2011C00610. Accessed 2 Aug 2013
12. European Commission (2012) Commission implementing regulation 579/2012 Off. J Eur Union L171:4–7
13. Government of Canada, Food and Drugs Act. Regulations Amending the Food and Drug Regulations, 1220-Enhanced Labelling for Food Allergen and Gluten Sources and Added Sulphites, 2011.; http://gazette.gc.ca/rp-pr/p2/2011/2011-02-16/html/sordors28-eng.html. Accessed 2 Aug 2013
14. Mierczynska-Vasilev A, Boyer A, Vasilev K, Smith PA (2017) A novel technology for the rapid, selective, magnetic removal of pathogenesis-related proteins from wines. Food Chem 232:508–514. https://doi.org/10.1016/j.foodchem.2017.04.050
15. Marangon M, Vincenzi S, Curioni A (2019) Wine fining with plant proteins. Molecules 24:2186. https://doi.org/10.3390/molecules24112186
16. Le Bourse D, Jégou S, Conreux A, Villaume S, Jeandet P (2010) Review of preparative and analytical procedures for the study of proteins in grape juice and wine. Anal Chim Acta 667(1–2):33–42. https://doi.org/10.1016/j.aca.2010.03.062
17. Neuhoff V, Arold N, Taube D, Ehrhardt W (1988) Improved staining of proteins in polyacrylamide gels including isoelectric focusing gels with clear background at nanogram sensitivity using Coomassie Brilliant Blue G-250 and R-250. Electrophoresis 9:255–262. https://doi.org/10.1002/elps.1150090603

Chapter 14

Isolating Synaptic Vesicles from Neurospheres for Proteomics

Caroline Brandão-Teles, Giuliana S. Zuccoli, Marcelo Ganzella, Victor Corasolla Carregari, Linda Olsthoorn, Érica de Almeida Duque, Carolina Demarchi Munhoz, Reinhard Jahn, Daniel Martins-de-Souza, and Fernanda Crunfli

Abstract

This chapter presents an optimized method for isolating synaptic vesicles (SVs) from neurospheres derived from human induced pluripotent stem cells (hiPSCs). The protocol begins with neurosphere cultivation to achieve mature neurons, which is essential for the functional studies of neuronal activity. Following this, neurosphere-derived synaptosomes are isolated, and SVs are enriched through differential centrifugation. The method culminates in the proteomic analysis of SVs using nano-liquid chromatography coupled with high-resolution tandem mass spectrometry (nanoLC-MS/MS), providing a detailed proteome profile of the isolated vesicles. This protocol can contribute to the understanding of SV molecular heterogeneity and the mechanisms of neurotransmitter uptake and release and be applied to the field of research in neurological and neuropsychiatric disorders.

Key words Synapse vesicles, Proteomics, Neurospheres, Neural stem cell, Vesicle isolation

1 Introduction

Generating human induced pluripotent stem cells (hiPSCs) from somatic human cells has revolutionized the study of human diseases, particularly by addressing a crucial aspect often lacking in animal models: the genetic background. This advantage is of even greater significance in research focused on neurological and neuropsychiatric diseases, such as autism, schizophrenia, and depression, all of which have a robust genetic component [1, 2]. Despite the limitations associated with in vitro models, hiPSC-derived cells offer a powerful tool for modeling human diseases and development [3].

Taufika Islam Williams (ed.), *Tissue Proteomics: Methods and Protocols*, Methods in Molecular Biology, vol. 2884, https://doi.org/10.1007/978-1-0716-4298-6_14,

Employing reprogramming factors allowed hiPSCs to be differentiated into neural stem cells (NSCs), capable of becoming neurons and glial cells later. Two primary cultivation methods are commonly employed to differentiate hiPSCs into neuronal cells: the adherent (2D) model and the formation of free-floating aggregates, known as neurospheres (3D). While the 2D model offers advantages in terms of single-cell analysis, neurospheres are considered a more naturalistic type of culture. This is attributed to their ability to better emulate the original tissue environment, encompassing tissue-specific architecture, mechanical and biochemical characteristics, cell-to-cell communication and signaling, and differentiation capabilities [4, 5].

Ultrastructural analysis of neurospheres has revealed a diverse population of cells in various states of differentiation [6], spanning from presumed stem/progenitor cells to fully differentiated neurons and glial cells [7]. The compelling benefit of having cells in different developmental stages is their unique capacity to facilitate the study of the development and functioning of live human neurons. Notably, not all neurons derived from hiPSCs are fully mature. This is apparent from the fact that only a fraction of the cells fire action potentials, some have immature synapses, and some express markers of immature neurons. The degree of neuron maturation observed in the neurospheres is highly dependent on the cultivation method applied [4, 5, 8].

Neurons attain maturity when they express a spectrum of ion channels, receptors, intracellular signaling elements, and vesicle-associated proteins. Some indications of advanced synaptic maturity have been observed within neurospheres, for example, in studies showing that most of the synapses appeared as entirely or nearly filled with synaptic vesicles (SVs) [8]. SVs are small vesicles, highly abundant in presynaptic nerve terminals of neurons, with a diameter of around 40 nm. These neurotransmitter reservoir organelles are a key component of neurotransmission. Neurotransmitter-filled SVs fuse with the plasma membrane, following an increase in cytosolic Ca^{2+} ions upon the arrival of an action potential (AP) at the presynaptic terminal. After fusion, the released neurotransmitter molecules bind to receptors on the postsynaptic side, thus initiating an intracellular signaling cascade. To sustain the fast stimulus-response coupling required for efficient neurotransmission, SVs are recycled through endo- and exocytosis mechanisms, are re-filled with neurotransmitters, and can then initiate another release cycle [9, 10].

Biochemically purified vesicles are essential for obtaining a better understanding of SV molecular heterogeneity, not only for the identification of crucial components of the molecular machinery responsible for calcium sensing, exo-, and endocytosis but also for unraveling the mechanisms of neurotransmitter uptake and storage [11–13]. The classical protocols for the isolation of SVs from rodent brains use synaptosomes as a starting point. Synaptosomes are pinched-off and spontaneously released presynaptic terminals formed by gentle homogenization of the tissue in a viscous solution. These synaptosomes are then subjected to hypoosmotic lysis to release SVs. SVs are further purified by sequential differential, rate-zonal, or isopycnic centrifugations. SVs prepared by these protocols are morphologically of impressive homogeneity and purity [14–17]. A better understanding of these vesicles' functional roles and associated molecular mechanisms can be gained through advanced analytical methods, such as proteomics, offering insights into their protein content.

By elucidating key proteins and their biological roles in synaptic vesicle function, proteomics brings the potential to drive the development of new therapeutic interventions and preventative strategies for neurological disorders [18].

In this chapter, an optimized method for the isolation of neurosphere-derived SVs is presented. Initially, the cultivation conditions of neurospheres were adjusted to achieve a high degree of maturity in neurons. The SV isolation protocol, as described below, represents a tailored and optimized approach specifically designed for neurospheres. It employs a series of differential centrifugation steps, beginning with the preparation of synaptosomes, followed by the enrichment of SVs. Lastly, a protocol for analyzing the proteomes of subcellular fractions, in this case focusing on SVs, using nano-liquid chromatography coupled to high-resolution tandem mass spectrometry (LC-MS/MS), is detailed.

2 Materials

2.1 Induction of Neural Stem Cells from Human Pluripotent Stem Cells and Neurosphere Differentiation

1. Human iPS cells.
2. 60-mm cell culture–treated dishes.
3. 6-well culture plates.
4. Serological sterile pipettes.
5. 37 °C humidified cell culture incubator with 5% CO_2.
6. Centrifuge.
7. CO_2 resistant shaker.
8. 15-mL and 50-mL sterile polypropylene conical tubes.
9. Ethylenediaminetetraacetic acid (EDTA) 0.5 mM: 0.02% in DPBS (0.5 mM), sterile-filtered.
10. Culture medium: StemFlex™ (or similar).
11. Neural induction medium (NIM): To prepare 100 mL, aseptically mix 98 mL of Neurobasal® Medium and 2 mL of Gibco® neural induction (NI) supplement.

12. Neural expansion medium (NEM): To prepare 100 mL, add 48.5 mL of Neurobasal® medium, 48.5 mL Advanced™ DMEM/F-12, 1 mL of GIBCO® neural induction supplement, and 1 mL of Pen/Strep to a final concentration of 1%.
13. Neurosphere medium: To prepare 100 mL, add 40 mL of DMEM/F12, 48 mL of Neurobasal® medium, 2 mL of 1% B27 (50×) to a final concentration of 1%, 1 mL of N2 (100×) to a final concentration of 1%, and 1 mL Pen/Strep to a final concentration of 1%.
14. Matrigel LDEV-Free (Corning).
15. Geltrex® LDEV-Free hESC-qualified reduced growth factor basement membrane matrix (Thermo Fisher Scientific).
16. ROCK inhibitor Y27632.
17. Phosphate-buffered saline (PBS) without calcium and magnesium.
18. Accutase.

2.2 Synaptic Vesicle Isolation and Protein Extraction

To ensure accurate results, use only high-purity (analytical grade) reagents. Prepare all solutions with Milli-Q water, and adjust the pH using pre-cooled solutions. Before use, filter the solutions through a 0.2-μm nitrocellulose membrane. These precautions are crucial for minimizing contamination and ensuring the accuracy, reliability, and reproducibility of experimental results.

1. 1.5-mL Teflon glass Dounce homogenizer
2. Beckman Optima TLX ultracentrifuge with TLA 110 rotor (or equivalent) and corresponding sealed polycarbonate ultracentrifuge tubes.
3. Homogenization buffer: Prepare a solution with 320 mM of sucrose and 5 mM of 2-[4-(2-hydroxyethyl)piperazin-1-yl] ethanesulfonic acid (HEPES). Adjust the pH to 7.4 using NaOH. The homogenization buffer should be freshly prepared before the experiment and stored at 4 °C.
4. HEPES solution: Prepare a 1 M HEPES stock solution by dissolving HEPES-free acid in deionized water. Pre-cool the solution and adjust the pH to 7.4 using KOH. Store aliquots of the solution at 4 °C for up to 6 months.
5. Protease inhibitor cocktail.
6. Plastic syringe 1 mL.
7. Hypodermic needle 22-gauge and 27-gauge.

2.3 Tryptic Digestion Procedure and Sample Preparation for NanoLC-MS/MS

1. Lysis buffer: Prepare lysis buffer by combining 100 mM Tris–HCl, 150 mM NaCl, 1 mM EDTA, 0.5% Triton X-100, and protease inhibitors. Adjust volumes based on the required buffer quantity and ensure thorough mixing.
2. BCA protein assay.
3. Microcon-10 kDa Centrifugal Filter Unit with Ultracel-10 membrane (centrifugal filter unit, with a 0.5 mL capacity and a 10 kDa nominal molecular weight limit (NMWL) membrane, facilitates the concentration of biomolecules).
4. 100 mM dithiothreitol (DTT): Add 15.4 mg of DTT powder to 8 mL of deionized water and use a stir bar with a stir plate (if available) to aid dissolving. Once the DTT is completely dissolved, use deionized water to bring the final volume of 10 mL. Store them at −20 °C for a longer shelf life. DTT degrades over time, especially at higher temperatures and with exposure to light and air.
5. 300 mM iodoacetamide (IAA): Prepare a 300 mM IAA solution by dissolving the required amount of IAA in distilled water. Protect the solution from light and use it immediately after preparation.
6. Sequencing grade modified porcine trypsin,
7. 0.1% formic acid (FA): Carefully prepare a 0.1% FA solution by adding the appropriate volume of formic acid to distilled water within a volumetric flask. Ensure to wear appropriate personal protective equipment throughout the process.

3 Methods

3.1 Induction of NSCs from iPS Cells

This protocol, adapted from Thermo Fischer Scientific® [19, 20], requires a sterile technique throughout all procedures and medium preparation. The following instructions are for a single well in a 6-well plate. For other culture ware, adjust volumes as needed.

1. Start with high-quality human iPS cells cultured in feeder-free conditions in StemFlex medium on Matrigel matrix (*see* **Note 1**).
2. When hiPSCs reach ~70–80% confluency, split them into 6-well culture plates (*see* **Note 2**).
3. To split the cells, start by aspirating the culture medium and add 1 mL of PBS.
4. Aspirate PBS and add 1 mL of EDTA 0.5 mM solution per well.
5. Incubate for 3–5 min at 37 °C in the incubator. If colonies remain attached, extend the incubation to 10 min (*see* **Note 3**).

6. Aspirate the EDTA solution with a serological pipette.
7. Add pre-warmed StemFlex medium containing 10 μM ROCK inhibitor to the plate (*see* **Note 4**).
8. Gently wash the well(s) with a medium to detach the cells. Harvest the detached colonies using a 5- or 10-mL serological pipette, avoiding bubble formation (*see* **Note 5**). Collect the cells in a 15-mL conical tube.
9. After generating a hiPSC suspension, transfer 1 mL of the cell suspension to a 15-mL conical tube for cell counting. It is crucial, to use a separate 15-mL conical tube (second tube) for **steps 9.1–9.4** to ensure an accurate estimation of the total cell number in the original suspension.
   9.1 Centrifuge the 15-mL conical tube with the cells at 200× *g* for 5 min to pellet the cells. Carefully aspirate the supernatant.
   9.2 Resuspend the cell pellet in 1 mL of pre-warmed Accutase dissociation reagent and incubate for 5 min at 37 °C.
   9.3 Gently pipette the cell suspension up and down several times using a 1-mL pipette to achieve a single-cell suspension. Avoid creating bubbles (*see* **Note 5**).
   9.4 Determine the total cell number using manual (hemocytometer) or automated counting (automated cell counters). These techniques are standard practice in cell culture laboratory settings.
10. After estimating the total cell number, plate the cells in each well of a 6-well plate at a density of 2.5×10^5–3×10^5 hiPSCs per well (*see* **Note 6**).
11. After plating, gently swirl the plates in several quick back-and-forth and side-to-side motions to disperse the cells across the surface. Place them gently in a CO_2 incubator (*see* **Note 7**).
12. On day 0 (typically 24 h after hiPSC splitting, when they reach 10–20% confluency), aspirate the medium to remove non-attached cells. Replace the medium with 2.5 mL of pre-warmed neural induction medium (NIM) per well in a 6-well plate and return the plates to the incubator (*see* **Note 8**).
13. At 48 h (day 2) after replacing to NIM, ensure the cell colonies exhibit uniform morphology (*see* **Note 9**). Replace the medium with 2.5 mL of pre-warmed complete NIM per well in a 6-well plate. Subsequently, refresh the medium daily (*see* **Note 10**).
14. By day 7 of neural induction, NSCs (passage number 0 = P0) are typically ready for harvesting and expansion (*see* **Note 11**).
15. Prepare Geltrex® matrix-coated vessels before proceeding with the next steps (*see* **Note 12**).

16. Prepare sufficient volumes of NEM supplemented with ROCK inhibitor Y27632 at a final concentration of 5 μM (*see* **Note 4**).
17. Gently rinse the cells once with 1 mL of PBS per well in a 6-well plate. Aspirate the PBS after rinsing.
18. Add 1 mL of pre-warmed Accutase cell dissociation reagent to each well of the 6-well plate and incubate at 37 °C for 5–8 min until most cells detach from the culture vessel surface. Use the microscope to determine if the cells are detaching. Gently transfer the detached cell clumps using a 5-mL pipette to a 15-mL conical tube (*see* **Note 13**).
19. Wash the wells with 1 mL of PBS per well in the 6-well plate to collect residual cells. Transfer the resulting cell suspension to the 15-mL conical tube containing the previously detached cells.
20. Gently pipette the cell suspension up and down several times using a 5-mL or 10-mL pipette to achieve a single-cell suspension (*see* **Note 14**). Avoid creating bubbles (*see* **Note 5**).
21. Centrifuge the cells at 200× *g* for 5 min at room temperature.
22. After removing the supernatant, mix the cells in NEM warmed and supplemented with ROCK inhibitor.
23. Add the diluted NSC suspension to each Geltrex® matrix-coated well and plate the cells at a density of 1×10^5 cells per well.
24. Gently swirl the culture plates in several quick back-and-forth and side-to-side motions to evenly distribute the cells. Carefully place them in a CO_2 incubator.
25. After overnight incubation, replace the medium with fresh NEM to remove the ROCK inhibitor Y27632. Subsequently, change the medium with NEM (without Y27632) every other day (*see* **Note 15**).
26. NSCs typically reach confluency within 4–6 days of plating in NEM. Once they are confluent, you have the option to either expand the NSCs further for neurosphere differentiation in fresh NEM or cryopreserve them for future use (*see* **Note 16**).
27. Neurospheres are typically generated from NSCs that have reached 90% confluency.
28. Gently aspirate the medium from the well/plate and add an appropriate volume (e.g., 1 mL for a 6-well plate) of pre-warmed Accutase cell dissociation reagent to detach the NSCs. Incubate the culture at 37 °C for 5 min.
29. Collect the cell supernatant in a 15-mL sterile conical tube.
30. Determine the total cell number using manual (hemocytometer) or automated counting (automated cell counters). These techniques are standard practice in cell culture laboratory settings.

31. Centrifuge the cells at 200 × *g* for 5 min.
32. After centrifugation, gently resuspend the NSC pellet in 1 mL of pre-warmed NEM. Count the cells using your preferred method (manual or automatic) and dilute the cell suspension further in NEM to achieve a final plating density of 2×10^6 cells per well of a 6-well plate.
33. Cells are incubated under constant gyratory shaking (88 rpm) at 37 °C and 5% CO_2.
34. At 48 h, if the NSCs have begun to form spherical structures (neurospheres), carefully replace the medium with a fresh neurosphere medium.
35. Continue culturing the cells in the shaker (88 rpm) for 15 days at 37 °C in a humidified atmosphere with 5% CO_2. Refresh the neurosphere medium every other day to ensure optimal nutrient supply.
36. After 8 weeks of differentiation, neurospheres typically exhibit uniform size and spherical morphology. Immunofluorescence staining with specific markers (e.g., neuronal or glial markers) can be used to assess differentiation status (Fig. 1).

3.2 Isolation of the Synaptic Vesicles

Handling Considerations for Synaptic Vesicle (SV) Isolation:

Detergent sensitivity: SVs are sensitive to detergents. Ensure all instruments and tools used throughout the isolation protocol are completely detergent-free. Even trace amounts of detergent can damage the vesicles. *Dedicated rotor tubes:* For optimal SV yield

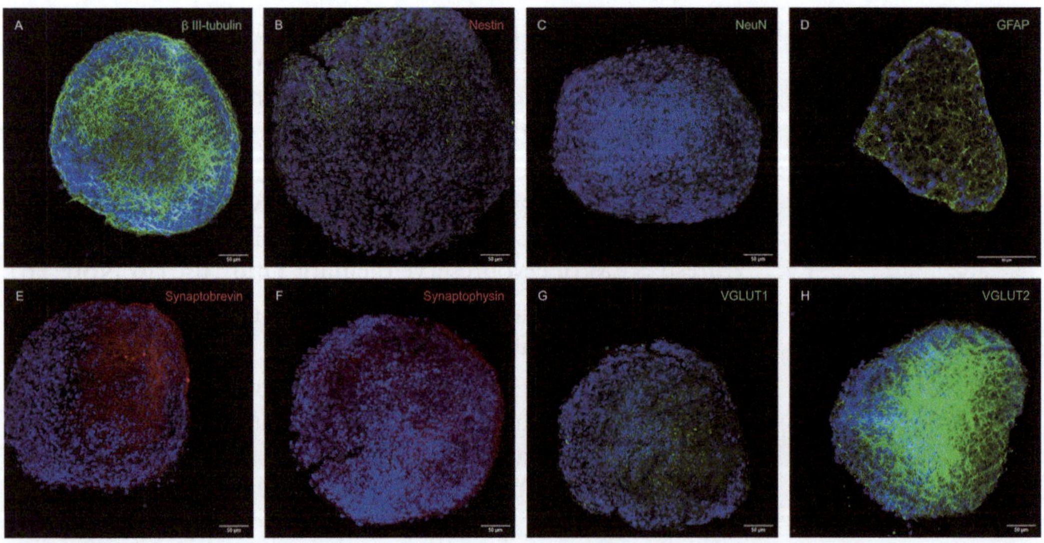

Fig. 1 Immunofluorescence assay of neurospheres derived from NSCs expressing: (**a**) β III-tubulin; (**b**) nestin; (**c**) NeuN; (**d**) GFAP; (**e**) synaptobrevin; (**f**) synaptophysin; (**g**) VGLUT1; (**h**) VGLUT2. Nuclei were stained with DAPI (blue)

and purity, it is strongly advised to use new rotor tubes specifically designated for this isolation protocol. These tubes should never have been previously exposed to detergents.

1. Carefully transfer the neurospheres from the 6-well plate to a new 15-mL conical tube and centrifuge for 5 min at 200× *g* (*see* **Note 17**).
2. Before homogenization, wash the neurospheres three times with an ice-cold homogenization buffer. To wash, centrifuge for 5 min at 200× *g* at 4 °C after each wash to pellet the neurospheres. Gently remove the supernatant after each centrifugation step (*see* **Note 18**).
3. Add 2 mL of homogenization buffer to the conical tube containing the neurospheres. Gently swirl the tube to ensure the neurospheres are completely covered by the buffer.
4. Gently transfer 1 mL of neurospheres using a wide-orifice tip (1.5 mm) in a tight Teflon glass Dounce homogenizer and apply 10 strokes for homogenization. Avoid creating air bubbles during this step (*see* **Note 5**). Transfer the homogenate to a 2-mL microcentrifuge tube. Repeat this process with the remaining neurospheres until all neurospheres have been homogenized.
5. Centrifuge the homogenate at 12,000× *g* for 15 min at 4 °C.
6. Following centrifugation, carefully discard the supernatant (S1) which likely contains mostly soluble proteins and free vesicles from the cytoplasm (*see* **Note 19**).
7. Gently resuspend the pellet (P1—enriched in synaptosomes) using a wide-orifice tip (1.5 mm) to avoid shearing. Use 100 μL of homogenization buffer for resuspension.
8. To hypostatically disrupt the pellet, add 900 μL of fresh Mili-Q Water supplemented with 5 mM HEPES and protease inhibitor.
9. To assist with disruption of the synaptosomes, carefully take up the sample with a 22-gauge needle coupled to a 1-mL syringe, change the needle to a 27-gauge, and slowly push it out, returning to the same microtube. Repeat this procedure 10 times.
10. Centrifuge the lysate at 17,000× *g* at 4 °C for 20 min.
11. Collect the supernatant and transfer it to polycarbonate ultracentrifuge tubes. Centrifuge at 200,000× *g* for 1 h in a Beckman Optima TLX ultracentrifuge equipped with a TLA 110 rotor (or equivalent) to pellet the SVs.
12. Following centrifugation, carefully discard the supernatant (S3). Gently resuspend the pellet (P3) using a pre-cut pipette tip to minimize shearing. Use 200 μL of lysis buffer for resuspension.

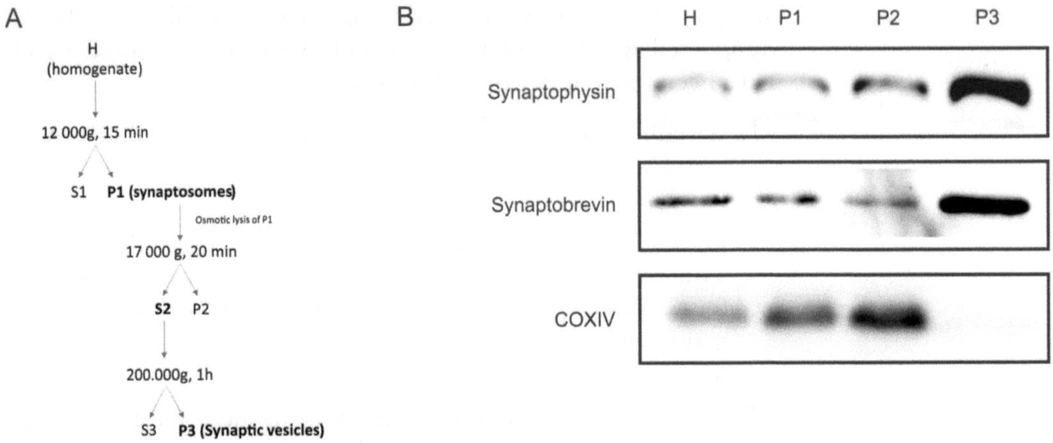

Fig. 2 Flowchart illustrating the purification protocol for synaptic vesicles from neurospheres (**a**). Representative enrichment profiles of synaptic vesicles from neurospheres monitored by Western blotting (**b**). The same amount of protein (10 μg) from each fraction was loaded per lane. The synaptic vesicles are enriched in the P3 fraction

13. The isolated SVs can be snap-frozen in liquid nitrogen and stored at −80 °C.
14. To verify the enrichment of SVs in each fraction, perform immunoblotting using antibodies against established SV resident proteins and potential contaminants (Fig. 2b). Proteins, such as synaptophysin and synaptobrevin, are ideal markers for detecting SV enrichment; thus, they should exhibit a parallel enrichment pattern, with the highest enrichment observed in the final SV fraction (P3). Conversely, markers for non-SV compartments, such as COXIV (mitochondrial marker), should ideally show depletion (or minimal enrichment) in the final fraction.

3.3 Proteomic Analysis

1. The SVs isolated from the neurospheres are homogenized in a lysis buffer and incubated on ice for 30 min for chemical disruption of the membrane vesicles.
2. After the lysis, determine the protein amount using the BCA protocol.
3. To minimize technical variability, sample contamination, and efficiently remove detergent and salt from the lysis buffer before LC-MS/MS analysis, we employed the filter-aided sample preparation (FASP) digestion protocol. This protocol used 30 micrograms of protein per sample for subsequent reduction, alkylation, and tryptic digestion [21, 22].
4. Adjust digested peptides in 0.1% of FA in water (solvent A) to a final concentration of 1 μg/μL.

5. Dilute peptides to a final concentration of 100 ng/μL and inject 500 ng into the column. Separate tryptic peptides using an UltiMate 3000 RSLCnano system (Thermo Fisher) equipped with a reverse-phase C18 column (3 μm, 75 μm × 150 mm) maintained at 40 °C. Employ a 120-min reversed-phase gradient at a flow rate of 250 nL/min, increasing the acetonitrile concentration from 5% to 40% over 90 min. Utilize water containing 0.1% formic acid (FA) as eluent A and acetonitrile containing 0.1% FA as eluent B. The separated peptides are analyzed by a high-definition Orbitrap Fusion Lumos (Thermo Fisher) spectrometer directly coupled to the chromatographic system. Acquire data in data-dependent acquisition (DDA) mode, scanning from 300 to 1500 *m/z* with a mass resolution of 120,000 at MS1 and a maximum injection time of 30 ms in positive ion mode. Set the intensity threshold for MS2 to 2.0e4 with a resolution of 15,000. Select ions with charges 2–5 for fragmentation, excluding previously analyzed ions for 60 s with a 10 ppm tolerance. Use two cycle times and an isolation window of 1.6 *m/z*. Apply HCD activation with a collision energy of 30 V.
6. Continuum nanoLC-MS/MS data from three replicate experiments for each sample are processed for qualitative and quantitative analyses using the software Proteome Discoverer version 2.4. The qualitative identification of proteins can be obtained by searching in the respective databank of the organism (UniProt). Search parameters are set as a tolerance of 10 ppm for precursor ions and 0.6 Da for product ions, a minimum of one fragment ion matched per peptide, a minimum of three fragment ions matched per protein, a minimum of one unique peptide matched per protein, two missed cleavage, carbamydomethylation of cysteines as fixed modification and oxidation of methionines as variable modifications, and false discovery rate (FDR) of the identification algorithm < 1%.
7. Label-free quantitative analysis is obtained using the normalized abundance intensity integrated with the Proteome Discoverer software, using all peptides for normalization (sum of the intensity of all peptides). The expression analysis is performed considering technical replicates available for each experimental condition following the hypothesis that each group is an independent variable. Filtered tables are generated to include proteins found at least in two out of three technical replicates and exclude proteins showing no statistical significance according to T.test ≥0.05.
8. Following nanoLC-MS/MS analysis, proteins that are found differently regulated are used to apply gene ontology and enrichment analysis (Fig. 3).

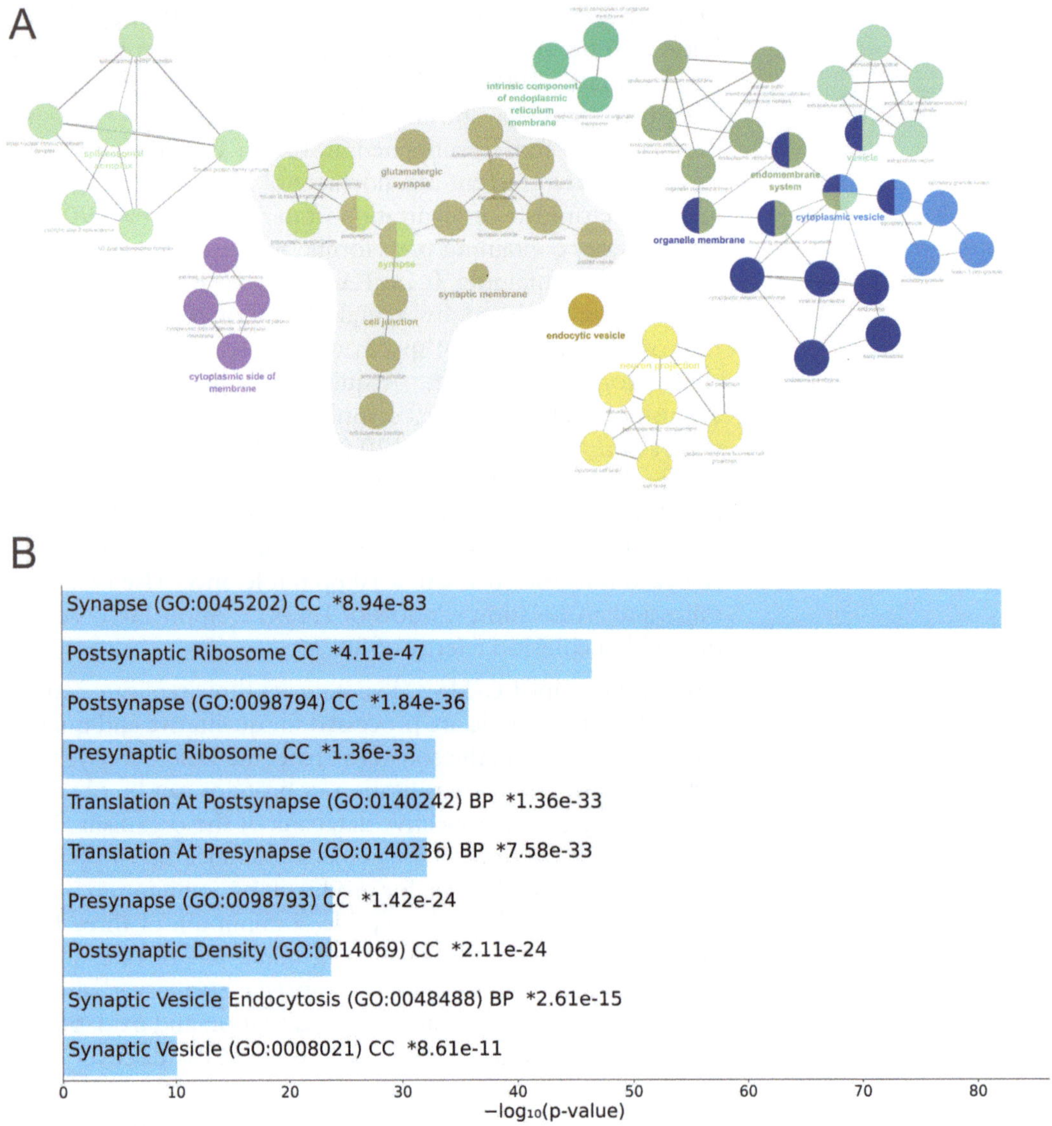

Fig. 3 Functional analysis of proteins identified in the synaptic vesicle enrichment. Cellular component terms were enriched by ClueGO [23] function analysis at a high level. The top 12 clusters were associated with spliceosomal complex, cytoplasmic side of membrane, synapse, glutamatergic synapse, synaptic membrane, cell junction, endocytic vesicle, intrinsic component of endoplasmic reticulum membrane, neuron projection, organelle membrane, endomembrane system, cytoplasmic vesicle, and vesicle (**a**). Bar chart of top enriched terms from the SynGO_2024 gene set library. The top ten enriched terms for the synaptic vesicle enrichment set are displayed based on the −log10(*p*-value), with the actual *p*-value shown next to each term. The term at the top has the most significant overlap with the input query gene set (**b**)

4 Notes

1. High-quality hiPSCs should be pluripotent, genetically stable, free of contaminants, highly viable, and maintain genomic integrity. They should form compact colonies with a uniform appearance, featuring tightly packed cells with well-defined, smooth edges. The individual cells are small and round, with a high nuclear-to-cytoplasmic ratio and prominent nucleoli. It is important to monitor for signs of differentiation, such as the presence of flat, elongated cells at the colony edges. While StemFlex is commonly used, other medium formulations can also support hiPSC culture, with the choice of the medium often depending on experimental requirements, cell line characteristics, and laboratory protocols. Additionally, vitronectin is an effective alternative to Corning Matrigel for culturing hiPSCs.
2. The most common way to assess confluence is by observing the cells under a microscope. At 70–80% confluence, the colonies should be well formed and nearly touching each other but with some space remaining between them. It is important to note that the ideal confluence percentage for passaging hiPSCs can vary depending on the specific cell line and culture conditions. Generally, passaging at 70–80% confluence is recommended to maintain optimal cell health and pluripotency. Regular monitoring and maintaining detailed records of cell culture conditions and growth rates will improve your accuracy in estimating confluence.
3. The incubation time may vary depending on the cell line or the non-enzymatic cell dissociation reagent. Observe the cells dissociating under a microscope until they begin to separate and round up, forming colonies with visible gaps. This indicates they are ready for removal from the vessel.
4. The ROCK inhibitor Y27632 is commonly used to improve the survival and proliferation of cells after dissociation. By inhibiting Rho-associated kinase, it prevents apoptosis and promotes cell attachment. Including ROCK inhibitors in the culture medium is often beneficial for maintaining cell viability and increasing cell numbers.
5. Bubbles in cell culture should be avoided because they can damage cells, disrupt gas exchange, interfere with microscopy, increase contamination risk, lead to uneven cell distribution, and alter pH levels. To prevent bubble formation, it is important to handle cells and media gently, avoid vigorous pipetting, and use proper pipetting techniques.
6. hiPSC lines exhibit varying growth rates and plating efficiencies. Therefore, providing a cell number range (2.5×10^5–

3×10^5 hiPSCs per well) allows for flexibility in accommodating these inter-line differences. This approach enhances experimental reproducibility and ensures optimal cell density for subsequent differentiation protocols.

7. For optimal hiPSC maintenance and neural induction, passage cells when they reach 10–20% confluence (*see* **Note 8**). Plate the cells in small clumps, avoiding single-cell plating as this can lead to increased cell death. The ideal split ratio (cells plated after passage compared to the starting number) will vary depending on the hiPSC line, depending on factors like growth rate, colony morphology, and passaging method. It is important to optimize this ratio for each cell line by experimenting and observing cell behavior over multiple passages. Neural induction protocols often begin around 24 h after passaging.
8. Observe the cells under a microscope, focusing on colony size, density, and empty space. At 10–20% confluence, colonies will be small and well spaced. The cells should be sparsely distributed, with large areas of the culture surface remaining uncovered. Cells appear as individual clusters of cells or small, defined colonies.
9. Uniform colony morphology, characterized by distinct colony borders and a homogenous cell distribution within the colony, is indicative of healthy hiPSC growth. Irregular or fuzzy borders may suggest senescence or differentiation, while variations in cell density within the colony can signal potential culture issues. Consistent monitoring of colony morphology is essential for maintaining optimal hiPSC culture conditions.
10. From day 4 onwards, double the volume of NIM (5 mL per well) to ensure optimal cell nutrition due to increasing cell density. Throughout this period (days 4–7), regular monitoring and detailed record-keeping of cell culture conditions and growth rates are essential. Minimal cell death is expected. If you observe a brownish cell color and many floating cells, it indicates a high starting PSC density. In this case, switch to daily medium changes with 5 mL of fresh NIM per well.
11. P0–P4 NSCs require overnight treatment with 5 μM ROCK inhibitor Y27632 upon plating to prevent cell death during expansion and differentiation into glial or neuronal cells. At this stage, you can freeze the cells to create a stock for future use.
12. Prepare a 1% Geltrex™ matrix solution by adding 1 mL of Geltrex™ matrix to 99 mL of pre-chilled (4 °C) DMEM/F-12 medium.
13. Accutase is the optimal dissociation method for the neural stem cell protocol due to its superior ability to maintain cell viability

and preserve cell surface proteins. It is gentler than traditional dissociation methods like trypsin, making it ideal for sensitive cell types such as stem cells.

14. Gently pipetting a cell suspension is essential for breaking up cell clumps, ensuring even distribution, and preventing cell damage. This, in turn, facilitates accurate cell counting and improves the success of downstream applications. It is crucial to emphasize that this step should be performed with the utmost care to minimize shear stress on the cells, ensuring their viability and functionality.
15. Prolonged exposure to ROCK inhibitors can interfere with normal cell physiology. Removing the ROCK inhibitor after overnight incubation allows for cellular recovery and adaptation to the culture environment. Regular medium changes with ROCK inhibitor-free medium are essential to maintain optimal growth conditions for neural stem cells.
16. To accurately assess the confluence of neural stem cells (NSCs), utilize phase contrast microscopy. This technique enhances the contrast between cells and the background, providing a clearer view of the distinct NSC morphology, characterized by the presence of neural rosettes. These rosettes are self-organized clusters of polarized neuroepithelial-like cells, resembling the ventricular zone (VZ) of a developing brain. For consistent and reliable visual estimations, assess confluence at the same time each day under uniform lighting conditions. Usually, the ideal confluence for passaging NSCs is within the 70–80% range to maintain optimal cell health and proliferation capacity. However, the specific cell line and culture conditions may necessitate adjustments to this guideline.
17. To obtain sufficient synaptic vesicles from neurospheres, start with at least one 6-well plate containing neurospheres.
18. To maximize synaptic vesicle yield, handle neurospheres gently during washes. Avoid breaking them by using a wide orifice tip (1.5 mm).
19. Enrichment profiles of target proteins can be monitored by immunoblotting subcellular fractions collected at each step of the protocol. To achieve this, aliquot an appropriate volume from each centrifugation pellet throughout the protocol steps.

Acknowledgments

We thank Prof. Emer Suavinho Ferro, who allowed us access to a Beckman Optima TLX ultracentrifuge, Mr. Paulo Baldasso (BSc) for the technical support provided, and Rodrigo Soares Caldeira

Brant and Michel Batista (Mass Spectrometry Platform, Carlos Chagas Institute–ICC Fio Cruz). We thank FAPESP (São Paulo Research Foundation) for research funding through grants 2017/25588-1, 2019/00098-7, 2022/12693-0, 2019/25957-2, and 2019/22398-2 and also the CNPq (the Brazilian National Council for Scientific and Technological Development) and the CAPES (Coordination of Superior Level Staff Improvement, Brazil).

Conflicts of Interest The authors declare no competing interests.

References

1. Dolmetsch R, Geschwind DH (2011) The human brain in a dish: the promise of iPSC-derived neurons. Cell 145:831–834
2. Galiakberova AA, Dashinimaev EB (2020) Neural stem cells and methods for their generation from induced pluripotent stem cells in vitro. Front Cell Dev Biol 8:815
3. Kim J, Koo B-K, Knoblich JA (2020) Human organoids: model systems for human biology and medicine. Nat Rev Mol Cell Biol 21:571–584
4. Kobolak J, Teglasi A, Bellak T et al (2020) Human induced pluripotent stem cell-derived 3D-neurospheres are suitable for neurotoxicity screening. Cells 9
5. Pauly MG, Krajka V, Stengel F et al (2018) Adherent vs. free-floating neural induction by dual SMAD inhibition for Neurosphere cultures derived from human induced pluripotent stem cells. Front cell. Dev Biol 6:3
6. Bez A, Corsini E, Curti D et al (2003) Neurosphere and neurosphere-forming cells: morphological and ultrastructural characterization. Brain Res 993:18–29
7. Del Valle L, Reiss K, Parker Struckhoff A (2022) Culture and phenotyping of glial cell cultures, gliospheres, and neurospheres. Methods Mol Biol 2422:217–232
8. Capetian P, Müller L, Volkmann J et al (2020) Visualizing the synaptic and cellular ultrastructure in neurons differentiated from human induced neural stem cells-an optimized protocol. Int J Mol Sci 21
9. Jahn R, Fasshauer D (2012) Molecular machines governing exocytosis of synaptic vesicles. Nature 490:201–207
10. Sudhof TC (2004) The synaptic vesicle cycle. Annu Rev Neurosci 27:509–547
11. Mori Y, Takamori S (2017) Molecular signatures underlying synaptic vesicle cargo retrieval. Front Cell Neurosci 11:422
12. Wolosker H, de Souza DO, de Meis L (1996) Regulation of glutamate transport into synaptic vesicles by chloride and proton gradient(∗). J Biol Chem 271:11726–11731
13. Holt M, Riedel D, Stein A et al (2008) Synaptic vesicles are constitutively active fusion machines that function independently of Ca2+. Curr Biol 18:715–722
14. Huttner WB, Schiebler W, Greengard P et al (1983) Synapsin I (protein I), a nerve terminal-specific phosphoprotein. III. Its association with synaptic vesicles studied in a highly purified synaptic vesicle preparation. J Cell Biol 96:1374–1388
15. Nagy A, Baker RR, Morris SJ et al (1976) The preparation and characterization of synaptic vesicles of high purity. Brain Res 109:285–309
16. Jahn R, Schiebler W, Ouimet C et al (1985) A 38,000-dalton membrane protein (p38) present in synaptic vesicles. Proc Natl Acad Sci USA 82:4137–4141
17. Ganzella M, Ninov M, Riedel D et al (2022) Isolation of synaptic vesicles from mammalian brain. In: Dahlmanns J, Dahlmanns M (eds) Synaptic vesicles: methods and protocols. Springer, New York, pp 131–145
18. Basu A, Ash PE, Wolozin B et al (2021) Protein interaction network biology in neuroscience. Proteomics 21:e1900311
19. Elkabetz Y, Studer L (2008) Human ESC-derived neural rosettes and neural stem cell progression. Cold Spring Harb Symp Quant Biol 73:377–387
20. Denham M, Dottori M (2009) Signals involved in neural differentiation of human embryonic stem cells. Neurosignals 17:234–241
21. Distler U, Kuharev J, Navarro P et al (2016) Label-free quantification in ion mobility-enhanced data-independent acquisition proteomics. Nat Protoc 11:795–812

22. Carregari VC (2022) Protein extraction and sample preparation methods for shotgun proteomics with central nervous system cells and brain tissue. In: Corasolla Carregari V (ed) Understanding PTMs in neurodegenerative diseases. Springer International Publishing, Cham, pp 1–15

23. Bindea G, Mlecnik B, Hackl H et al (2009) ClueGO: a Cytoscape plug-in to decipher functionally grouped gene ontology and pathway annotation networks. Bioinformatics 25:1091–1093

Chapter 15

Isolation and Identification of Brain Tissue Extracellular Vesicles for Translational Proteomics

Michelle V. Lihon, Nathan A. Tuchscherer, and W. Andy Tao

Abstract

Extracellular vesicles (EVs) are small membrane-bound structures that play important roles in intercellular communication and the transfer of biomolecules between cells. EVs have become a topic of interest for research in translational proteomics for disease biomarker discovery due to their ability to reflect changes in the cellular proteome, including diseases affecting the brain. Utilizing the proteome analysis of EVs to its fullest potential requires proper isolation and purity. In this chapter, we describe a detailed method for the isolation and identification of brain tissue EVs for translational proteomics using our in-house chemical affinity magnetic bead–based (non-antibody) method, the EVtrap. We also discuss various methods for quantification, characterization, and functional analysis of isolated brain tissue EVs, including western blotting, and proteomic profiling of post-translational modifications (PTMs) involved in neurodegenerative diseases, such as protein N-terminal acetylation. This protocol provides a valuable resource for studies conducted on brain tissue EVs and their potential as biomarkers and therapeutic targets for neurological diseases.

Key words Extracellular vesicles (EVs), Exosomes, Tissue, Brain tissue, EVtrap, Mass spectrometry, DDA, Label-free proteomics, Relative quantitative proteomics

1 Introduction

EVs are lipid-bound vesicles secreted by cells into the extracellular space [1]. Furthermore, EVs are nanosized, membrane-bound vesicles that play a critical role in intercellular communication and biomolecule transport between cells [2]. EVs have gained significant interest in translational proteomics research due to their potential as disease biomarkers and therapeutic targets, particularly in the field of neurodegenerative diseases [3]. By analyzing the proteome of EVs, researchers can identify biomarkers that reflect changes in the cellular proteome, including those occurring in the brain. Moreover, as EVs can cross the blood-brain barrier, they hold promise as a non-invasive diagnostic tool, especially for neurodegenerative diseases [3]. The method described here offers

Taufika Islam Williams (ed.), *Tissue Proteomics: Methods and Protocols*, Methods in Molecular Biology, vol. 2884,
https://doi.org/10.1007/978-1-0716-4298-6_15,

researchers a direct approach to isolate and analyze EVs from brain tissue for the identification of potential biomarkers that could aid in the diagnosis and monitoring of neurodegenerative diseases.

In addition to the proteome, post-translational modifications (PTMs) of proteins, such as N-terminal acetylation, play a crucial role in the regulation of protein folding, protein localization, protein function, and cellular signaling [4]. Identifying and characterizing this modification in EVs can provide valuable insights into the underlying biological processes in neurological diseases. N-terminal acetylation is one of the most abundant PTMs identified in vivo [4]. It is an irreversible PTM in which an acetyl group is covalently bound to the amino-terminal of a protein by an enzyme classified as an N-terminal acetyltransferase [4]. Disruption in the homeostasis of acetylation has been identified in several neurodegenerative disorders, including Alzheimer's disease (AD), Parkinson's disease (PD), and Huntington's disease (HD) [5–8]. The disturbance in acetylation homeostasis has shown accumulations in proteins such as tau, α-synuclein, and Huntingtin protein implicated in AD, PD, and HD, respectively [9]. Therefore, precise quantification of N-terminal acetylation levels in EVs may provide a non-invasive tool for the early detection and monitoring of neurodegenerative diseases and allow a better understanding of their underlying mechanisms.

In this protocol chapter, we provide a comprehensive method for isolating and characterizing extracellular vesicles (EVs) from brain tissue samples using a mouse model system. This includes the description of our group's novel chemical affinity magnetic bead, known as extracellular vesicle total recovery and purification (EVtrap) [10, 11]. EVtrap is designed with a unique combination of lipophilic and hydrophilic groups, enabling it to interact specifically with the lipid bilayer of EVs for efficient isolation [10, 11].

Additionally, this protocol also provides a method for selectively enriching protein N-terminal acetylated tryptic peptides using strong cation exchange (SCX). In general, PTMs often occur at low stoichiometry compared to the total protein pool. Enrichment of PTMs narrows the dynamic range by removing unmodified peptides, improving the detection and quantification of low-abundance modified peptides. Ultimately, this targeted enrichment approach enables the determination of changes in the abundances of these modifications when comparing control vs. experimental samples, providing valuable insights into disease mechanisms. This method employs SCX silica resin, which selectively binds positively charged ions (cations) from a solution to negatively charged functional groups on the resin, enabling separation and purification based on their differing affinities for the resin. This enrichment strategy separates peptides according to their charge and orientation, with tryptic peptides in acidic conditions having a charge of +2 due to Lys or Arg amino acid on one end of the peptide and the N-terminus on the other end of the peptide, and protein Nt-Ac peptides are singly charged. This difference in charge between internal protein peptides and Nt-Ac peptides will allow for their separation by the SCX column solid phase.

This method was designed to be fully compatible with LC-MS-based EV proteome analysis, providing the high recovery and purity levels necessary. In this chapter, we detail the procedure for EV isolation and proteome analysis from brain tissue using a mouse model system via EVtrap, phase-transfer surfactant (PTS) lysis and protein extraction, protein N-term acetylation enrichment via SCX, and DDA mass spectrometry (MS) analysis (*see* Fig. 1 for the overall workflow).

2 Materials

1. Collagenase III: ≥ 100 CDU/mg (CDU, collagenase digestion unit).
2. Serum-free Dulbecco's Modified Eagle Medium (DMEM).
3. 1× phosphate-buffered saline (PBS): 137 mM sodium chloride (NaCl), 2.7 mM potassium chloride (KCl), 10 mM sodium phosphate dibasic (Na_2HPO_4), and 1.8 mM monobasic potassium phosphate (KH_2PO_4); pH 7.4.
4. EVtrap functionalized magnetic beads (Tymora Analytical Operations).
5. EVtrap loading buffer (Tymora Analytical Operations).
6. EVtrap washing buffer (Tymora Analytical Operations).
7. EVtrap elution solution: 200 mM triethylamine (TEA) in HPLC-grade water. This solution is recommended to be prepared fresh.
8. HPLC-grade water.
9. Triethylammonium bicarbonate (TEAB).
10. Sodium deoxycholate (SDC).
11. Sodium lauroyl sarcosinate (SLS).
12. Tris(2-carboxyethyl) phosphine hydrochloride (TCEP).
13. Chloroacetamide (CAA).
14. Phase-transfer surfactant (PTS) lysis solution: 12 mM SDC, 12 mM SLS, 10 mM TCEP, 30 mM CAA in 50 mM TEAB. Vortex after each added ingredient. This solution must be prepared fresh and used immediately.
15. 50 mM TEAB in HPLC-grade water.
16. Proteomics grade trypsin (Millipore Sigma, Promega).

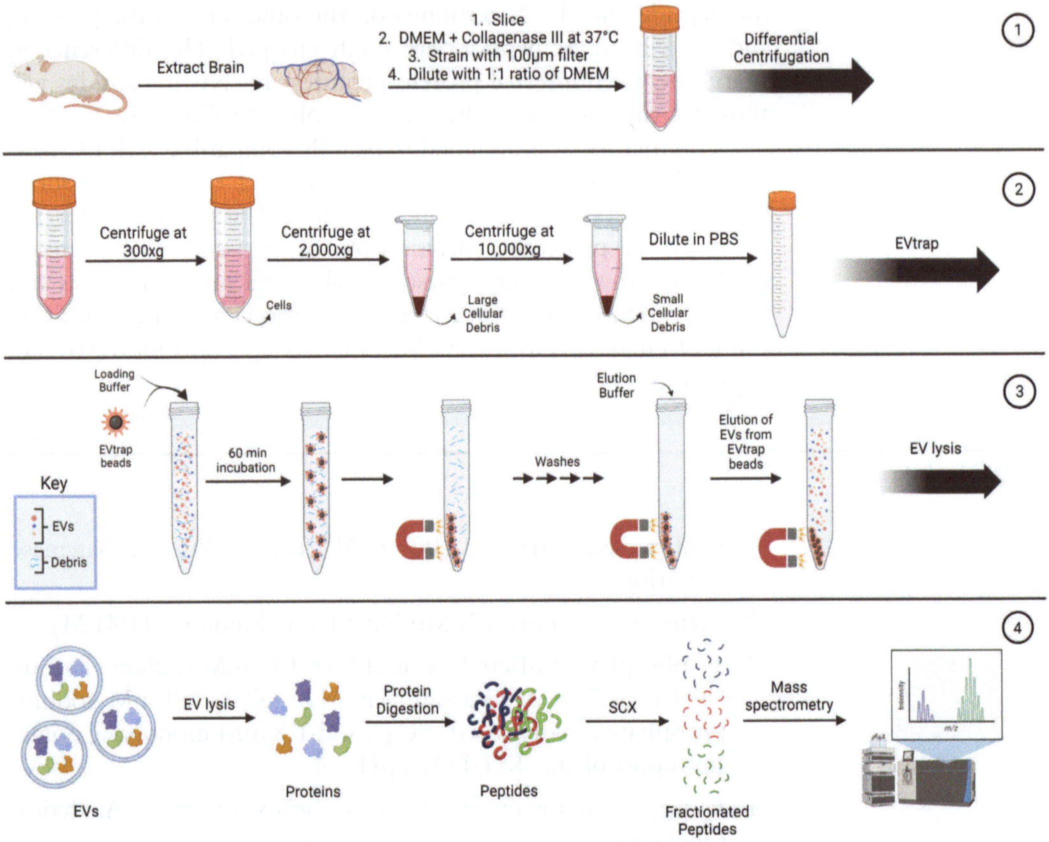

Fig. 1 Workflow overview. Panel 1, the top panel, displays the extraction and release of EVs from the extracellular space. Panel 2, the middle panel, displays the differential centrifugation steps prior to EVtrap. Centrifuging at 300× *g*, 2000× *g*, and 10,000 × *g* pellets the cells, large cellular debris, and small cellular debris, respectively. The 10,000 × *g* supernatant is then diluted 25× with 1× PBS prior to EVtrap. Panel 3, the bottom panel, displays the steps in EVtrap. The sample is rotated with EVtrap beads and a loading buffer to bind the EVs to the EVtrap beads. Then, using a magnet, the EVtrap beads are held undisturbed to the side of the tube, while the remaining fluid is removed and analyzed as the flowthrough. Next, the EVtrap beads are washed twice with washing buffer and twice with 1× PBS. The washed beads are then shaken with 0.2 M TEA to release the EVs from the EVtrap beads. This solution is then separated from the EVtrap beads and analyzed as the elution. This sample containing EVs can then be dried and analyzed via western blot and mass spectrometry

17. Acetonitrile (ACN).
18. Trifluoroacetic acid (TFA).
19. Ethyl acetate.
20. C18 loading solution: 0.1% TFA, 3% acetonitrile in HPLC-grade water. Loading solution can be stored at room temperature for up to 6 months.

21. C18 elution solution: 0.1% TFA, 60% acetonitrile in HPLC-grade water. Loading solution can be stored at room temperature for up to 6 months.
22. LC-MS loading solution: 0.05% TFA, 3% acetonitrile or 0.1% formic acid, 3% acetonitrile in HPLC-grade water.
23. Centrifuge capable of 16,000× *g* speed and 4 °C temperature.
24. Thermal shaker capable of reaching 95 °C.
25. Sep-Pak C18 column (Waters). Alternatively, any other C18 column vendor may also be suitable for this purpose.
26. Centrifugal vacuum concentrator.
27. Magnetic separator rack.
28. Ammonium formate.
29. Strong cation exchange resin (Polysulfoethyl A TopTip) or any other SCX column vendor may also be suitable for this purpose.
30. 15-mL and 50-mL conical tubes.
31. 1.7-mL nonstick microcentrifuge tubes.
32. Bicinchoninic acid (BCA) protein assay kit (Life Technologies, Millipore Sigma).
33. Pierce Quantitative Peptide Assays and Standards.
34. Razor blade.
35. Water bath capable of achieving 37 °C.
36. 100-μM filter.
37. End-over-end rotator.
38. pH paper.
39. Microcentrifuge tube shaker.
40. Pierce quantitative peptide assay (Thermo Scientific).
41. Pierce BCA protein assay (Thermo Scientific).
42. Ammonium bicarbonate (ABC).
43. *Optional*: EasynLC 1000 system (Thermo Fisher Scientific).
44. EasynLC 1000 system buffer A: 0.1% formic acid.
45. EasynLC 1000 system buffer B: 0.1% formic acid in 80% (vol/vol) acetonitrile.
46. *Optional*: hybrid high-resolution LTQ-Orbitrap Velos Pro mass spectrometer (Thermo Fisher Scientific).
47. *Optional*: column measuring 45 cm in length and packed in-house (360 μm OD × 75 μm ID) with C18 resin (2.2 μm, 100 Å; Michrom Bioresources).

3 Methods

3.1 Extraction of Tissue Extracellular Vesicles (EVs)

Note: This extraction of tissue EV protocol has been adapted from Vella et al., 2017.

1. Thinly slice the brain with a razor blade for a greater surface area to volume ratio. If brain tissue is frozen, thaw on ice then thinly slice the brain with a razor blade.
2. Weigh sliced tissue and transfer it to a 15-mL conical tube containing 75 U/mL of collagenase type III enzyme in serum-free Dulbecco's Modified Eagle Medium (DMEM). Add 800 μL of enzyme per 100 mg of brain tissue (*see* **Note 1**).
3. Gently mix the sample by inverting the tube several times.
4. Incubate in a water bath at 37 °C for a total of 20 min. After the first 5 min of incubation, gently invert the tube and return to the water bath. After 10 min of incubation, gently pipette up and down to further break down the tissue.
5. After incubation time, immediately place the sample on ice and dilute the sample with serum-free DMEM 1:1 to dilute down collagenase type 3 (*see* **Note 2**).
6. Allow the tissue to settle to the bottom of the tube, carefully remove the excess solution using a sterile pipette, and transfer it into a clean 15-mL conical tube (*see* **Note 3**).
7. Place a 100-μM filter on a 50-mL conical tube and add the sample from the previous step (**step 6**) onto the filter to remove the excess tissue and other debris. Then, wash the strainer with 1 mL of DMEM and collect the rinse in the same 50-mL conical tube.
8. Centrifuge this filtrate at 300× *g* for 10 min at 4 °C (the pellet contains brain cells, and the supernatant contains EVs).
9. Transfer the supernatant into a clean conical tube and centrifuge at 2000× *g* for 10 min at 4 °C (the pellet contains large cell debris, and the supernatant contains EVs).
10. Transfer the supernatant into clean 1.7-mL nonstick microcentrifuge tubes and centrifuge at 10,000× *g* for 30 min at 4 °C (the pellet contains small cell debris, and the supernatant contains EVs).
11. Collect the supernatant (that had been centrifuged at 10,000× *g*) and dilute it 25× with 1× PBS.

3.2 Enrichment of EVs

1. Take the desired amount of diluted 10,000× *g* supernatant and add EVtrap loading buffer (for every 1 mL of sample, add 100 μL of loading buffer).
2. Add EVtrap functionalized magnetic beads (20 μL for every 1 mL of the original sample volume).

3. Incubate in an end-over-end rotator for 1 h at room temperature.
4. Place the tubes on a magnetic rack, wait for the beads to settle toward the magnet, and remove the supernatant from the microcentrifuge tube (*see* **Note 4**).
5. Wash the beads twice with 1 mL of 1× PBS. Mix the tubes gently by inverting them several times in between each wash. Place the tubes back onto a magnetic rack and discard the supernatant.
6. Elute the EVs from the beads twice by adding 50 μL of 20 0 mM TEA. Place the sample in a shaker for 10 min. Place the sample on a generic microcentrifuge tube shaker for 5 min in between each elution. Place tubes back on the magnetic rack after each 5-min shaking and transfer the elution to a clean tube without disturbing the beads.
7. The eluted EVs can be dried using a speed vacuum (*see* **Notes 5** and **6**).

3.3 EV Lysis

1. Prepare fresh PTS lysis solution (*see* **Note** 7).
2. Resuspend the dried EVs in 100 μL appropriate volume of PTS solution.
3. Incubate the samples at 95 °C for 5 min to lyse the EVs.
4. Cool the sample to room temperature and dilute fivefold with fresh 50 mM TEAB or 50 mM ammonium bicarbonate (ABC; ensure that the pH is ~7.8–8.2 by using pH paper).
5. Prepare protein standards following the Thermo Scientific BCA protein assay kit's instructions and perform the BCA assay to determine the protein concentration of the lysed EVs (*see* **Note 8**).

3.4 Protein Digestion and Surfactant Removal

1. Add the appropriate amount of trypsin into the sample at a 1: 50 ratio of trypsin to protein w/w (e.g., 1 μg trypsin per 50 μg protein to be digested). Digest overnight at 37 °C with shaking at 900 rpm for 12–16 h.
2. Acidify the sample with 10% TFA to pH < 3 (e.g., ~1% final TFA concentration) and vortex the sample.
3. Add equal parts of ethyl acetate (according to the sample volume) and shake the sample vigorously for 2 min. Then, spin down the sample at 16,000× *g* for 3 min.
4. Carefully pipette out the organic phase on top and discard it (*see* **Note 9**).
5. Dry the sample completely using a centrifugal vacuum concentrator to remove any remaining organic solvent.

3.5 Peptide Desalting

1. The sample can be desalted with a disposable C18 column/tip. The desalting step described in this section is specific for using the Sep-Pak C18 column (Waters) (*see* **Note 10**).
2. Condition the column with 1X column volume of 100% methanol and let it drain completely, followed by 2X column volumes of 0.1% TFA (*see* **Note 11**).
3. Load the acidified sample onto the column with a slow drip by gravity or by slow spin down in a centrifuge (see manufacturer's instructions).
4. Wash 3X column volumes of 0.1% TFA (*see* **Note 11**).
5. Elute the sample with 1X column volume of 0.1% TFA in 60% ACN into a 1.7-mL nonstick microcentrifuge tube. Do this step with a slow drip or low-speed centrifugation and collect the eluent.
6. Dry the sample completely using a centrifugal vacuum concentrator before proteome analysis (*see* **Notes 12** and **13**).

3.6 Strong Cation Exchange Processing

1. Resuspend the peptides in 100 μL of 5 mM ammonium formate (AmFo), 70% ACN, pH 3 (*see* **Note 14**).
2. Equilibrate the SCX tip column with 2X column volumes of 5 mM AmFo, 70% ACN, and pH 3 and centrifuge at 500× *g* for 1 min.
3. Load the sample onto the tip column and centrifuge at 150× *g* for 5 min. Repeat if there is still a remaining sample in the tip column.
4. Elute the samples into clean microcentrifuge tubes following the sequence below, centrifuging the samples at 150× *g* for 4 min to collect each fraction separately:

   Fraction 1: 1X column volume of 5 mM AmFo, 70% ACN.

   Fraction 2: 1X column volume of 10 mM AmFo, 70% ACN.

   Fraction 3: 1X column volume of 15 mM AmFo, 70% ACN.

   Fraction 4: 1X column volume of 20 mM AmFo, 70% ACN.
5. Split each fraction into two aliquots 20% and 80%, and then, dry the sample completely using a centrifugal vacuum concentrator. The 20% aliquot will be used to determine the peptide concentration, and the 80% aliquot will be used for N-term acetylation proteome analysis.
6. Resuspend the 20% aliquot from each fraction in 10 μL of ultrapure water, and then, measure the peptide concentration using Pierce quantitative peptide assay. The 80% aliquot of each fraction is ready for nanoLC-MS analysis via data-dependent acquisition (DDA) (*see* **Notes 12** and **15**).

3.7 NanoLC/MS

The method described here utilizes an EasynLC 1000 system coupled with an LTQ-Orbitrap Velos Pro mass spectrometer, providing a robust platform for DDA mass spectrometry analysis. However, it is important to note that there are various other systems and methods available for DDA mass spec analysis, including data-independent acquisition (DIA) analysis. Researchers may choose alternative systems and methods based on their specific experimental requirements and analytical preferences.

1. Dissolve the dried enriched peptide samples from Subheading 3.6, **step 6** in an appropriate volume of 0.25% formic acid with 3% (vol/vol) ACN to inject a maximum amount of enriched peptides.
2. Inject 4 μL of each prepared sample into an EasynLC 1000 system (Thermo Fisher Scientific).
3. Perform peptide separation with a column measuring 45 cm in length and packed in-house (360 μm OD × 75 μm ID) with C18 resin (2.2 μm, 100 Å; Michrom Bioresources).
4. Use 0.1% formic acid in HPLC-grade water (referred to as buffer A), while the eluting buffer consisted of 0.1% formic acid in 80% (vol/vol) acetonitrile (referred to as buffer B) for the mobile phase buffer.
5. From minute 0 to minute 60, use a linear gradient ranging from 3% to 30% buffer B; from minute 60 to minute 70, continue with a linear gradient increasing from 30% to 50% buffer B; then, from minute 70 to minute 71, further increase the gradient to 95% buffer B. Finally, maintain a constant 95% buffer B from minute 71 to minute 75, all at a flow rate of 300 nL/min.
6. Analyze mass spectrometry in data-dependent mode, starting with a full-scan MS over the range of m/z 300 to 1500, with a resolution of 30,000 at m/z 400. Perform MS/MS fragmentation on the 10 most intense ions detected during the full-scan MS analysis.

3.8 Data Interrogation

After acquiring mass spectrometry data, it is crucial to perform thorough data interrogation to extract meaningful insights. Various software tools are available for analyzing proteomics data, each offering unique functionalities and capabilities. Here, we outline the basic steps for data interrogation:

1. *Data analysis/processing:* Begin by processing raw mass spectrometry data using software tools such as MaxQuant, Proteome Discoverer, or equivalent. Please note that mass spectrometry raw data typically come in formats like .raw, . mzXML, and .mzML, and make sure that the software used for data analysis is compatible with such formats.

2. *Parameter setup:* Configure the parameters specific to your experiment. This includes selecting the appropriate instrument settings, defining search parameters such as enzyme specificity, maximum missed cleavages, PTMs, precursor and fragment ion mass tolerances, and choosing the database against which the spectra will be searched (e.g., UniProt). Specify search parameters for accurate peptide/protein identification.
3. *Database search and identification of PTMs:* the software of your choice will perform a database search to identify peptides in the mass spectrometry raw data. Some software programs use an algorithm such as Andromeda for this purpose. The search includes matching experimental spectra against theoretical spectra generated from the protein sequence database. At the same time, these software are capable of identifying PTMs, such as acetylation. It does this by searching for mass shifts in the peptide ions that correspond to the presence of PTMs. These software programs also offer advanced algorithms for accurate quantification based on tandem mass spectrometry data.
4. *Statistical analysis:* After protein identification, conduct statistical analysis to identify significant changes in protein expression levels between experimental conditions. Software tools like Perseus and MSstatsShiny provide comprehensive statistical analysis capabilities, including t-tests, ANOVA, and multiple testing corrections.
5. *Gene Ontology analysis:* Once statistical analysis has been performed, perform Gene Ontology (GO) analysis to gain insights into their biological functions, cellular components, and molecular processes (Fig. 2). GO analysis helps in understanding the functional significance of the identified modified proteins or lack of modified proteins within biological systems. There are several software programs available for this such as:

   DAVID (Database for Annotation, Visualization, and Integrated Discovery).

   Panther (Protein ANalysis THrough Evolutionary Relationships).

   GOrilla (Gene Ontology enRIchment anaLysis and visuaLizAtion tool).

   STRING (Search Tool for the Retrieval of Interacting Genes/Proteins)—includes GO enrichment analysis functionality.
6. *Visualization and interpretation:* Visualize and interpret the results obtained from data analysis to draw meaningful conclusions. Utilize visualization tools within the chosen software platforms to generate plots, heatmaps, and other visual representations of the data, aiding in result interpretation. This can be done using software such as:

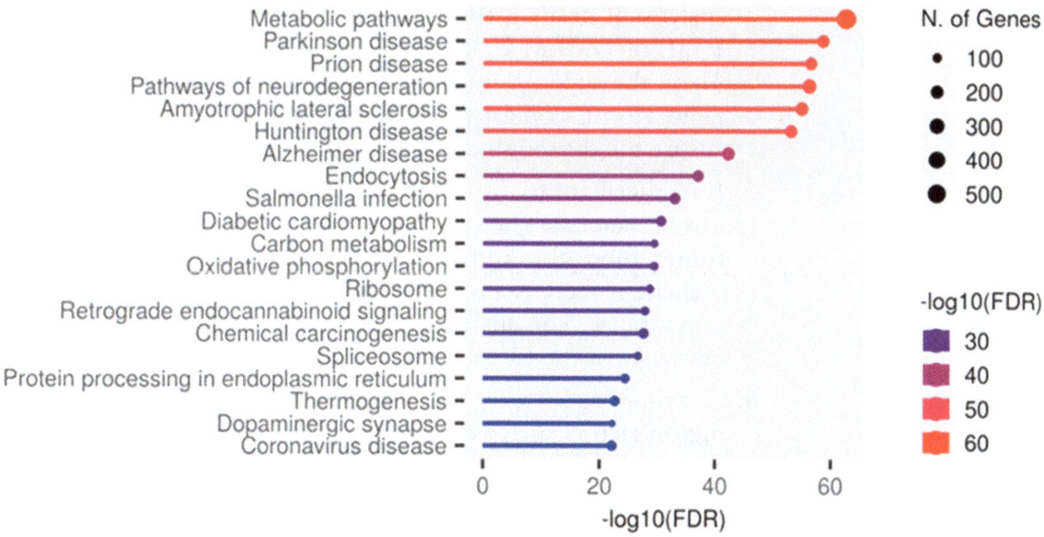

Fig. 2 Pathways associated with the gene segments of identified proteins of isolated EVs from EVtrap elution. The proteins identified from the isolated EVs from brain tissue identified protein markers correlated with neurodegenerative disease pathways such as Parkinson's disease, Alzheimer's disease, and Huntington's disease. The identification of proteins in these neurodegenerative disease pathways provides evidence to support the utilization of this method for further research on these disease pathways

R Studio.

Mass Dynamics.

Perseus.

Scaffold.

Progenesis QI.

Skyline.

ProteoWizard.

4 Notes

1. It is essential to use collagenase type III enzyme to gently digest the tissue, facilitating the release of extracellular vesicles (EVs) while minimizing cell disruption [12, 13]. This is extremely important since cell lysis and release of protein into the sample mixture can affect EV isolation, EV content characterization, and other downstream analyses.
2. This step is crucial for minimizing tissue digestion by diluting the collagenase III enzyme, ensuring controlled processing of the sample.
3. It is important not to disturb the undigested tissue settled at the bottom of the tube during transfer. This is to ensure no contamination of tissue debris that may affect downstream

analysis. If needed, the undigested tissue can be washed with 1 mL of DMEM by gently inverting the tube several times. Then, the supernatant can be removed and transferred to the same 15-mL conical tube used in **step 6**. These washes can ensure the thorough collection of EVs from the tissue pellet.

4. It is crucial not to disturb the beads during this step to prevent loss of material. The supernatant removed from the microcentrifuge tube, also known as the flowthrough, can be saved to evaluate if there is any loss of material (EVs) during the isolation process. This analysis can provide valuable insights into the efficiency of the EVtrap method.

5. The time required for the eluted EVs to dry may vary depending on factors such as the volume of the sample and the temperature of the speed vacuum. If necessary, drying the EVs at a lower temperature, such as 4 °C, can expedite the process, although it is not always necessary. Over-drying should be avoided to prevent potential damage to the EVs. Once the sample is dry, proceed with the next steps of the protocol or store the dried EVs at −80 °C. It is important to note that the drying process is essential to remove the organic solvent (TEA) used to elute EVs from the beads, preventing it from affecting subsequent steps.

6. After EV isolation, it is advisable to perform western blot analysis before proceeding with proteomics or other omics analyses. This additional step helps confirm the efficiency of EV isolation from the tissue sample, assess low contamination from brain tissue and blood, and confirm if cells were not lysed during tissue digestion, leading to the release of proteins into the matrix, such as endoplasmic reticulum proteins. Known EV markers like HSPA8 (heat shock protein family A member 8), Alix (ALG-2-interacting protein X), and CD9 (cluster of differentiation 9) have been used to confirm the purity of EV isolation (Fig. 3). These proteins are commonly found on the

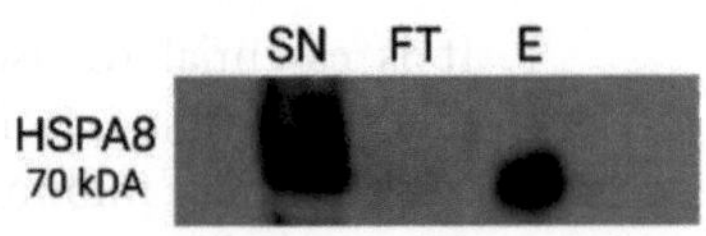

Fig. 3 Western blot analysis from the workflow of the 10,000× *g* supernatant diluted 25× in 1× PBS. The western blot shown was performed using HSPA8, a positive control for the presence of EVs. The samples from left to right are the 10,000× *g* supernatant of the sample diluted 25× in 1× PBS (SN), EVtrap flowthrough (FT), and EVtrap elution (E). As displayed in the western blot, this workflow used to isolate EVs from the brain tissue produces a high recovery in the elution with minimal EVs wasted in the flowthrough of the EVtrap

surface of EVs. Additionally, non-EV markers can also be blotted, like calnexin, an endoplasmic reticulum membrane protein, and myelin proteins are indicative of contamination from cellular compartments or brain tissue.

7. Typically, 100 μL of this solution per sample is sufficient for lysing isolated EVs. To prepare 100 μL of PTS lysis solution, follow the recipe below. Add the following stock solutions in the order given to 54 μL of ultrapure water immediately before use, and vortex after each added ingredient:

   10 μL of 500 mM TEAB

   10 μL of 120 mM SDC stock solution. Stock solution can be stored at room temperature for up to 6 months.

   10 μL of 120 mM SLS stock solution. Stock solution can be stored at room temperature for up to 6 months.

   5 μL of 200 mM TCEP stock solution. Stock solution can be prepared ahead of time and stored for up to 6 months at −20 °C. Freeze–thaw is not recommended; discard any remaining TCEP stock solution.

   10 μL of 300 mM CAA stock solution. The stock solution must be prepared fresh and used immediately.

8. The PTS lysis solution used for lysing EVs can interfere with the BCA protein assay. Therefore, it is essential to dilute the sample 5X with 50 mM TEAB before performing the BCA assay. Additionally, it is important to include a blank containing the PTS solution diluted 5X as a blank to subtract the background from the samples. Alternatively, there are other reliable total protein quantification assays that can be applied to obtain the total protein amount. This step is crucial for determining the appropriate amount of trypsin enzyme to add to the sample.

9. Ensure not to disturb the intermediate layer, which contains a portion of the peptides. After culminating **step 3**, there will be two phases, with a white disc visible between them indicating the presence of peptides. The presence of the white disc may sometimes be indicative of the amount of material resulting from the digestion. However, sometimes, this layer may not be visible. Regardless, always be careful not to disrupt the intermediate layer.

10. There are several C18 column or tip vendors like Waters, that can be used for desalting. Additionally, C18 tips can be packed in-house as described by Rappsilber et al., 2007 [14]. It is important to note that these columns have loading capacities, and it is essential to use the appropriate column/tip size that matches the binding capacity for the material being processed.

Typically, loading approximately 80% of the capacity is recommended to avoid overloading or loss of material due to underloading or overloading the column.

11. All conditioning steps can be done quickly by pushing the solution through the column using a pipette bulb or vacuum manifold, but take extra care to avoid drying out the column.
12. The time required for the sample to dry may vary depending on factors such as the volume of the sample and the temperature of the speed vacuum. To dry the sample faster, freeze the eluted samples at −80 °C for 5–10 min, and then, place the sample in the centrifugal vacuum. Alternatively, the sample can be dried by using a lyophilizer. Over-drying should be avoided, and once the sample is dry, proceed with the next steps of the protocol or store the dried peptides at −80 °C.
13. After desalting, the sample is prepared for proteome analysis of EVs by LC-MS. Before proceeding with PTM enrichment steps, it is advisable to reserve 1% of the desalted sample for proteome analysis of EVs. This serves as a control to assess the efficiency of EV isolation and content before proceeding to PTM enrichment (Fig. 4). Researchers can utilize resources such as Vesiclepedia (http://www.microvesicles.org), a web-based compendium of RNA, proteins, lipids, and metabolites identified in EVs from both published and unpublished studies, to further evaluate the efficiency of EV isolation. This platform can provide valuable insights into the expected EV content and aid in the assessment of isolation efficiency. Additionally, this reserved sample can aid in evaluating the efficiency of PTM enrichment. Analyzing a portion of the sample prior to enrichment enables the assessment of any changes or losses in

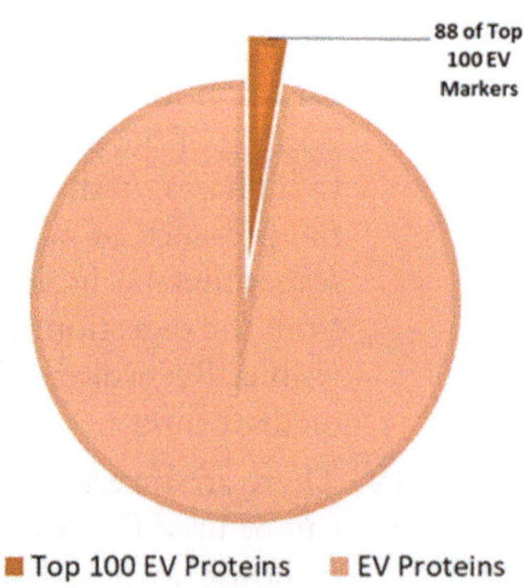

Fig. 4 EV marker identification. The mass spectrometric analysis of the EVs from the EVtrap elution yielded a large number of identified proteins, including 88 of the top 100 EV markers [15]

protein content throughout the protocol. Comparing the proteome of the initial EV sample to that of the enriched sample allows for the evaluation of the specificity and efficiency of the PTM enrichment process.

14. When making buffers for strong cation exchange processing, it is crucial to adjust the pH with TFA to prevent the need for re-desalting our samples, as all reagents used at this step need to be volatile. Additionally, it is essential to check the pH of solutions before using them, as an incorrect pH can disrupt the fractionation process. Ensure that buffers are fresh (within 1 month) to maintain their effectiveness.
15. The optimal resuspension solution is 0.05% TFA in 3% ACN or 0.1% formic acid (FA) in 3% ACN.

References

1. Doyle L, Wang M (2019) Overview of extracellular vesicles, their origin, composition, purpose, and methods for exosome isolation and analysis. Cells 8
2. Zaborowski MP, Balaj L, Breakefield XO, Lai CP (2015) Extracellular vesicles: composition, biological relevance, and methods of study. Bioscience 65. Preprint at https://doi.org/10.1093/biosci/biv084
3. Banks WA et al (2020) Transport of extracellular vesicles across the blood-brain barrier: brain pharmacokinetics and effects of inflammation. Int J Mol Sci 21
4. Ree R, Varland S, Arnesen T (2018) Spotlight on protein N-terminal acetylation. Experimental and Molecular Medicine 50. Preprint at https://doi.org/10.1038/s12276-018-0116-z
5. Iyer A, Sidhu A, Subramaniam V (2022) How important is the N-terminal acetylation of alpha-synuclein for its function and aggregation into amyloids? Front Neurosci 16. Preprint at https://doi.org/10.3389/fnins.2022.1003997
6. Watson MD, Lee JC (2019) N-terminal acetylation affects α-Synuclein fibril polymorphism. Biochemistry 58
7. Bell R et al (2023) Effects of N-terminal acetylation on the aggregation of disease-related α-synuclein variants: N-terminal acetylation of α-synuclein variants. J Mol Biol 435
8. Gottlieb L, Guo L, Shorter J, Marmorstein R (2021) N-alpha-acetylation of huntingtin protein increases its propensity to aggregate. J Biol Chem 297
9. Kabir F, Atkinson R, Cook AL, Phipps AJ, King AE (2023) The role of altered protein acetylation in neurodegenerative disease. *Frontiers in aging*. Neuroscience 14. Preprint at https://doi.org/10.3389/fnagi.2022.1025473
10. Iliuk A et al (2020) Plasma-derived extracellular vesicle Phosphoproteomics through chemical affinity purification. J Proteome Res 19
11. Chen IH et al (2017) Phosphoproteins in extracellular vesicles as candidate markers for breast cancer. Proc Natl Acad Sci USA 114
12. Vella LJ et al (2017) A rigorous method to enrich for exosomes from brain tissue. J Extracell Vesicles 6
13. Zhang Z et al (2023) Comprehensive characterization of human brain-derived extracellular vesicles using multiple isolation methods: implications for diagnostic and therapeutic applications. J Extracell Vesicles 12
14. Rappsilber J, Mann M, Ishihama Y (2007) Protocol for micro-purification, enrichment, pre-fractionation and storage of peptides for proteomics using StageTips. Nat Protoc 2
15. Pathan M et al (2019) Vesiclepedia 2019: a compendium of RNA, proteins, lipids and metabolites in extracellular vesicles. Nucleic Acids Res 47

Chapter 16

Cross-Linking Mass Spectrometry to Capture Protein Network Dynamics of Cell Membranome

Lucia Santorelli, Michele Costanzo, Sara Petrosino, Michele Santoro, Marianna Caterino, Margherita Ruoppolo, and Paolo Grumati

Abstract

Interactions among proteins are fundamental in driving functions and activities that regulate cell biology, mechanotransduction, and cell-to-cell communication/recognition. Recently, cross-linking mass spectrometry (XL-MS) has emerged as a powerful tool for interaction discovery and characterization, driving the enlightenment of novel binding partners otherwise undetected. Covalent linkages of two amino acid residues of proteins (or within complexes) in close proximity can be identified by MS, thus providing structural insights such as distance restraints or unraveling interaction dynamics.

The XL-MS workflow described here is applied to map the plasma membrane protein (PMP) networks since they play important roles in the modulation of diverse molecular processes, including transport, signal transduction, endocytosis, and secretion. The strategy includes cross-linking of PMP-enriched fractions, label-free nanoLC-MS/MS, and bioinformatics data analysis. "Membranome" interconnections constitute around 30% of the mammalian proteome and 60% of all drug targets. Exploring such networks under different biological conditions is a promising and unbiased approach to depicting regulatory pathways that govern cell behavior.

Key words Cross-linking mass spectrometry, XL-MS, Membranomics, Interactomics, Protein-protein interactions, Regulatory networks, Structural proteomics, Protein dynamics

1 Introduction

Cross-linking mass spectrometry (XL-MS) has emerged as a powerful and versatile technique in the field of structural biology, providing unprecedented insight into the three-dimensional organization of macromolecular complexes. This innovative approach enables the elucidation of protein-protein interactions (PPIs), protein conformational changes, and the mapping of intricate networks within biological systems. By covalently linking proximal

Lucia Santorelli, Michele Costanzo and Sara Petrosino contributed equally with all other contributors.

Taufika Islam Williams (ed.), *Tissue Proteomics: Methods and Protocols*, Methods in Molecular Biology, vol. 2884, https://doi.org/10.1007/978-1-0716-4298-6_16,

amino acid residues, XL-MS offers a unique glimpse into the spatial arrangements of proteins, overcoming the limitations of traditional structural biology methods [1–3].

XL-MS has gained widespread recognition due to its ability to capture dynamic interactions in native-like environments, allowing researchers to study macromolecular assemblies in physiologically relevant conditions. It represents a valuable approach to dissecting the intricacy of cellular connections, allowing the mapping and characterization of PPIs [4–6]. Moreover, this technique captures interactome dynamics from a small-to-large scale. Classical structural biology methodologies employed to define molecular structure, such as X-ray crystallography (XRC), cryo-electron microscopy (cryo-EM), and nuclear magnetic resonance (NMR) spectroscopy, can be integrated with XL-MS, thus obtaining complementary and transversal insight [7–9].

Cross-linking approaches involve the insertion of cross-linkers, which are molecules with different chemical functionalities that covalently attach via a spacer to functional groups on proteins. These covalent bonds are established between spatially close amino acid residues, within one protein, or between interacting proteins/protein complexes. Subsequent MS analysis of the cross-linked products allows for the identification of linked peptides, providing important information in regard to distance constraints, guiding the generation of low-resolution structures, and/or refining high-resolution models [10, 11].

Examples of cross-linkers are photoreactive ones, characterized by the presence of UV-inducible amino acids, such as photo-methionine or photo-leucine [12]. On the other hand, chemical cross-linkers may be provided with two identical or different reactive groups, defined as homobifunctional or heterobifunctional, respectively [13]. In addition, they also carry a linker (the spacer arm) that connects two adjacent amino acids whose length correlates with the distance between them [14, 15]. Other categories of cross-linkers are characterized by a third functional group that is exploited for affinity enrichment purposes (e.g., using a biotin moiety [4] or click-based affinity tags [16]). The use of enrichable cross-linkers is often mandatory because cross-linked peptides are usually low abundant when confronted with their linear counterparts [17]; consequently, enrichment or affinity purification strategies (e.g., affinity chromatography, size exclusion chromatography, or ion-exchange chromatography) are required [18].

Chemical cross-linking reagents have the ability to link reactive groups of amino acids in close proximity, including primary amines, sulfhydryl groups, or, as in the case of bifunctional cross-linkers, carboxyl and amine groups [5, 17]. Primary amines are ideal targets of cross-linking reactions because of their high reactivity and for being recurrently located at the solvent-accessible surface of proteins owing to lysine polarity [19].

In addition to categorizing cross-linkers as homo- or hetero-bifunctional or by the length of their spacer arm, they can also be categorized according to their fragmentation behavior. In fact, non-MS-cleavable linkers result in intact cross-linkers that do not undergo fragmentation [5, 20] [e.g., disuccinimidyl suberate, bis-(sulfosuccinimidyl) suberate (BS3), bis-(sulfosuccinimidyl) glutarate (BS2G), and diethyl suberthioimidate (DEST)]. Conversely, MS-cleavable cross-linkers are specifically designed to break up in the mass spectrometer's (MS) collision cell, owing to labile bonds that are easily fragmented [5, 20], such as disuccinimidyl dibutyric urea (DSBU), disuccinimidyl sulfoxide (DSSO), and 1,1'-carbonyldiimidazole (CDI).

Therefore, chemical diversity and linker lengths offer a variety of strategies with possible applications to interrogate biological problems [21]. After the cross-linking reaction, proteins or protein complexes undergo hydrolysis through specific proteases (like trypsin) or a combination of them (trypsin + Lys-C) [22]. This enzymatic cleavage results in linear and cross-linked peptides, and the latter can be easily enriched on affinity beads [23] and analyzed by high-resolution (HR) nano-liquid chromatography-tandem mass spectrometry (nanoLC-MS/MS), following standard proteomic procedures [24–28].

To be sure, data analysis and peptide identification pose a challenge in MS-based cross-linking experiments. The identification of peptide pairs, linked by intact cross-linkers, relies on MS/MS spectra that contain fragment ions that originate from two peptides and is complicated by the fact that the mass of the linker is added to the precursor masses. Moreover, the fragmentation of one peptide by collision-induced dissociation (CID) or high-energy collision dissociation (HCD) is often more favored than the other, and tryptic peptides frequently result in more *y*-ions than *b*-ions [29, 30]. This unbalanced product ion distribution influences the accurate and confident assignment of cross-linked peptides.

To reduce data complexity, the adoption of MS-cleavable cross-linkers represents the best choice. Peptides connected by similar types of molecules generate a characteristic reporter ion profile at the MS2 level, allowing for the analysis of such peptides in mass spectrometers without further MS fragmentation steps [31]. Nevertheless, the computational challenges in the analysis of XL-MS data arise from the considerable number of potential cross-linked peptide combinations, a factor influenced by sample complexity, fragmentation efficiency, specific database requirements, and the significantly increased search space [32].

In this chapter, we explain and comment on an experimental workflow used to capture protein network dynamics of the cellular membranome subjected to different states. Specifically, we applied the following detailed protocol on total protein lysates of standard HeLa cells in which proteins are cross-linked with DSBU as the cross-linker agent, before (native) and after (disrupted) plasma membrane (PM) dissolution. We detected the best experimental settings in terms of cross-linker concentration, cross-linking reaction time, PM enrichment, MS acquisition, and elaboration. Finally, we show the building of an intricate PM PPIs network, capturing already known and de novo interacting protein partners.

2 Materials

Required buffers and reagents are reported in Table 1.

Prepare all solutions using ultra-pure water (HPLC grade) or MS-compatible solvents. All the reagents used should be provided with the highest available purity to avoid contamination.

3 Methods

3.1 Cross-Linking of Whole Cell Proteome

1. Scrape cells (*see* **Note 1**) in ice-cold phosphate-buffered saline (PBS) and then spin down (800× *g*, 4 °C, 5 min) for cell pelleting (*see* **Note 2**) (Fig. 1).

Native (NT) Condition (a)

2. (a) Gently resuspend the cells (*see* **Note 3**) in ice-cold cross-linking buffer (CB) complemented with protease and phosphatase inhibitors according to cell amount (*see* **Note 4**). Add 1.5 mM DSBU in dimethyl sulfoxide (DMSO) anhydrous and incubate for 1 h at 4 °C on the wheel. Stop the reaction by adding quenching buffer (QB) in a 50:1 QB/DSBU molar ratio and incubating for 30 min at 4 °C on the wheel.
3. (a) Centrifuge at 800× *g* for 5 min at 4 °C. Wash the pellet twice with lysis buffer (LB) (*see* **Note 5**) and resuspend it in LB. Mechanically destroy the cell membrane with a 24-G needle syringe.

Table 1
List of buffers used in the cross-linking phase with their chemical composition

Buffer/reagent
Lysis buffer:12.5 mM Tris, 115 mM NaCl, 1% NP 40 pH 7.5
CB (cross-linking buffer):20 mM Hepes, 100 mM KCl pH = 7.5
QB (quenching buffer): 0.5 M Tris–HCl, 1.25% NP40 pH = 7.5
HB (homogenizer buffer): Plasma membrane protein extraction kit-ab6540

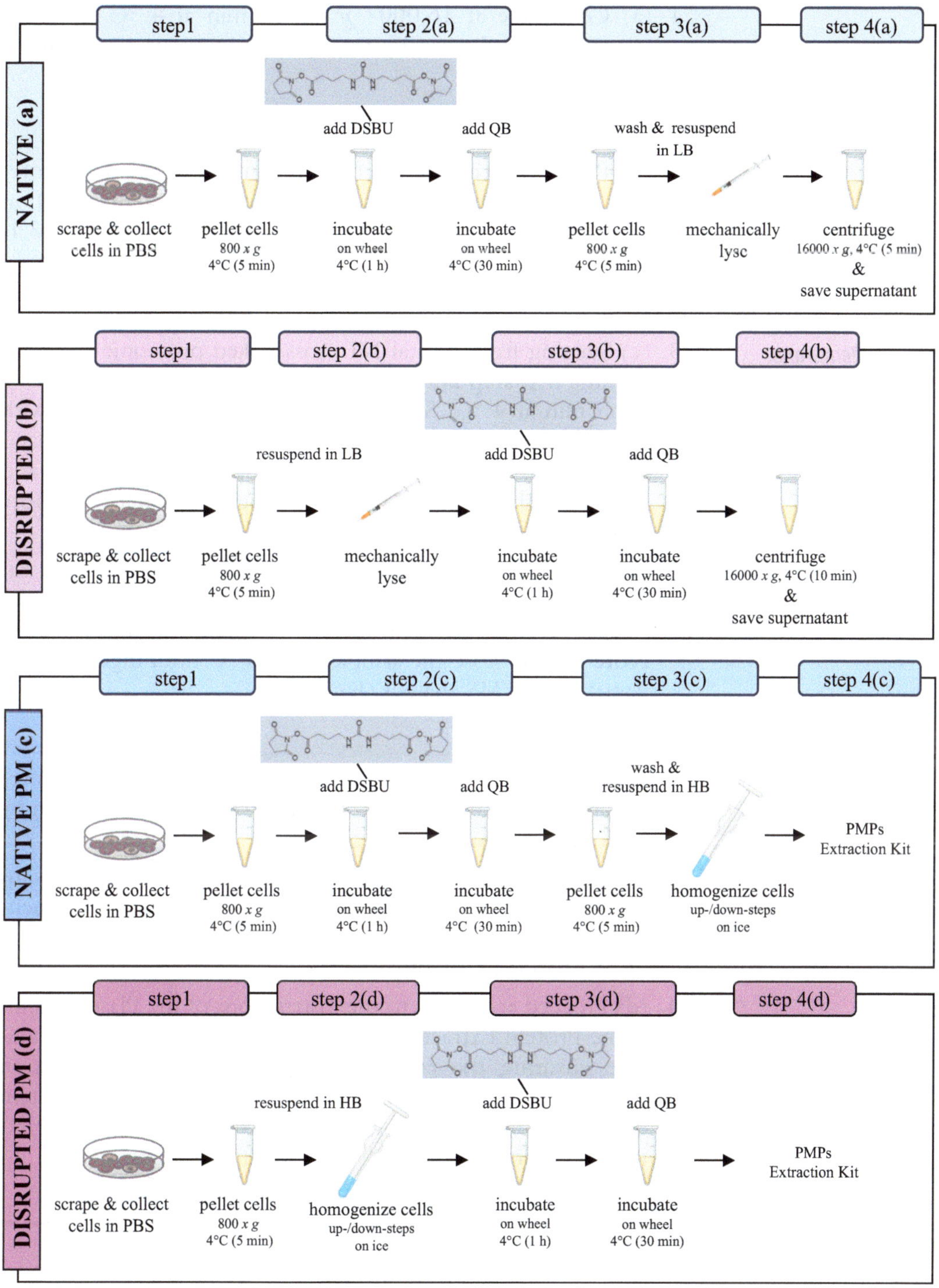

Fig. 1 Cross-linking mass spectrometry sample preparation of HeLa whole proteome and plasma membrane–enriched fraction. Proteins were cross-linked with DSBU ex vivo, before (native, **a**) and after (disrupted, **b**) cell lysis. The plasma membrane cross-linked fraction was enriched from cells cross-linked in the native state (**c**). The figure was created with BioRender.com. Abbreviations: *PBS* phosphate-buffered saline, *DSBU* disuccinimidyl dibutyric urea, *CB* cross-linking buffer, *QB* quenching buffer, *LB* lysis buffer, *HB* homogenizer buffer, *PM* plasma membrane

4. (a) Centrifuge at 16,000× *g* for 10 min at 4 °C. Save the supernatant and quantify total protein content (*see* **Note 6**) (Fig. 1).

Disrupted (DR) Condition (b)

2. (b) Resuspend cells in ice-cold LB complemented with protease and phosphatase inhibitors and perform protein extraction by exploiting a syringe, as described in Subheading 3.1, **step 3(a)**.

3.2 Cross-Linking of Plasma Membrane Proteins

Native PM (c)

3. (c) Starting from a total NT cross-linked proteome [*see* Subheading 3.1, **step 2(a)**], remove QB by centrifuging at 800 × *g* for 5 min at 4 °C. Subsequently, wash the pellet twice with homogenizer buffer (HB), complemented with protease and phosphatase inhibitors. Resuspend the cells in 2 mL of HB and gently lyse them with a Dounce homogenizer on ice (*see* **Note 2**) with 50 up/down steps (*see* **Note** 7).
4. (c) Proceed using a Plasma Membrane Protein Extraction Kit according to the manufacturer's instructions [33]. Once the PM is extracted, perform total PM protein quantification (*see* **Note 6**) and store the sample at −20 °C until digestion and nanoLC-MS/MS analysis (Fig. 1).

Disrupted PM (d)

2. (d) Resuspend cells in ice-cold LB complemented with protease and phosphatase inhibitors and perform cell disruption by exploiting a Dounce homogenizer.
3. (d) After the cell lysis in HB, crosslink the extracted proteins with 1.5 mM DSBU dissolved in anhydrous DMSO and incubate for 1 h at 4 °C on the wheel. Add QB to stop the reaction by incubating at 30 min at 4 °C.
4. (d) Proceed according to the instructions of the Plasma Membrane Protein Extraction Kit or any other commercially evaluable ensuring the enrichment/purification of PM fraction [32]. Once the PM is extracted, perform total PM protein quantification (*see* **Note 6**) and store the sample at −20 °C until digestion and nanoLC-MS/MS analysis (Fig. 1).

4 Quality Check of PM Sample Preparation

1. To assess the efficiency of the sample preparation workflow, check the protein pattern of the DSBU-linked proteins by mono-dimensional sodium dodecyl-sulfate polyacrylamide gel electrophoresis (1D-SDS-PAGE) followed by Coomassie blue (CB) staining. In the case presented here, first, a DSBU

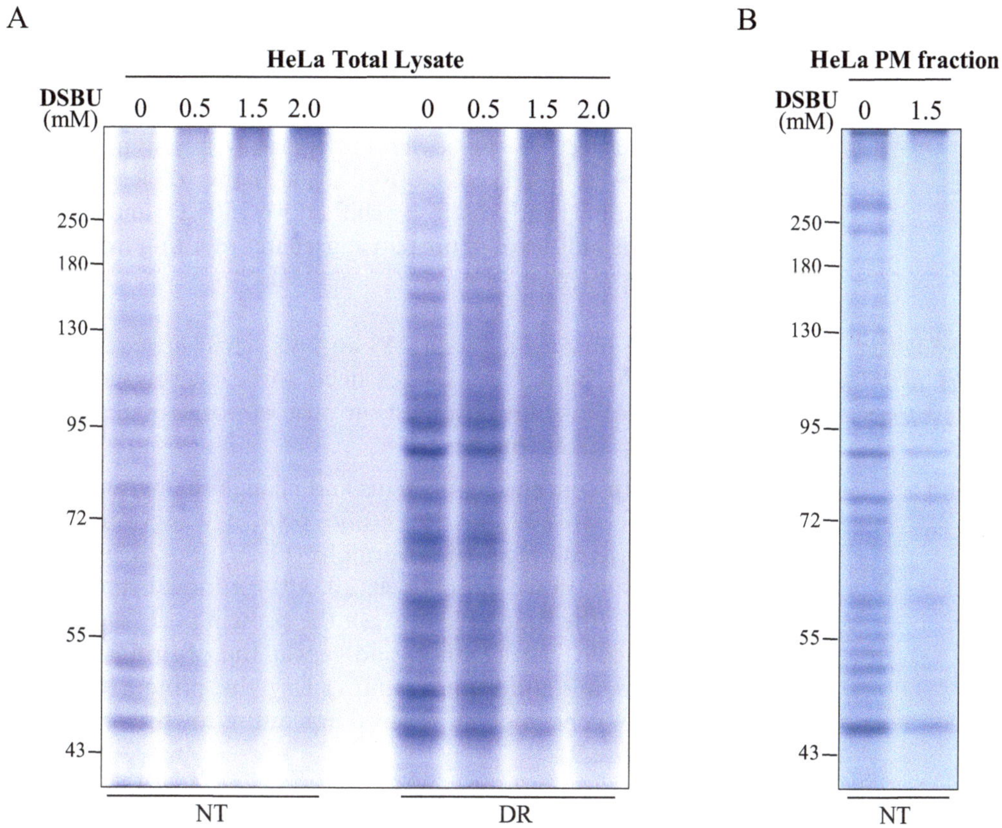

Fig. 2 Cross-linked and non-cross-linked protein profiles from total lysates and plasma membrane enrichment in HeLa cells. (**a**) DSBU titration assay performed on the total cell proteome cross-linked in native and disrupted status; control samples (0 mM DSBU) were treated with DMSO, the vehicle of DSBU. (**b**) PM fraction enriched from cross-linked and non-cross-linked native cells. For each sample, 15 μg of protein were loaded onto the 7.5% acrylamide gel and run by SDS-PAGE with a constant voltage of 80 V. Coomassie blue protein staining was performed, and molecular weight markers (kDa) are shown on the left of the gels

titration (0, 0.5, 1.5, 2 mM DSBU) was carried out to ascertain the best cross-linker/protein ratio under NT and DR conditions (Fig. 2a). Then, the samples were run on a 7.5% polyacrylamide gel with a constant voltage of 80 Volts for 2 h. The gel shows peculiar profiles, whereas the pattern of bands is reasonably specific to cross-linking for the higher DSBU concentrations for both groups. According to the gel profile and MS elaboration (*see* Subheading 3), the optimal concentration of DSBU cross-linker was determined to be 1.5 mM in the NT state. In the low molecular weight region, fewer bands are observed. In the high molecular weight region (containing cross-linked proteins), there is increased intensity with fewer sharp bands, which is indicative of protein aggregates resulting from over-cross-linking. Upon obtaining the best DSBU/protein ratio, we enriched for PM proteins in the NT condition

(Fig. 2b). The PM fractions from cross-linked (1.5 mM DSBU) and untreated (0 mM DSBU) native cells were subjected to SDS-PAGE and CB stain as above. The cross-linked PM protein profile presents a peculiar 1D pattern. Specifically, with respect to the lower molecular weight regions, fewer bands were observed. Conversely, at higher molecular weight areas, some bands disclose a shift above 250 kDa and at 55 kDa, which is indicative of protein aggregates resulting from a cross-linking reaction.

4.1 In-Solution Digestion of Cross-Linked Proteins

1. Before proceeding with MS analysis, cross-linked proteomes and PM protein complexes need to be hydrolyzed. Apply a standard in-solution digestion protocol or something similar [34, 35] (*see* **Note 8**).
2. To both concentrate proteins and change the buffer, add 4/1 (v/v) ice-cold MS-grade acetone/protein lysate volume and let the proteins precipitate overnight (O/N) at −20 °C. The day after, spin down proteins at 20,000× *g* for 20 min at 4 °C. Remove the supernatant, wash the protein pellet with 500 μL of ice-cold 70% ethanol, and repeat the centrifugation step, maintaining the same settings. Dry the protein pellet in a vacuum centrifuge, at room temperature for the needed time (*see* **Note 9**).
3. Dissolve the dried pellet in 6 M guanidine hydrochloride (GnHCl), 50 mM Tris–HCl pH 8.5 complemented with 10 mM Tris(2-carboxyethyl)phosphine hydrochloride (TCEP) and 40 mM chloroacetamide (CAA). Boil the samples for 10 min at 95 °C.
4. Dilute the samples 1:6 with 50 mM Tris pH 8.5 and add a mix of Lys-C/trypsin hydrolases These enzymes can be added individually or can be bought in a ready-to-use mixture. Add each enzyme in an enzyme-to-protein ratio of 1:100 (w/w) or 1:50 for the hydrolases mix, and incubate/agitate overnight (O/N) at 37 °C (*see* **Notes 10** and **11**). On the day after, quench the digestion process by adding 1% trifluoracetic acid (TFA).
5. Collect and desalt the peptides by the C18 StageTip. The StageTip workflow is from the method described by Rappsilber et al. [36]. Briefly, prepare as many tips as the number of samples. Wash them with methanol, and then, elute them with an 80% acetonitrile (ACN)/0.1% formic acid (FA) solution. Load the samples onto separate tips, allowing for peptide adsorption onto the C18 phase. Elute the peptide mixture into a new fresh tube using a 0.1% FA solution (*see* **Note 12**). Dry the volume in a vacuum centrifuge, and store at −20 °C until nanoLC-MS/MS analysis.

4.2 NanoLC-MS/MS Analysis

1. Dissolve dried peptides in 0.1% FA and inject them into the nanoLC system (*see* **Note 13**).
2. Perform nanoLC-MS/MS analysis. For this protocol, we adapted the nanoLC separation and MS acquisition according to the methods published by Iacobucci et al. [37] (*see* **Note 14**). Our nanoLC-MS/MS setup consists of an EASY-nLC 1200 coupled via an electrospray ionization (ESI) source to a quadrupole-based Q Exactive HF benchtop MS. More generally, any other nanoLC coupled to a high-resolution (HR) tandem mass spectrometer can be properly used. Here, the nanoLC gradient applied was 100 min long with a flow rate of 300 nL/min, using a binary buffer system consisting of solvent A (0.1% FA) and solvent B (80% ACN, 0.1% FA). The gradient involved the following steps: 7 to 44% B in 89 min, 44 to 95% B in 5 min, 95% B for 4 min, and 5% B for 1 min for column wash and equilibration. MS data acquisition was performed in a data-dependent acquisition (DDA) format, acquiring tandem MS data on the 15 most abundant (peptide) ions eluting from the nanoLC column at a given point in time on the nanoLC gradient. Stepped collision-induced dissociation (CID) mode was applied, using 15 variable windows (mass range = 300–1700 m/z) and normalized collision energy (NCE) of 25%, 27.5%, and 30%. The resolution was set to 120,000 for MS1 and 45,000 for MS2.

4.3 Identification of Cross-Linked Peptides

1. The identification of cross-links can be performed using different software available to researchers. In our workflow, we employ the MeroX tool [38], as listed in the following steps. MeroX is a freely available software, continuously improved and updated, specifically designed for analyzing highly complex cross-linking datasets.
2. Prepare a protein database (for the organism under investigation) in FASTA format (*see* **Note 15**).
3. Upload the *.raw* data into the MeroX software (*see* **Note 16**).
4. Specify search parameters, e.g., protease sites, number of missed cleavages, blocked residues by XL reaction, cross-linker agent, instrument settings, and so on (*see* **Note 17**).
5. Once started, the data analysis process can take hours. The elaboration time may depend on both the number of experimental spectra acquired per sample and the computational resources available (*see* **Note 18**).
6. At the end of the analysis, a result file is generated with the *.zhrm* extension (openable in MeroX), from which diverse information can be exported, including the type of cross-linker reaction products and their distribution on the proteins. Indeed, after the cross-linking treatment, diverse reaction

products with specific links combination could form (1) intramolecular cross-links, occurring within the same molecule; (2) intermolecular cross-links, linking two amino acids belonging to two different proteins; (3) dead-links, regarding those peptides modified with a hydrolyzed crosslinker portion; and (4) homotypic (or loop-link) links, describing the cross-linking reaction between two close amino acids in the same peptide. These last two are generally considered side effects of an out-numbered presence of cross-linker molecules compared with the total protein amount. In addition, the distance restraints among the proteins derived from the cross-linker and experimentally determined by the XL-MS analysis are also provided, typically summarized in an exportable table format. These values can be validated, by comparing them with the conformational status of proteins reported in the Protein Data Bank (PDB). For an immediate and comprehensive interpretation of the topological data, they also can be loaded in xVis, a web server for schematic visualization of cross-linked-derived spatial restraints [39].

7. Evaluate the whole proteome of the XL-MS experiment, checking the cross-linked and non-cross-linked peptides of the DSBU titration curve in terms of protein identification (ID): in our protocol, over 1500 different protein species were identified for each experimental point [false discovery rate (FDR) of peptide-spectrum matches (PSM) at 1% and at least one unique peptide identification] (Fig. 3a). As expected, membrane disruption before the cross-linking reaction showed higher ID numbers compared to the cross-linking reaction in NT. Interestingly, DSBU conjugation appears to have an important impact; in fact, the increase in cross-linker concentration is inversely correlated with the number of protein IDs. This can be explained by the fact that samples with a higher DSBU concentration present a great portion of cross-linked peptides that remain unmatched in a classical proteomics identification workflow [40, 41], in which the search for cross-linked modifications is not contemplated.
8. Perform principal component analysis (PCA) to control profile similarities of protein intensities (Fig. 3b). Here, PCA was performed [42–44] and revealing a characteristic protein profile associated with NT or DR state, corroborating the evidence that cross-linking before or after the cell lysis leads to enrichment of different protein populations (Fig. 3b, c). The difference observed in ID terms according to the cross-linking treatment before or after the PM dissolution could be explained by the limited membrane permeability typical of some cross-linkers, including the DSBU.

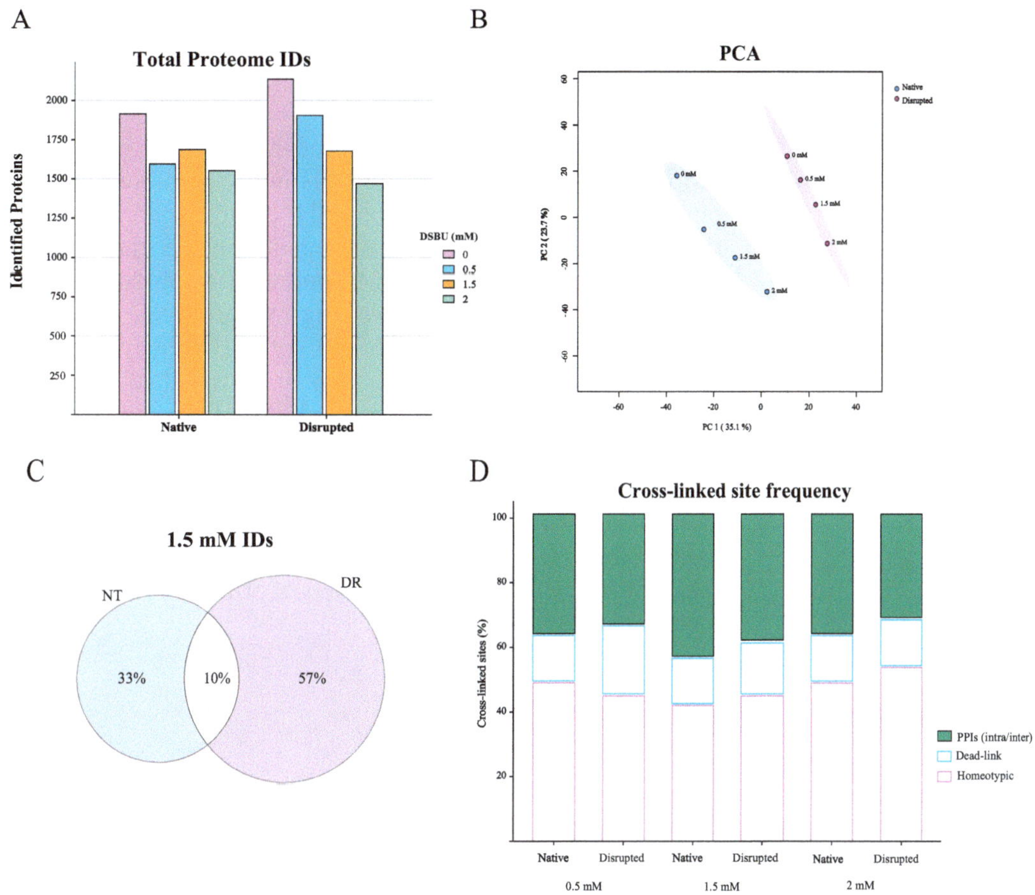

Fig. 3 XL-MS analysis of HeLa total cell lysate. (**a**) Total proteome ID distribution of DSBU titration curve. (**b**) PCA analysis of quantitative data showing NT and DR separation. (**c**) Overlap of cross-linked identified proteins between NT and DR groups treated with 1.5 mM DSBU. (**d**) Cross-linked sites detected in NT and DR groups treated with different DSBU concentrations; PPIs and dead link and homotypic link percentages are shown

9. Evaluate the distribution of all the DSBU reaction products to determine the recurrence frequency of the residue cross-links in the peptide population. With its major percentage of inter–/intra-PPIs and lower presence of collateral products (dead links and homotypic), cross-linking with 1.5 mM in NT state proved to be the best protocol (Fig. 3d).

   However, choosing a better strategy (NT or DR) remains an open issue, dependent on the biological question that the XL-MS experiment should answer. For instance, if the interest is directed to describe the PPIs network related to processes that involve membranes or the general cell architecture, a NT cross-linking approach might be the best choice. In all other cases, where it is not necessary to preserve the integrity of cell structures, the DR cross-linking could be applied. Eventually, if the required starting material is available, it is recommended to use both approaches in parallel in order to get a more complete picture of existing PPIs.

4.4 Identification of Membranome Proteins and PPIs

1. Check the quality of PM enrichment, since this protocol is proposed to efficiently capture PPIs in the membranome. To obtain an overview of the subcellular compartments in which the identified proteins are implicated, the MS data can be matched against a specific PM annotated database to count the number of IDs. In establishing this protocol, the whole cell proteome contained around 10% of PMPs regardless of the state and DSBU concentration (Fig. 4a). Instead, after the PM enrichment (at NT and after 1.5 mM DSBU), the percentage of PMP IDs rose up to 60% compared to the total of detected IDs (Fig. 4b).
2. The quality of PM fractionation can also be easily accessed via western blot monitoring of the enrichment in PM markers (e.g., Glut1) with respect to markers of other cellular compartments (Fig. 4c).

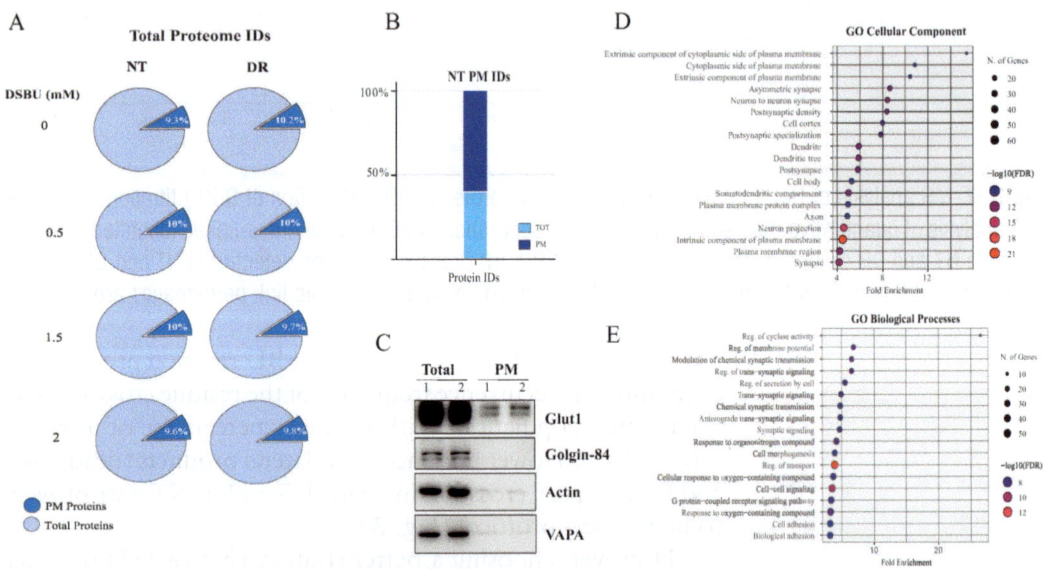

Fig. 4 XL-MS analysis of HeLa membranome. Percentage of proteins annotated as resident in PM (**a**) before and (**b**) after the enrichment. (**c**) Immunoblot of total cell lysate and PM fraction samples to check for the fractionation efficiency. Positive control: Glut1, protein annotated as PM member. Negative controls: Golgin-84, actin, and VAPA, markers of other cellular compartments (Golgi apparatus, cytoskeleton, endoplasmic reticulum). GO enrichment analysis of proteins detected in the PM sample enriched in NT state, with classification according to (**d**) cellular component and (**e**) biological process. The GO was performed using the ShinyGo tool [50], while the SRplot platform was employed for the major part of graphical representations [51]

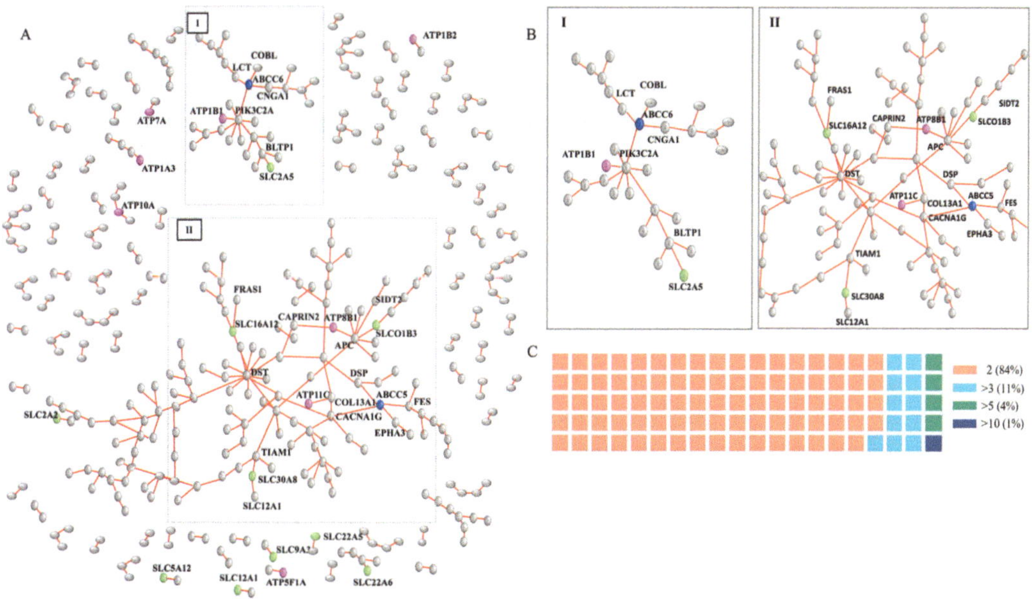

Fig. 5 Plasma membrane interactome of HeLa cells. The PPIs network identified in the XL-MS dataset shows PM resident proteins. (**a**) The global network is composed of unique residue-to-residue pairs originating from more than 300 proteins assigned to the PM compartment. Comparison to known PPIs revealed an overlap of 25%, confirming the accuracy of our dataset. With respect to PM-associated proteins, we identified the highest number of PPIs for specific PM member families: ATP-binding cassette (ABC) transporters (blue dots), solute carrier (SLC) transporter family (green dots), ATPase Na^+/K^+, Cu^{2+}, and phospholipid transporters (purple dots). The remaining cross-links were identified for proteins which are currently not described as interacting with each other, presenting potentially de novo partners. (**b**) The two biggest networks (I-II) were expanded. (**c**) The global network is composed of hundreds of interactions: a large portion was represented by binary interactions, but also, large node interconnections were found (>3, >5, >10 interactors)

3. Moreover, performing a Gene Ontology (GO) analysis can confirm the belonging of the enriched proteins to the PM. Thus, the analysis of GO cellular components (Fig. 4d) or biological processes (Fig. 4e) should provide terms connected with PM localization or function (e.g., secretion, transport, signaling transmission) as significantly enriched.
4. Then, use the XL-MS data derived from the PM of NT cross-linked cells to build an interactome network in MeroX (Fig. 5a, b).
5. Data can be categorized according to the number of interactors in diverse groups (Fig. 5c).

5 Notes

1. For the whole-cell proteome cross-linking preparation, about 2×10^6 cells are required, while 10×10^8 cells are necessary for the PM protein cross-linking protocol.

2. Carry out all steps at 4 °C. It is necessary to maintain ice-cold conditions during all procedures to prevent sample degradation.
3. It is recommended to cut the pipette tip to gently resuspend the cell pellet to avoid damaging the cell membranes.
4. It is not recommended to decrease the ratio between the volume of CB and cell pellet since the reaction could be less efficient. Hence, 500 μL of CB is recommended for 10×10^8 cells, while 200 μL of CB is recommended for 2×10^6 cells.
5. If it is necessary to change the LB, we strongly suggest maintaining the Tris–HCl concentration lower than 75 mM to avoid quenching of the DSBU-mediated reaction.
6. It is recommended to avoid Bradford's total protein quantification since the DSBU presence can negatively affect the results, inducing an underestimation of protein concentration. Alternative colorimetric (e.g., BCA assay, DC protein assay) or fluorometric methods should be preferred.
7. Homogenization should be performed based on the cell type. The number of up/down steps needs to be experimentally established. To verify homogenization efficiency, we recommend a microscope inspection of the cell lysate to exclude cells that remain intact, despite the mechanical disruption.
8. The cross-linked proteins may be separated by electrophoresis and in situ digested or directly hydrolyzed using in-solution or column-based methods. In the case of in situ digestion, the main advantage is represented by the fact that cross-linked species are separated from non-cross-linked ones before digestion. In the case of other protocols for proteolysis, any in-solution or column-based methods, such as PreOmics in-StageTip (iST), suspension trapping (S-Trap), filter-aided sample preparation (FASP), single-pot, and solid-phase-enhanced sample preparation (SP3), can be considered as they are largely applied in linear proteomics studies [45–47]. However, with regard to XL-MS workflows, we recommended a preliminary setup phase to verify their efficiency in the processing of cross-linking samples.
9. It is possible to air-dry the pellet, but we highly recommend the drying step be carried out in a vacuum centrifuge to maximize efficiency, limit time consumption, and avoid contamination.
10. It is important to dilute 6 M GnHCl up to 1 M before adding digestive enzymes since its presence in high concentrations could compromise the reaction efficiency of these proteases.
11. It could be that trypsin cleaves with a lower frequency at modified lysine residues, generated as a consequence of the DSBU reaction. A combination of hydrolytic enzymes, such as trypsin/Lys-C, trypsin/Glu-C, and trypsin/Asp-N, with their diverse consensus motifs, could be helpful.

12. Every laboratory uses its own setup for the concentration and desalting of peptide mixtures; usually, peptides are loaded onto a reversed-phase C18 tip. Due to the low abundance of the cross-linked peptides compared with the linear counterparts, we suggest optimizing their enrichment by adding other steps of fractionation. Diverse are the strategies developed to enrich the cross-linked species, based on their larger size and higher isoelectric point compared with the linear peptides. Size exclusion chromatography (SEC), as well as strong cation exchange (SCX), might be applied for a better enrichment of cross-linked products [1, 48].
13. Adjust the volume for dissolving dried peptides—especially in the case of PM proteins—according to the sensitivity of the mass spectrometer to obtain sufficient signal.
14. The nanoLC and MS methods can be adapted to many working sets published in the literature according to the instrumentation and platform available.
15. To greatly reduce the computational cost and time, we highly recommend generating an experimental database to use as a reference. For example, this database could be derived from the MS of the non-cross-linked counterpart dataset.
16. To allow MeroX to properly perform, it is necessary to convert MS raw files into the *.mzML* extension (e.g., utilizing the ProteoWizard MSConvert tool [49]) and upload those converted files into the MeroX software.
17. When setting the search parameters for cross-linked identification in MeroX, pay special attention to the FDR estimation. For good reliability of the detected cross-links, we recommend not setting the FDR cut-off higher than 20%.
18. Applying our proposed membranomics XL-MS workflow, we recorded an average analysis time of about 5 h to process each single nanoLC-MS data file. MeroX was run on a virtual machine with 56 GB of memory space and 68 processors. The .fasta file employed consisted of about 4000 protein entries.

References

1. Sinz A (2006) Chemical cross-linking and mass spectrometry to map three-dimensional protein structures and protein–protein interactions. Mass Spectrom Rev 25:663–682. https://doi.org/10.1002/mas.20082
2. Leitner A, Faini M, Stengel F, Aebersold R (2016) Crosslinking and mass spectrometry: an integrated technology to understand the structure and function of molecular machines. Trends Biochem Sci 41:20–32. https://doi.org/10.1016/j.tibs.2015.10.008
3. Chavez JD, Bruce JE (2019) Chemical cross-linking with mass spectrometry: a tool for systems structural biology. Curr Opin Chem Biol 48:8–18. https://doi.org/10.1016/j.cbpa.2018.08.006
4. Zhang H, Tang X, Munske GR et al (2009) Identification of protein-protein interactions

and topologies in living cells with chemical cross-linking and mass spectrometry. Mol Cell Proteomics 8:409–420. https://doi.org/10.1074/mcp.M800232-MCP200

5. Santorelli L, Caterino M, Costanzo M (2022) Dynamic interactomics by cross-linking mass spectrometry: mapping the daily cell life in postgenomic era. Omi A J Integr Biol 26: 633–649. https://doi.org/10.1089/omi.2022.0137
6. Gonzalez Melo M, Fontana AO, Viertl D et al (2021) A knock-in rat model unravels acute and chronic renal toxicity in glutaric aciduria type I. Mol Genet Metab 134:287–300. https://doi.org/10.1016/j.ymgme.2021.10.003
7. Dobson CM (2019) Biophysical techniques in structural biology. Annu Rev Biochem 88:25–33. https://doi.org/10.1146/annurev-biochem-013118-111947
8. García-Nafría J, Tate CG (2021) Structure determination of GPCRs: cryo-EM compared with X-ray crystallography. Biochem Soc Trans 49:2345–2355. https://doi.org/10.1042/BST20210431
9. Roviello GN, Roviello G, Musumeci D et al (2014) Synthesis and supramolecular assembly of 1,3-bis(1′-uracilyl)-2- propanone. RSC Adv 4:28691–28698. https://doi.org/10.1039/c4ra03713h
10. Liu F, Rijkers DTS, Post H, Heck AJR (2015) Proteome-wide profiling of protein assemblies by cross-linking mass spectrometry. Nat Methods 12:1179–1184. https://doi.org/10.1038/nmeth.3603
11. Rogawski R, Sharon M (2022) Characterizing endogenous protein complexes with biological mass spectrometry. Chem Rev 122:7386–7414. https://doi.org/10.1021/acs.chemrev.1c00217
12. Suchanek M, Radzikowska A, Thiele C (2005) Photo-leucine and photo-methionine allow identification of protein-protein interactions in living cells. Nat Methods 2:261–268. https://doi.org/10.1038/nmeth752
13. Tinnefeld V, Sickmann A, Ahrends R (2014) Catch me if you can: challenges and applications of cross-linking approaches. Eur J Mass Spectrom 20:99–116. https://doi.org/10.1255/ejms.1259
14. Staros JV, Wright RW, Swingle DM (1986) Enhancement by N-hydroxysulfosuccinimide of water-soluble carbodiimide-mediated coupling reactions. Anal Biochem 156:220–222. https://doi.org/10.1016/0003-2697(86)90176-4
15. Bitan G, Teplow DB (2004) Rapid photochemical cross-linking a new tool for studies of metastable, amyloidogenic protein assemblies. Acc Chem Res 37:357–364. https://doi.org/10.1021/ar000214l
16. Chowdhury SM, Du X, Tolić N et al (2009) Identification of cross-linked peptides after click-based enrichment using sequential collision-induced dissociation and electron transfer dissociation tandem mass spectrometry. Anal Chem 81:5524–5532. https://doi.org/10.1021/ac900853k
17. Barth M, Schmidt C (2021) Quantitative cross-linking of proteins and protein complexes. Methods Mol Biol 2228:385–400
18. Jiang P, Wang C, Diehl A et al (2022) A membrane-permeable and immobilized metal affinity chromatography (IMAC) enrichable cross-linking reagent to advance in vivo cross-linking mass spectrometry. Angew Chemie Int Ed 61:e202113937. https://doi.org/10.1002/anie.202113937
19. Patil US, Qu H, Caruntu D et al (2013) Labeling primary amine groups in peptides and proteins with N-hydroxysuccinimidyl ester modified Fe3O4@SiO2 nanoparticles containing cleavable disulfide-bond linkers. Bioconjug Chem 24:1562–1569. https://doi.org/10.1021/bc400165r
20. Sinz A (2017) Divide and conquer: cleavable cross-linkers to study protein conformation and protein–protein interactions. Anal Bioanal Chem 409:33–44. https://doi.org/10.1007/s00216-016-9941-x
21. Mattson G, Conklin E, Desai S et al (1993) A practical approach to crosslinking. Mol Biol Rep 17:167–183. https://doi.org/10.1007/BF00986726
22. Ser Z, Cifani P, Kentsis A (2019) Optimized cross-linking mass spectrometry for in situ interaction proteomics. J Proteome Res 18: 2545–2558. https://doi.org/10.1021/acs.jproteome.9b00085
23. Rey M, Dupré M, Lopez-Neira I et al (2018) eXL-MS: an enhanced cross-linking mass spectrometry workflow to study protein complexes. Anal Chem 90:10707–10714. https://doi.org/10.1021/acs.analchem.8b00737
24. Costanzo M, Caterino M, Salvatori I et al (2022) Proteome data of neuroblastoma cells overexpressing Neuroglobin. Data Br 41: 107843. https://doi.org/10.1016/j.dib.2022.107843
25. Manganelli V, Salvatori I, Costanzo M et al (2021) Overexpression of neuroglobin promotes energy metabolism and autophagy

induction in human neuroblastoma SH-SY5Y cells. Cells 10:3394. https://doi.org/10.3390/cells10123394
26. Santorelli L, Barigazzi E, Pitto M, Raimondo F (2020) Investigation of the N-glycoproteome in the urinary exosomes: technical challenges. In: NATO science for peace and security series a: chemistry and biology. Springer, pp 257–258
27. Barigazzi E, Santorelli L, Morello W et al (2020) New insight into idiopathic nephrotic syndrome: strategy based on urinary exosomes. In: NATO science for peace and security series a: chemistry and biology. Springer, pp 217–218
28. Costanzo M, Fiocchetti M, Ascenzi P et al (2021) Proteomic and bioinformatic investigation of altered pathways in neuroglobin-deficient breast cancer cells. Molecules 26:2397. https://doi.org/10.3390/molecules26082397
29. Trnka MJ, Baker PR, Robinson PJJ et al (2014) Matching cross-linked peptide spectra: only as good as the worse identification. Mol Cell Proteomics 13:420–434. https://doi.org/10.1074/mcp.M113.034009
30. Trnka MJ, Burlingame AL (2010) Topographic studies of the GroEL-GroES chaperonin complex by chemical cross-linking using diformyl ethynylbenzene. Mol Cell Proteomics 9:2306–2317. https://doi.org/10.1074/mcp.M110.003764
31. Stieger CE, Doppler P, Mechtler K (2019) Optimized fragmentation improves the identification of peptides cross-linked by MS-cleavable reagents. J Proteome Res 18:1363–1370. https://doi.org/10.1021/acs.jproteome.8b00947
32. Yılmaz Ş, Shiferaw GA, Rayo J et al (2018) Cross-linked peptide identification: a computational forest of algorithms. Mass Spectrom Rev 37:738–749. https://doi.org/10.1002/mas.21559
33. Shuid AN, Safi N, Haghani A et al (2015) Apoptosis transcriptional mechanism of feline infectious peritonitis virus infected cells. Apoptosis 20:1457–1470. https://doi.org/10.1007/s10495-015-1172-7
34. Reggio A, Buonomo V, Berkane R et al (2021) Role of FAM134 paralogues in endoplasmic reticulum remodeling, ER-phagy, and collagen quality control. EMBO Rep 22:e52289. https://doi.org/10.15252/embr.202052289
35. Costanzo M, Caterino M, Cevenini A et al (2020) Dataset of a comparative proteomics experiment in a methylmalonyl-CoA mutase knockout HEK 293 cell model. Data Br 33:106453. https://doi.org/10.1016/j.dib.2020.106453
36. Rappsilber J, Ishihama Y, Mann M (2003) Stop and go extraction tips for matrix-assisted laser desorption/ionization, nanoelectrospray, and LC/MS sample pretreatment in proteomics. Anal Chem 75:663–670. https://doi.org/10.1021/ac026117i
37. Iacobucci C, Götze M, Ihling CH et al (2018) A cross-linking/mass spectrometry workflow based on MS-cleavable cross-linkers and the MeroX software for studying protein structures and protein–protein interactions. Nat Protoc 13:2864–2889. https://doi.org/10.1038/s41596-018-0068-8
38. Götze M, Pettelkau J, Schaks S et al (2012) StavroX—A software for analyzing crosslinked products in protein interaction studies. J Am Soc Mass Spectrom 23:76–87. https://doi.org/10.1007/s13361-011-0261-2
39. Grimm M, Zimniak T, Kahraman A, Herzog F (2015) xVis : a web server for the schematic visualization and interpretation of crosslink-derived spatial restraints. Nucleic Acids Res 43:W362–W369. https://doi.org/10.1093/nar/gkv463
40. Tyanova S, Temu T, Cox J (2016) The MaxQuant computational platform for mass spectrometry based shotgun proteomics. Nat Protoc 11:2301–2319. https://doi.org/10.1038/nprot.2016.136
41. Orsburn BC (2021) Proteome discoverer—a community enhanced data processing suite for protein informatics. Proteomes 9:15. https://doi.org/10.3390/proteomes9010015
42. Costanzo M, Caterino M (2023) Targeted lipidomics data of COVID-19 patients. Data Br 48:109089. https://doi.org/10.1016/j.dib.2023.109089
43. Pang Z, Chong J, Zhou G et al (2021) MetaboAnalyst 5.0: narrowing the gap between raw spectra and functional insights. Nucleic Acids Res 49:W388. https://doi.org/10.1093/nar/gkab382
44. Scarcella M, Scerra G, Ciampa M et al (2024) Metabolic rewiring and autophagy inhibition correct lysosomal storage disease in mucopolysaccharidosis IIIB. iScience 27:108959. https://doi.org/10.1016/j.isci.2024.108959
45. Kulak NA, Pichler G, Paron I et al (2014) Minimal, encapsulated proteomic-sample processing applied to copy-number estimation in eukaryotic cells. Nat Methods 11:319–324. https://doi.org/10.1038/nmeth.2834
46. Haile Mariam M, Eguez RV, Singh H et al (2018) S-trap, an ultrafast sample-preparation approach for shotgun proteomics. J Proteome

Res 17:2917–2924. https://doi.org/10.1021/acs.jproteome.8b00505

47. Hughes CS, Moggridge S, Müller T et al (2019) Single-pot, solid-phase-enhanced sample preparation for proteomics experiments. Nat Protoc 14:68–85. https://doi.org/10.1038/s41596-018-0082-x

48. Schmidt R, Sinz A (2017) Improved single-step enrichment methods of cross-linked products for protein structure analysis and protein interaction mapping. Anal Bioanal Chem 409:2393–2400. https://doi.org/10.1007/s00216-017-0185-1

49. Adusumilli R, Mallick P (2017) Data conversion with ProteoWizard msConvert. Methods Mol Biol 1550:339–368. https://doi.org/10.1007/978-1-4939-6747-6_23

50. Ge SX, Jung D, Yao R (2020) ShinyGO: a graphical gene-set enrichment tool for animals and plants. Bioinformatics 36:2628–2629. https://doi.org/10.1093/bioinformatics/btz931

51. Tang D, Chen M, Huang X et al (2023) SRplot: a free online platform for data visualization and graphing. PLoS One 18:e0294236. https://doi.org/10.1371/journal.pone.0294236

Chapter 17

In-Depth Analysis of Tissue Phosphoproteomics with TiO_2 Enrichment and Fractionation with High-pH Reversed-Phase Chromatography

Xiulan Chen, Shasha Wei, and Fuquan Yang

Abstract

Protein phosphorylation is an important post-translational modification that regulates almost all cellular processes, such as cellular metabolism, growth, differentiation, signal transduction, and gene regulation. Mass spectrometry, which acts as an automated and sensitive method, enables global analysis of protein phosphorylation. However, several technical challenges need to be addressed when analyzing protein phosphorylation in a global manner. Low-abundant phosphopeptides need to be enriched before analysis with LC-MS/MS, so specific enrichment of phosphopeptides is central for a successful analysis of the phosphoproteome. Due to the complexity of phosphoproteome, fractionation of phosphopeptides before LC-MS/MS is essential to increase it coverage. Here, we present a detailed protocol for in-depth analysis of tissue phosphoproteome, including collection of tissue samples, extraction of tissue proteins, proteolytic digestion of proteins into peptides, enrichment of phosphopeptides with TiO_2 using lactic acid as non-phosphopeptide excluder, fractionation of phosphopeptides with TEA-based high-pH reversed-phase (HpH-RP) chromatography, and identification of phosphopeptides with LC-MS/MS. We also outline the essential steps for data processing.

Key words Protein phosphorylation, Post-translational modification (PTM), Tissue phosphoproteomics, TiO_2 enrichment, High-pH reversed-phase chromatography (HpH-RP) fractionation, Mass spectrometry (MS)

1 Introduction

Protein phosphorylation is an essential post-translational modification (PTM). It controls downstream signaling pathways that regulate many cellular processes, such as cell proliferation, differentiation, and metabolism. Abnormal protein phosphorylation can contribute to the genesis and progression of a range of diseases [1, 2].

Authors Xiulan Chen and Shasha Wei have equally contributed to this chapter.

Taufika Islam Williams (ed.), *Tissue Proteomics: Methods and Protocols*, Methods in Molecular Biology, vol. 2884,
https://doi.org/10.1007/978-1-0716-4298-6_17,

Mass spectrometry–based phosphoproteomics has become a powerful tool for analyzing protein phosphorylation in a global manner [3]. However, several challenges need to be addressed when analyzing protein phosphorylation. Phosphoproteins are often low abundant in complex samples, and phosphorylation sites are at low stoichiometry, which results in low signals of phosphopeptides when analyzed in the presence of high-abundant non--modified peptides. So selective enrichment of phosphopeptides from complex samples is central for a successful phosphoproteomic analysis. In the past two decades, many different enrichment methods, such as affinity chromatography using TiO_2 or ion metal affinity chromatography (IMAC), antibody and protein domain–based enrichment of tyrosine phosphorylated peptides, or chemical derivatization using beta-elimination, have been developed to enrich phosphopeptides from complex samples [4–6]. These enrichment methods differ in selectivity, specificities, sensitivity, and capacity. For example, antibody and protein-based enrichment methods selectively enrich tyrosine phosphorylated peptides, while chemical derivatization using beta-elimination is used for the study of serine and threonine phosphorylated peptides. Besides that, TiO_2 chromatography enriches mono-phosphorylated peptides more efficiently than multiply phosphorylated peptides, whereas IMAC has a preference for multiply phosphorylated peptides, but it has low capacity and selectivity, which limits its application in highly complex samples. Indeed, the most widely used enrichment techniques are based on IMAC or TiO_2.

It is estimated that nearly 30% of all human proteins may be phosphorylated [4], which poses a significant technical challenge to get a deep coverage of the phosphoproteome. A combination of fractionation and phosphopeptide enrichment to reduce sample complexity prior to LC-MS/MS analysis is essential to solve the problem. Traditionally, the strategy of "fractionation before enrichment" is applied to increase the coverage of phosphoproteome. In this strategy, complex peptide mixtures are separated into several fractions with chromatographic methods, such as strong cation exchange (SCX) [7, 8] or high-pH reversed-phase (HpH-RP) chromatography [9], prior to phosphopeptide enrichment of the fractionated peptide fractions. However, this strategy requires a large amount of starting materials (a few milligrams), and it is quite labor intensive, and might result in poor enrichment of phosphopeptides due to suboptimal protein-to-resin ratios [5]. Recently, the strategy of "fractionation after enrichment", in which enrichment of phosphopeptides is performed as the first step and the purified phosphopeptides are fractionated using other methods, such as HILIC [10] or HpH-RP [11], has been adopted in large-scale phosphoproteomic analyses. This strategy is simple and needs less starting material for enrichment. Here, we introduce a detailed protocol that combines TiO_2 enrichment with

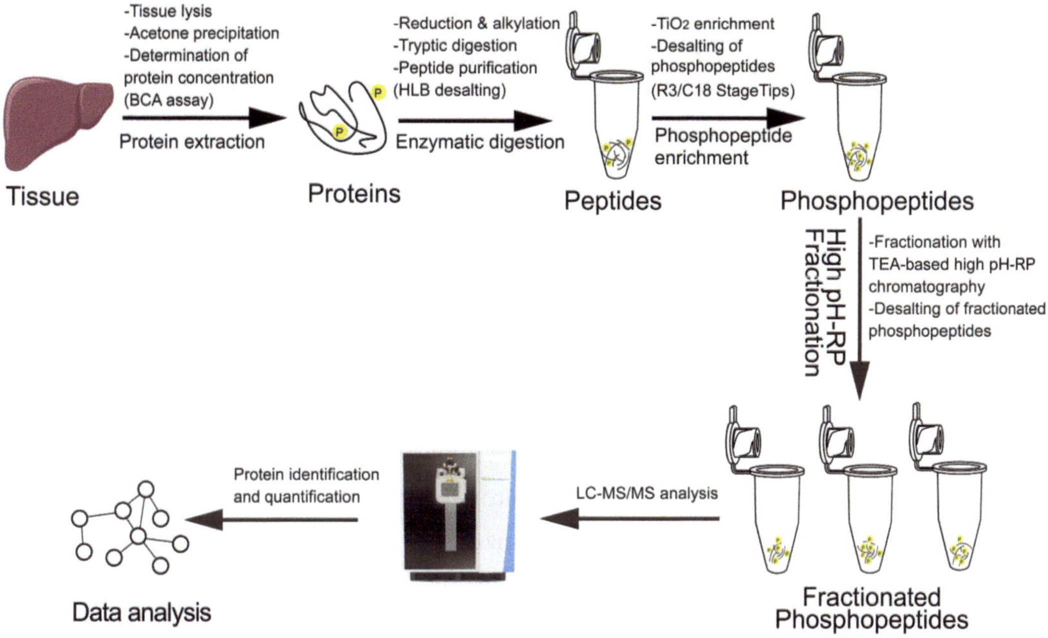

Fig. 1 Comprehensive phosphoproteomic workflow combining TiO_2 enrichment and fractionation with triethylamine (TEA)-based HpH-RP chromatography. The phosphoproteomic workflow consists of several parts: ① protein extraction: proteins are extracted from tissues, precipitated with acetone, and protein concentration is determined with BCA assay; ② enzymatic digestion: proteins are reduced, alkylated, and digested into peptides, and then, peptides are desalted with HLB columns; ③ phosphopeptide enrichment: phosphopeptides are enriched with TiO_2, and the enriched phosphopeptides are desalted with R3/C18 StageTips; ④ fractionation of phosphopeptides: enriched phosphopeptides are fractionated with TEA-based high-pH RP chromatography, and the fractionated phosphopeptides are desalted with R3/C18 StageTips; ⑤ LC-MS/MS analysis: the fractionated phosphopeptides are analyzed with LC-MS/MS analysis; ⑥ data analysis: MS data are analyzed to get protein identification and quantification results

fractionation using triethylamine (TEA)-based HpH-RP chromatography for the comprehensive and deep phosphoproteomic analysis of tissue samples (Fig. 1).

2 Materials

2.1 Collection of Tissue Samples

1. Animals used for collection of tissues (here male adult C57BL/6J mice).
2. Anesthesia machine.
3. Anesthesia (isoflurane).
4. Povidone-iodine solution.
5. Surgical instruments: dissection scissors, straight surgical forceps, curved surgical forceps.
6. 2 mL cryogenic vial with silicone washer.

7. Sterile saline solution (0.9% (w/v) NaCl).
8. EDTA-free protease inhibitor cocktail.
9. Phosphatase inhibitor cocktail.
10. 10-cm cell culture dish.
11. Dust-free wiping paper.
12. 5-mL dropper.
13. Liquid nitrogen.

2.2 Tissue Lysis, Protein Extraction, and Protein Precipitation

1. Tissue samples to be lysed (here mouse liver).
2. Lysis buffer: 8 M (w/v) urea, 100 mM (w/v) Tris–HCl, pH 8.5, freshly prepared in H_2O (*see* **Notes 1** and **2**).
3. EDTA-free protease inhibitor cocktail.
4. Phosphatase inhibitor cocktail.
5. Precellys homogenizer (Bertin Technologies).
6. Soft tissue homogenizing kit: 2-mL tube and 1.4-mm ceramic beads (Bertin Technologies).
7. Acetone (ice-cold).
8. Bicinchoninic acid (BCA) protein assay.

2.3 In-Solution Digestion of Proteins

1. Reducing reagent: 1 M (w/v) dithiothreitol (DTT), freshly prepared in H_2O.
2. Alkylation regent: 1 M (w/v) iodoacetamide (IAM), freshly prepared in H_2O.
3. Mass spectrometry (MS)-grade lysyl endopeptidase (Lys-C).
4. Dilution buffer: 100 mM (w/v) Tris–HCl, pH 8.0.
5. Sequencing grade modified trypsin.
6. HPLC-grade formic acid (FA).

2.4 Desalting of Tryptic Peptides with HLB Cartridges

1. Oasis® HLB (hydrophilic-lipophilic balanced) extraction cartridge, 1 cc/10 mg sorbent per cartridge, 30 μm particle size.
2. Supelco Visiprep SPE vacuum manifold.
3. HPLC-grade acetonitrile (ACN).
4. HLB equilibration and desalting buffer: 0.1% (v/v) FA in H_2O.
5. HLB elution buffer 1: 20% (v/v) ACN, 0.1% (v/v) FA in H_2O.
6. HLB elution buffer 2: 40% (v/v) ACN, 0.1% (v/v) FA in H_2O.
7. HLB elution buffer 3: 60% (v/v) ACN, 0.1% (v/v) FA in H_2O.
8. Quantitative colorimetric peptide assay.

2.5 Enrichment of Phosphopeptides with TiO_2

1. TiO_2 beads (5 μm, GL Sciences).
2. 18-gauge blunt-tipped needle and plunger.
3. Empore Octyl C8 47-mm extraction disk (3 M).
4. 200-μL low-binding pipette tip.
5. 200-μL pipette tip adapter.
6. 1.5-mL low-binding microcentrifuge tube.
7. Lactic acid.
8. HPLC-grade trifluoroacetic acid (TFA).
9. Sample loading buffer: 70% (v/v) ACN, 5% (v/v) TFA, 20% (v/v) lactic acid, 5% (v/v) H_2O.
10. Washing buffer 1: 30% (v/v) ACN, 0.5% (v/v) TFA.
11. Washing buffer 2: 80% (v/v) ACN, 0.4% (v/v) TFA.
12. Elution buffer 1: 4% (v/v) $NH_3.H_2O$ (pH ~11.3) (*see* **Note** 3).
13. Elution buffer 2: 4% (v/v) $NH_3.H_2O$, 50% (v/v) ACN.
14. 10% FA in water.

2.6 Desalting of Phosphopeptides with R3/C18 StageTips

1. 18-gauge blunt-tipped needle and plunger.
2. Empore Octadecyl C18 47-mm extraction disk (3 M).
3. 200-μL low-binding pipette tip.
4. 200 μL pipette tip adapter.
5. 1.5-mL low-binding microcentrifuge tube.
6. OLIGO R3 reversed-phase (RP) Resin (Thermo Fisher Scientific).
7. HPLC-grade methanol.
8. StageTip desalting buffer: 0.1% (v/v) FA in H_2O.
9. StageTip elution buffer 1: 20% (v/v) ACN, 0.1% (v/v) FA in H_2O.
10. StageTip elution buffer 2: 40% (v/v) ACN, 0.1% (v/v) FA in H_2O.
11. StageTip elution buffer 3: 60% (v/v) ACN, 0.1% (v/v) FA in H_2O.

2.7 Fractionation of Phosphopeptides with HpH-RP StageTip

1. Durashell C18 resin (Agela Technologies, 3 μm particle size, 150 Å).
2. Empore Octyl C8 47-mm extraction disk (3 M).
3. 0.1% TFA.
4. Buffer A: 0.1% triethylamine (TEA) (v/v) in H_2O.
5. Elution buffers: X% ACN (v/v), 0.1% TEA in H_2O (X = 2, 5, 8, 15, 40).
6. 10% FA in H_2O.

2.8 Nanoflow LC-MS/MS Analysis

1. Mobile phase A: 0.1% (v/v) FA in H_2O, LC-MS grade.
2. Mobile phase B: 80% (v/v) ACN, 0.1% (v/v) FA in H_2O.
3. Nanoflow liquid chromatography: Easy-nLC 1200 HPLC system with a fused silica trap column (100 μm ID * 2 cm) in-house packed with reversed-phase silica (ReproSil-Pur C18 AQ, 5 μm, Dr. Maisch GmbH) and an analytical column (75 μm ID * 20 cm) packed with reversed-phase silica (ReproSil-Pur C18 AQ, 3 μm, Dr. Maisch GmbH).
4. Mass spectrometer: Orbitrap Exploris 480 with a high-field asymmetric waveform ion mobility spectrometry (FAIMS) device or any other high-resolution mass spectrometric system.

2.9 Raw Data Processing and Data Analysis

1. Protein sequence data in FASTA format for the analyzed species, such as data downloaded from the public comprehensive protein sequence database UniProt (https://www.uniprot.org/) supplemented with common contaminants.
2. Protein identification and quantification software for raw data processing, such as commercially available Proteome Discoverer 2.4 (PD 2.4) or freely available MaxQuant (https://www.maxquant.org/).
3. Data analysis software, such as Perseus (https://maxquant.net/perseus/) or R (https://www.r-project.org/).

3 Methods

3.1 Collection of Tissue Samples

1. Anesthetize the mice by isoflurane inhalation, collect blood by eyeball extirpating, and euthanize the mice by cervical dislocation (*see* **Note 4**).
2. Shave the abdomen using an electric shaver to allow for cleaner exposure and disinfect the area with a povidone-iodine solution.
3. Identify linea alba (midlines white fascia) of the abdominal musculature and dissect the mice to collect the liver tissue (*see* **Note 5**).
4. Put the liver in a 10-cm cell culture dish on the ice, and wash the liver with precooled saline solution containing protease and phosphatase inhibitor cocktails using 5-mL droppers (*see* **Note 6**).
5. Remove the liquid on the liver surface with dust-free wiping paper.
6. Transfer the liver to a new 10-cm cell culture dish on the ice and cut the tissue into several big pieces.

7. Immediately pack the tissue into a cryogenic vial, snap freeze in liquid nitrogen for 10–30 min, and store at −80 °C for further analysis.

3.2 Tissue Lysis, Protein Extraction, and Protein Precipitation

1. Chill the Precellys homogenizer to 4 °C with dry ice.
2. For soft tissue homogenization, add 1.4-mm ceramic beads into 2-mL tubes to cover the conical bottom of the tubes.
3. Label 2-mL tubes and put them on ice to keep them cool.
4. Add ~50 mg of frozen tissues into the labeled tubes.
5. Add 500 μL ice-cold lysis buffer containing protease and phosphatase inhibitor cocktails into the labeled tubes (*see* **Note** 7).
6. Vortex briefly, and spin to cover the tissues in lysis buffer.
7. Place the tubes on the chilled Precellys homogenizer. Homogenize with Precellys homogenizer for nine cycles of 15 s on/30 s off at 7800 rpm.
8. Repeat **step** 7 twice.
9. Centrifuge the homogenized tissue lysate for 10 min at 20,000 × *g*, 4 °C.
10. Transfer the supernatant to a new 1.5-mL microcentrifuge tube carefully and avoid the lipid layer on top of the cleared lysate. For big pellets at the bottom of the tube, it is recommended to sonicate with a sonicator on ice to homogenize the tissues completely.
11. Take 150 μL of clear lysate and add 600 μL ice-cold acetone (4 volumes of ice-cold acetone to 1 volume of protein solution) to precipitate proteins and remove small-molecule interferences (*see* **Note 8**).
12. Mix thoroughly and keep the samples at −20 °C for at least 2 h (*see* **Note 9**).
13. Centrifuge for 15 min at 15,000× *g*, 4 °C.
14. Discharge supernatant carefully and retain the protein pellets.
15. Dry the pellets briefly by tilting the tube on tissue paper to eliminate any acetone residue (*see* **Note 10**).
16. Re-solubilize protein pellets with 200 μL ice-cold lysis buffer (*see* **Note 11**).
17. Sonicate samples on ice for 2 min to increase the solubility of proteins.
18. Determine protein concentration with BCA protein assay.
19. Take 10 μg of protein lysate and run sodium dodecyl-sulfate polyacrylamide gel electrophoresis (SDS-PAGE) to check the efficiency of protein extraction.

3.3 In-Solution Digestion of Proteins

1. Take an aliquot of lysate containing 1 mg of protein according to the BCA protein assay (or if less material is available, *see* **Note 12**).
2. Add DTT to a final concentration of 10 mM to reduce disulfide bonds for 1 h at 30 °C with agitation at 400 rpm (*see* **Note 13**).
3. Add IAM to a final concentration of 40 mM to alkylate-free thiols for 45 min at room temperature in the dark (*see* **Note 14**).
4. Add DTT to a final concentration of 40 mM to remove excess IAM at 30 °C for 30 min.
5. Add Lys-C protease to the reduced and alkylated samples (enzyme to protein ratio = 1:100, w/w) and digest at 37 °C for 3 h with agitation at 600 rpm (*see* **Note 15**).
6. Dilute the samples with 50 mM Tris–HCl (pH 8.0) to reduce the concentration of urea to less than 2 M (*see* **Note 16**).
7. Add trypsin protease to the diluted samples (enzyme to protein ratio = 1:50, w/w) and digest at 37 °C overnight with agitation at 600 rpm.
8. Quench the digestion by adding FA to a final concentration of 0.5% (v/v). Check that the pH with pH indicator paper is about 3.
9. Centrifuge the digested samples at 20,000× *g* for 10 min at 4 °C to remove undigested proteins and collect the supernatant.
10. Desalt the supernatant with HLB cartridges (*see* Subheading 3.4).

3.4 Desalting of Tryptic Peptides with HLB Cartridges

1. Set up the Oasis HLB 1 cc Vac cartridges (10 mg sorbent, maximum binding capacity of 2 mg) onto the Supelco Visiprep SPE vacuum manifold.
2. Condition the cartridges with 1 mL of ACN. Let the liquid run through under gravity (*see* **Note 17**).
3. Repeat **step 2** twice.
4. Equilibrate the cartridges with 1 mL of HLB equilibration and desalting buffer three times.
5. Load the digested samples onto the cartridges (*see* **Note 18**) and collect the flow-through (FT).
6. Reload the FT onto the cartridges to recover peptides as much as possible.
7. Repeat **step 6** once.
8. Wash the cartridges with 1 mL of HLB equilibration and desalting buffer three times.
9. Elute peptides retained on the C18 material into low-binding microcentrifuge tubes sequentially with 400 μL HLB elution

buffer 1, 400 μL HLB elution buffer 2, and 400 μL HLB elution buffer 3.

10. Mix the eluates with vortexing and dry them with a SpeedVac.
11. Dissolve the desalted peptides with 1.5 mL 0.1% FA, and determine peptide concentration with BCA colorimetric peptide assay.
12. Take 10 μg peptides to check protein digestion efficiency before phosphopeptide enrichment and perform protein-level quantification with LC-MS/MS analysis (*see* **Note 19**).
13. Take 1 mg peptide digest and dry with a SpeedVac for phosphopeptide enrichment (*see* Subheading 3.5).

3.5 Enrichment of Phosphopeptides with TiO_2

Here, phosphopeptides are enriched from 1 mg of peptides with TiO_2 beads using lactic acid as a non-phosphopeptide excluder according to a protocol published previously [11]. The protocol is subdivided into three parts: preparation of TiO_2 beads (*see* Subheading 3.5.1), preparation of C8 tips (*see* Subheading 3.5.2), and phosphopeptide enrichment (*see* Subheading 3.5.3).

3.5.1 Preparation of TiO_2 Beads

1. Weigh TiO_2 beads according to the total quantity of peptides to be enriched. The beads-to-peptide ratio used is 10.5:1 (w/w) for the three incubations mentioned below (*see* **Note 20**). Here, 12 mg TiO_2 beads are used.
2. Place TiO_2 beads in a 1.5-mL low-binding microcentrifuge tube.
3. Wash TiO_2 beads with 0.5 mL washing buffer 2 and place the tube on the shaker (1000 rpm) at room temperature for 5 min.
4. Centrifuge briefly to pellet TiO_2 beads and discard the supernatant.
5. Repeat **steps 3** and **4** twice.
6. Equilibrate the beads with 0.5 mL loading buffer and place the tube on the shaker (1000 rpm) at room temperature for 5 min.
7. Centrifuge briefly to pellet TiO_2 beads and discard the supernatant.
8. Repeat **steps 6** and 7 twice.
9. Add 0.8 mL loading buffer to 12 mg beads to make the concentration of TiO_2 beads slurry 15 mg/mL.
10. Split TiO_2 beads into 6, 3, and 1.5 mg to three low-binding microcentrifuge tubes for phosphopeptide enrichment from 1 mg of peptides (*see* **Note 21**).
11. Centrifuge to pellet the beads and remove the supernatant.

3.5.2 Preparation of C8 Tips

1. Take a C8 disk with an 18-gauge blunt-tipped needle and plunger to the bottom of a 200 μL low-binding pipette tip to make C8 tips. Prepare a few tips in excess.
2. Place the C8 tips with the appropriate adapter into 1.5-mL low-binding microcentrifuge tubes.
3. Place collection tubes, together with adapter and tips, in a centrifuge.
4. Add 100 μL of washing buffer 2 into each tip and centrifuge for 1 min at 400× *g*.
5. Check the liquid levels of washing buffer 2 in the tips and keep the tips with comparable liquid levels.
6. Centrifuge the tips and keep a small amount of liquid on the tips for phosphopeptide enrichment.

3.5.3 Phosphopeptide Enrichment

1. Dissolve 1 mg of dried peptides with 0.5 mL freshly prepared sample loading buffer to make the final concentration of peptides 2 μg/μL.
2. Mix it thoroughly by shaking the peptide solution for 10 min at room temperature.
3. Add peptide solution into the low-binding microcentrifuge tubes containing 6 mg of TiO_2 beads prepared in Subheading 3.5.1 (*see* **Note 22**).
4. Incubate the peptide solution and TiO_2 beads at room temperature for 15 min on a shaker at 1000 rpm. Adjust the shaking speed to keep the beads suspended in the peptide solution.
5. Centrifuge the samples in a table centrifuge for 1 min to pellet the TiO_2 beads.
6. Transfer the supernatant to a low-binding microcentrifuge tube containing 3 mg of TiO_2 beads prepared in Subheading 3.5.1 and incubate it for 15 min at room temperature on a shaker at 1000 rpm (*see* **Note 23**).
7. Centrifuge to pellet TiO_2 beads and transfer the supernatant to another low-binding microcentrifuge tube containing 1.5 mg TiO_2 beads prepared in Subheading 3.5.1 (*see* **Note 24**) and incubate for 15 min at room temperature on a shaker at 1000 rpm.
8. Centrifuge to pellet the beads and collect the supernatant as the TiO_2-FT fraction (*see* **Note 25**).
9. Pool the TiO_2 beads from the three incubations using 150 + 50 μL of loading buffer and transfer them to the C8 tips prepared in Subheading 3.5.2. Make sure the TiO_2 beads are completely transferred to the C8 tips by rinsing the three tubes with another round of 50 μL of loading buffer.

10. Centrifuge at 450× *g* for 3–5 min and transfer the FT to the TiO_2-FT tube. Adjust centrifugation speed to pass the liquid through the tips within 5 min.
11. Wash TiO_2 beads on C8 tips with 150 μL of loading buffer.
12. Centrifuge at 450× *g* for 3–5 min and transfer the FT to the TiO_2-FT tube.
13. Repeat **steps 11** and **12** for 10–15 times (*see* **Note 26**).
14. Wash the TiO_2 beads on C8 tips with 150 μL washing buffer 1.
15. Centrifuge at 450× *g* for 3–5 min and transfer the FT to the TiO_2-FT tube.
16. Repeat **steps 14** and **15** three times.
17. Wash the TiO_2 beads on C8 tips with 150 μL washing buffer 2.
18. Centrifuge at 450× *g* for 3–5 min and transfer the FT to the TiO_2-FT tube.
19. Repeat **steps 17** and **18** twice.
20. Elute phosphopeptides with 200 μL elution buffer 1. Centrifuge at 1000× *g* for 3 min (*see* **Note 27**).
21. Elute phosphopeptides with 200 μL elution buffer 2. Centrifuge at 1000× *g* for 3 min.
22. Immediately acidify the eluates using 10% FA in water. Check the pH of the solution by spotting a small amount on the pH indicator paper. Adjust the pH to be below 3 (*see* **Note 28**).
23. Lyophilize the eluted phosphopeptides with a SpeedVac.
24. Desalt phosphopeptides with R3/C18 StageTips (*see* Subheading 3.6).

3.6 Desalting of Phosphopeptides with R3/C18 StageTips

Empirically, the yield of phosphopeptides is 1–2% of the original peptide input. According to this estimation, 10–20 μg phospho peptides can be purified from 1 mg peptides. Two homemade R3/C18 StageTips are used to desalt phosphopeptides purified from 1 mg peptides. The purification is performed sequentially, meaning that all the phosphopeptides go through the first C18/R3 StageTip and the FT goes directly onto a second C18/R3 StageTip. Both tips are washed with desalting buffer and eluted with elution buffer (*see* **Note 29**).

1. Take the C18 disk with an 18-gauge blunt-tipped needle and plunger to the bottom of a 200-μL low-binding pipette tip to make a C18 tip. Prepare a few tips in excess.
2. Weigh a desired amount of OLIGO R3 resins.
3. Wash the resins with 1 mL of methanol for 5 min with shaking at 1000 rpm.

4. Centrifuge at 3000× *g* for 1 min to pellet the resins and remove the supernatant.
5. Repeat **steps 3** and **4** once.
6. Add 1 mL of methanol to the resins to make R3 slurry.
7. Add R3 slurry to C18 tips to make the length of C18 StageTips 1 cm.
8. Place the homemade R3/C18 StageTips, adapter, and 1.5-mL tubes onto a table centrifuge.
9. Activate StageTips by adding 150 μL of methanol into each tip and centrifuge for 3 min.
10. Repeat **step 9** once.
11. Perform a quality check of the tips as described in Subheading 3.5.2 and keep tips with comparable elution speed.
12. Equilibrate the selected tips with 100 μL StageTip desalting buffer and centrifuge for 2 min (*see* **Note 30**).
13. Repeat the equilibration step (**step 12**) twice.
14. Reconstitute the dried phosphopeptides with 100 μL StageTip desalting buffer and sonicate on ice for 2 min to increase the solubility of peptides.
15. Take two equilibrated R3/C18 StageTips and load the sample onto the first R3/C18 StageTip.
16. Centrifuge for 3 min and collect the flow-through (FT).
17. Load the FT onto a second R3/C18 StageTip.
18. Reload the FT onto two StageTips sequentially once.
19. Add 100 μL of StageTip desalting buffer into the two tips for one sample and centrifuge for 3 min to wash out the unbound materials.
20. Repeat **step 19** once.
21. Elute both tips with 80 μL of StageTip elution buffer 1, 80 μL of StageTip elution buffer 2, and 80 μL of StageTip elution buffer 3 sequentially into one 1.5-mL low-binding microcentrifuge tube by centrifugation for 3 min.
22. Lyophilize the desalted phosphopeptides with a SpeedVac.
23. Store the samples at −80 °C for further processing (HpH-RP fractionation or direct LC-MS/MS analysis).

3.7 Fractionation of Phosphopeptides with HpH-RP StageTips

To increase the depth of phosphoproteomic analysis, phosphopeptides enriched from 1 mg peptides are separated into several fractions with HpH-RP fractionation using TEA as an additive to the mobile phase of HpH-RP chromatography (*see* **Notes 31** and **32**). The protocol is subdivided into three parts: preparation of C18 resins (*see* Subheading 3.7.1), preparation of HpH-RP StageTips (*see* Subheading 3.7.2), and HpH-RP fractionation (*see* Subheading 3.7.3).

3.7.1 Preparation of C18 Resins

1. Weigh the desired amount of Durashell C18 resin (here 20 mg) (*see* **Note 33**) and place it in a 1.5-mL tube.
2. Wash the C18 resin with 1 mL methanol, shaking for 10 min at room temperature.
3. Centrifuge at 3000× *g* for 2 min to pellet the resin and remove the supernatant.
4. Repeat **steps 2** and **3** once.
5. Add 500 μL of methanol to make a 40 mg/mL C18 resin slurry.

3.7.2 Preparation of HpH-RP StageTips

1. Prepare C8 tips as described in Subheading 3.5.2.
2. Take 75 μl of Durashell C18 resin slurry (approximately 3 mg) onto a C8 tip (*see* **Note 34**).
3. Centrifuge at 3000× *g* for 2 min to remove the methanol.
4. Add 150 μL of ACN to activate the HpH-RP StageTip and centrifuge at 3000× *g* for 2–3 min to remove the ACN.
5. Repeat **step 4** once.
6. Equilibrate HpH-RP StageTips with 150 μL of 0.1% TFA and centrifuge at 3000× *g* for 3–4 min.
7. Repeat **step 6** once.

3.7.3 HpH-RP Fractionation

1. Reconstitute the purified phosphopeptides enriched from 1 mg of peptides with 100 μL of 0.1% TFA (*see* **Note 35**).
2. Load the samples onto the equilibrated HpH-RP StageTips and centrifuge the tips at 3000× *g* for 3 min.
3. Wash HpH-RP StageTips with 200 μL of H_2O (*see* **Note 36**).
4. Elute phosphopeptides sequentially with 80 μL of elution buffer with a stepwise gradient of ACN (2%, 5%, 8%, 15%, and 35%) in 0.1% TEA and collect the five eluates (*see* **Note 37**).
5. Acidify the five eluates immediately with 10% FA. Check for pH lower than 3.
6. Combine the five eluates into three fractions: combine 2% ACN/0.1% TEA fraction and 15% ACN/0.1% TEA fraction as Fraction 1, keep 5% ACN/0.1% TEA fraction as Fraction 2, and combine 8% ACN/0.1% TEA and 35% ACN/0.1% TEA fraction as Fraction 3 (*see* **Note 38**).
7. Dry the three fractions by a SpeedVac.

8. Desalt the three fractions with homemade OLIGO R3/C18 StageTips as described in Subheading 3.6. One C18/R3 StageTip is used for each fraction.

3.8 LC-MS/MS Analysis

1. Reconstitute the dried (fractionated) phosphopeptides or peptides for proteome analysis in an appropriate volume of mobile phase A, e.g., 12 μL.
2. Sonicate peptide solution on ice for 2 min.
3. Measure the concentration of peptides/phosphopeptides with a NanoDrop spectrophotometer (Thermo Scientific) or equivalent.
4. Take 1 μg of peptides/phosphopeptides according to the Nanodrop results and add mobile phase A to make the volume of samples 10 μL.
5. Centrifuge the samples at 20,000× *g* for 10 min.
6. Transfer the supernatant to an autosampler vial for loading onto the trap column.
7. Perform peptide separation on an analytical column at a flow rate of 300 nL/min with a 132 min gradient (0–5% B, 6 min; 5–20% B, 69 min; 20–30% B, 39 min; 30–99% B, 9 min; 99% B, 9 min).
8. In the described workflow, LC-MS/MS analysis is performed on an Orbitrap Exploris 480 (Thermo Fisher Scientific) with a high-field asymmetric waveform ion mobility spectrometry (FAIMS) device (Thermo Fisher Scientific) coupled to an Easy-nLC 1200 HPLC system.
9. For MS/MS (MS2) acquisition of peptides, a resolution of 15,000 and a maximum injection time of 22 ms are used to sequence as many peptides as possible to get a high coverage of the proteome. For MS2 acquisition of phosphopeptides, higher resolution and longer injection time (30,000 resolution and 54 ms) are used to improve the fragment spectra of phosphopeptides.

3.9 Raw Data Processing and Data Analysis

1. Only the essential steps for raw data processing and data analysis are covered. The described data analysis workflow is adapted to the materials used and different bioanalytical questions investigated.
2. LC-MS/MS raw data are processed with PD 2.4 software for protein identification and quantification using SequestHT searching engine against the mouse UniProt/Swiss-Prot database supplemented with common contaminants (*see* **Note 39**).
3. Set database searching parameters as follows: enzyme specificity for trypsin and up to two missed cleavages are allowed and minimum peptide length is 6. Set mass tolerance for precursor

and fragment ions as 10 ppm and 0.02 Da, respectively. Set cysteine carbamidomethylation as a fixed modification. Include methionine oxidation, acetylation at the N-terminal of proteins, deamidation at asparagine and glutamine, and phosphorylation at serine, threonine, and tyrosine as variable modifications with a maximum number of three modifications per peptide. Calculate the false discovery rate (FDR) using the Percolator algorithm in PD 2.4 and set FDR on peptide and protein levels as 1%. The precursor detector node in PD 2.4 is included to reduce the influence of chimeric spectra. Set PhosphoRS localization probability greater than 0.75.

4. Filter out contaminants and keep phosphopeptides with fully localized sites (*see* **Note 40**).
5. Perform further data analysis: statistical analysis, functional annotation, GO enrichment, and protein interaction network analysis using Perseus or R.

3.10 Anticipated Results

1. Check phosphopeptide enrichment specificity by comparing the number of phosphopeptides identified with the total number of peptides identified. For phosphopeptide enrichment, a good specificity value should be above 80%. If the enrichment specificity is significantly lower, check the TiO_2 beads-to-peptide ratio and the washing steps.
2. For single-shot LC-MS/MS analysis with a 2-h LC gradient, about 9000–10,000 phosphopeptides can be identified. After HpH-RP fractionation, about 15,000–20,000 phosphopeptides can be identified from 1 mg starting material for phosphopeptide enrichment.

4 Notes

1. It is recommended to use freshly prepared lysis buffer or lysis buffer stored at −20 °C for no longer than 3 months when urea is used in lysis buffer to solubilize proteins. As in aqueous solution, a small amount of urea can decompose into cyanic acid, which can react with the side chains of lysine and arginine and the N-terminal amino groups of peptides to form carbamylated residues. This urea-induced carbamylation can be minimized in two ways. First, use freshly prepared lysis buffer as urea degrades in an aqueous solution. Second, perform sample preparation at a lower temperature (30 °C), as the decomposition of urea increases at high temperatures.
2. 100 mM of Tris–HCl, instead of 100 mM NH_4HCO_3, is used in the lysis buffer to extract proteins from tissues, as Tris–HCl

buffer can reduce artificial deamidation on Asn during in-solution digestion [12].

3. Prepare elution buffers freshly and check the pH (~11.3) prior to use.
4. All animal experiments should be performed in accordance with the guidelines of laws about animal care and approved by the Ethics Committee of the Institute or University.
5. To reduce the interferences from the blood for tissue proteomic/phosphoproteomic analysis, perfuse from the portal vein with pre-cooled saline solution containing protease and phosphatase inhibitor cocktails if there is too much blood left in the liver after laparotomy (the color of the liver is observed to be red). However, some cells (Kuff cells, hepatic sinus endothelial cells, etc.) may come out with the lavage fluid, so it is necessary to pay attention to the flow rate of the perfusion fluid.
6. Protease inhibitor cocktails and phosphatase inhibitor cocktails should be added to the saline solution or lysis buffer (*see* Subheading 3.2) just before use.
7. The ratio between the volume of lysis buffer (μL) and the wet weight of tissue (mg) is about 10:1. Adjust the volume of lysis buffer according to the weight of tissue.
8. Acetone precipitation can remove substances which may interfere with protein digestion and/or downstream phosphopeptide enrichment [13].
9. If the protein concentration is very low, it is recommended to keep the samples at −20 °C overnight.
10. Do not dry the protein pellets completely, as it is difficult to solubilize the pellets when completely dried.
11. Adjust the volume of lysis buffer to make protein concentration greater than 5 mg/mL, as too low of a protein concentration during the tryptic digestion would lead to higher missed cleavage rates for the peptides.
12. Phosphopeptides enriched from 200 μg of protein are enough for single-shot LC-MS/MS analysis. However, to increase the depth of phosphoproteomic analysis, 1 mg of protein should be used for the enrichment of phosphopeptides, and the enriched phosphopeptides can be fractionated into several fractions using HpH-RP fractionation.
13. Here, a lower temperature (30 °C) is used during the reduction step, as the lysis buffer contains high amounts of urea. 30 °C is sufficient for reduction and minimizes unwanted side modifications (urea-induced carbamylation).
14. IAM is sensitive to light, so its exposure to light during alkylation should be minimized.

15. Lys-C protease can operate at higher concentrations of urea to digest proteins into large peptides and increase cleavage efficiency compared with single tryptic digestion [14].
16. To keep high activity of trypsin, the concentration of urea during tryptic digestion should be diluted to lower than 2 M.
17. Never let the column material dry out, and always leave a liquid layer above the column material during the whole desalting process.
18. Dilute the digested samples to 1 mL with 0.1% FA when the sample volume is less than 0.3 mL; otherwise, it will lead to significant peptide loss during the desalting process.
19. Analyzing a small aliquot of digested peptides by LC-MS/MS can serve two purposes: (1) check the digestion efficiency before phosphopeptide enrichment and (2) assess protein-level changes and decipher the changes that are truly based on changes of phosphorylation but not changes in protein abundance, as the differential phosphopeptide abundances reflect not only changes in phosphorylation statue but also the expression of substrate proteins [15].
20. The TiO_2 beads-to-peptide ratio is an important factor for specific phosphopeptide enrichment [16]. So the amount of TiO_2 beads used should be adjusted according to the quantity of peptides to be enriched.
21. It is difficult to weigh 6, 3, and 1.5 mg of TiO_2 beads, so just measure the total amount of beads needed and split into three fractions for three incubations in the phosphopeptide enrichment steps.
22. In the first incubation, the TiO_2 beads-to-peptide ratio is 6:1. Add 0.6 mg TiO_2 beads per 100 μg of peptide (solution).
23. In the second round of incubation, half of the amount of TiO_2 beads is used as in the first incubation.
24. The incubation of peptides with TiO_2 beads can be repeated a total of three times to recover as much phosphopeptide as possible. 1/4 of the amount of TiO_2 beads used in the first incubation is used in the third incubation.
25. The TiO_2-FT fraction can be lyophilized and desalted for proteomic analysis or used to enrich other PTMs, such as acetylation [17].
26. The times for washing with loading buffer depend on the amount of starting material for phosphopeptide enrichment. If a lower amount of starting materials is used, the number of washing steps can be decreased.
27. This process can be repeated once to elute multi-phosphorylated peptides binding strongly to TiO_2 beads. The

centrifugation speed can be increased to 1500× g to reduce the elution time.

28. As the phosphorylated moiety is not stable at high pH, 10% FA is used to rapidly neutralize the high pH of the elution buffer.
29. Commercially available purification tips, such as ZipTip, can be applied to purify phosphopeptides. However, it cannot be used to purify peptides greater than 5 μg (the maximum capacity of a ZipTip is 5 μg). The StageTip protocol is more effective when large amounts of enriched samples are processed.
30. Avoid letting the disks dry in between the equilibration and sample loading. Check if the solution passed through the tip before going further.
31. It is recommended that HpH-RP fractionation is carried out for phosphopeptides enriched from at least 1 mg of peptides. If a lesser amount of starting peptides is used to enrich phosphopeptides, a limited amount of phosphopeptides can be enriched for further fractionation.
32. HpH-RP chromatography can be used to fractionate phosphopeptides enriched from TMT-labeled or iTRAQ-labeled peptides. However, the ACN gradient used for elution should be adjusted, as TMT labeling could increase the hydrophobicity of peptides [18].
33. Other high-pH-resistant C18 reagents, such as Oligo R3 resins, can also be used for HpH-RP fractionation.
34. Empirically, 3–4 mg of C18 beads can be used for fractionation of 20–30 μg phosphopeptides.
35. Reconstitute samples in acidic buffer as some phosphopeptides, especially multi-phosphorylated peptides, may not bind with RP materials at high pH values. The reason is that multi-phosphorylated peptides attain a high number of negative changes at high pH values, which reduces their interaction with RP materials.
36. As samples are loaded onto HpH-RP StageTips in an acidic buffer, washing HpH-RP StageTips with H_2O can adjust the pH to neutral for fractionation with a high pH elution buffer.
37. Phosphopeptides are naturally more hydrophilic than non-phosphorylated peptides, so the ACN gradient starts from 2%. If peptides are fractionated with HpH-RP chromatography, ACN gradient can start from 5%.
38. It is ok to analyze the five fractions separately. However, compared with analyzing three concatenated fractions, analyzing five fractions could not increase the depth of phosphoproteome coverage by too much but required almost twice as

much MS time. There is a balance between MS analysis time and the depth of phosphoproteomic coverage.

39. Other database-searching software, such as MaxQuant with the Andromeda search algorithm, can be used for protein identification and quantification.
40. If there are multiple (two or three) phosphosites within one phosphopeptide and at least one phosphosite in the phosphopeptide is ambiguous, this phosphopeptide is not regarded as a localized phosphopeptide.

Acknowledgements

This work was supported by grants from the Natural Science Basic Research Program of Shaanxi (Program No. 2022JQ-899) for Shasha Wei.

References

1. Cohen P (2001) The role of protein phosphorylation in human health and disease. The Sir Hans Krebs Medal Lecture. Eur J Biochem 268(19):5001–5010. https://doi.org/10.1046/j.0014-2956.2001.02473.x
2. Wesseling H, Mair W, Kumar M, Schlaffner CN, Tang S, Beerepoot P, Fatou B, Guise AJ, Cheng L, Takeda S, Muntel J, Rotunno MS, Dujardin S, Davies P, Kosik KS, Miller BL, Berretta S, Hedreen JC, Grinberg LT, Seeley WW, Hyman BT, Steen H, Steen JA (2020) Tau PTM profiles identify patient heterogeneity and stages of Alzheimer's disease. Cell 183(6):1699–1713.e1613. https://doi.org/10.1016/j.cell.2020.10.029
3. Franciosa G, Locard-Paulet M, Jensen LJ, Olsen JV (2023) Recent advances in kinase signaling network profiling by mass spectrometry. Curr Opin Chem Biol 73:102260. https://doi.org/10.1016/j.cbpa.2022.102260
4. Low TY, Mohtar MA, Lee PY, Omar N, Zhou H, Ye M (2021) Widening the bottleneck of phosphoproteomics: evolving strategies for phosphopeptide enrichment. Mass Spectrom Rev 40(4):309–333. https://doi.org/10.1002/mas.21636
5. Taewook Kang SB, Edwards A, Larsen MR (2022) Phosphoproteomics: methods and challenges. In: Reference module in life sciences. Elsevier. https://doi.org/10.1016/B978-0-12-821618-7.00031-6
6. Thingholm TE, Jensen ON, Larsen MR (2009) Analytical strategies for phosphoproteomics. Proteomics 9(6):1451–1468. https://doi.org/10.1002/pmic.200800454
7. Yue XS, Hummon AB (2013) Combination of multistep IMAC enrichment with high-pH reverse phase separation for in-depth phosphoproteomic profiling. J Proteome Res 12(9):4176–4186. https://doi.org/10.1021/pr4005234
8. Hou J, Tobe BT, Lo F, Blethrow JD, Crain AM, Wolf DA, Snyder EY, Singec I, Brill LM (2013) Combined total proteomic and phosphoproteomic analysis of human pluripotent stem cells. Methods Mol Biol 1029:163–189. https://doi.org/10.1007/978-1-62703-478-4_12
9. Batth TS, Francavilla C, Olsen JV (2014) Off-line high-pH reversed-phase fractionation for in-depth phosphoproteomics. J Proteome Res 13(12):6176–6186. https://doi.org/10.1021/pr500893m
10. Engholm-Keller K, Birck P, Storling J, Pociot F, Mandrup-Poulsen T, Larsen MR (2012) TiSH--a robust and sensitive global phosphoproteomics strategy employing a combination of TiO2, SIMAC, and HILIC. J Proteome 75(18):5749–5761. https://doi.org/10.1016/j.jprot.2012.08.007
11. Li J, Wang J, Yan Y, Li N, Qing X, Tuerxun A, Guo X, Chen X, Yang F (2022) Comprehensive evaluation of different TiO2-based phosphopeptide enrichment and fractionation methods for Phosphoproteomics. Cells 11(13):2047

12. Hao P, Ren Y, Datta A, Tam JP, Sze SK (2015) Evaluation of the effect of trypsin digestion buffers on artificial deamidation. J Proteome Res 14(2):1308–1314. https://doi.org/10.1021/pr500903b
13. Potel CM, Lin MH, Heck AJR, Lemeer S (2018) Defeating major contaminants in Fe (3+)- immobilized metal ion affinity chromatography (IMAC) phosphopeptide enrichment. Mol Cell Proteomics 17(5): 1028–1034. https://doi.org/10.1074/mcp.TIR117.000518
14. Glatter T, Ludwig C, Ahrné E, Aebersold R, Heck AJ, Schmidt A (2012) Large-scale quantitative assessment of different in-solution protein digestion protocols reveals superior cleavage efficiency of tandem Lys-C/trypsin proteolysis over trypsin digestion. J Proteome Res 11(11):5145–5156. https://doi.org/10.1021/pr300273g
15. Wu R, Dephoure N, Haas W, Huttlin EL, Zhai B, Sowa ME, Gygi SP (2011) Correct interpretation of comprehensive phosphorylation dynamics requires normalization by protein expression changes. Mol Cell Proteomics 10(8):M111.009654. https://doi.org/10.1074/mcp.M111.009654
16. Li QR, Ning ZB, Tang JS, Nie S, Zeng R (2009) Effect of peptide-to-TiO2 beads ratio on phosphopeptide enrichment selectivity. J Proteome Res 8(11):5375–5381. https://doi.org/10.1021/pr900659n
17. Melo-Braga MN, Ibáñez-Vea M, Kulej K, Larsen MR (2021) Comprehensive protocol to simultaneously study protein phosphorylation, acetylation, and N-linked sialylated glycosylation. Methods Mol Biol 2261:55–72. https://doi.org/10.1007/978-1-0716-1186-9_5
18. Chen X, Sun Y, Zhang T, Shu L, Roepstorff P, Yang F (2021) Quantitative proteomics using isobaric labeling: a practical guide. Genomics Proteomics Bioinformatics 19(5):689–706. https://doi.org/10.1016/j.gpb.2021.08.012

Chapter 18

Mass Spectrometry–Based Glycomics and Proteomics Profiling of On-Slide Digested Tissue from Complex Biological Samples

Sayantani Chatterjee, Joseph Zaia, and Manveen K. Sethi

Abstract

Mass spectrometry–based investigation of the heterogeneous glycoproteome from complex biological specimens is a robust approach to mapping the structure, function, and dynamics of the glycome and proteome. Sampling whole wet tissues often provides a large amount of starting material; however, there is a reasonable variability in tissue handling prior to downstream processing steps, and it is difficult to capture all the different biomolecules from a specific region. The on-slide tissue digestion approach, outlined in this protocol chapter, is a simple and cost-effective method that allows comprehensive mapping of the glycoproteome from a single spot of tissue of 1 mm or greater diameter. It provides a selection of target areas on tissue slides appropriate for tissue volumes of 10 nL or greater, corresponding to a 1 μL droplet of enzyme solution applied to a 1-mm diameter target on a 10-μm-thick tissue slice. Sequential enzymatic digestions and desalting of the biomolecules without any prior derivatization from the surface of fresh frozen or formalin-fixed paraffin-embedded tissue slides enable the simultaneous identification of glycosaminoglycan disaccharides such as hyaluronan, chondroitin sulfate and heparan sulfate, asparagine or *N*-linked glycans, and intact (glyco)peptides using liquid chromatography-tandem mass spectrometry. The in-depth information obtained from this method including the disaccharide compositions, glycan structures, peptide abundances, and site-specific glycan occupancies provides a detailed profiling of a single spot of tissue which has the potential to be disseminated to biomedical laboratories.

Key words Formalin-fixed paraffin-embedded tissue, Fresh frozen tissue, Glycomics, Glycoproteomics, Liquid chromatography, Mass spectrometry, On-slide tissue digestion, Proteomics

1 Introduction

Protein glycosylation, the addition of complex carbohydrates (glycans) to polypeptide chains, is a dynamic post-translational modification that influences various biological functions [1, 2]. Asparagine or *N*-linked and mucin-type *O*-linked glycosylation are the two common and well-studied types of mammalian protein glycosylation [3, 4]. Proteoglycans (PGs), another form of

Taufika Islam Williams (ed.), *Tissue Proteomics: Methods and Protocols*, Methods in Molecular Biology, vol. 2884, https://doi.org/10.1007/978-1-0716-4298-6_18,

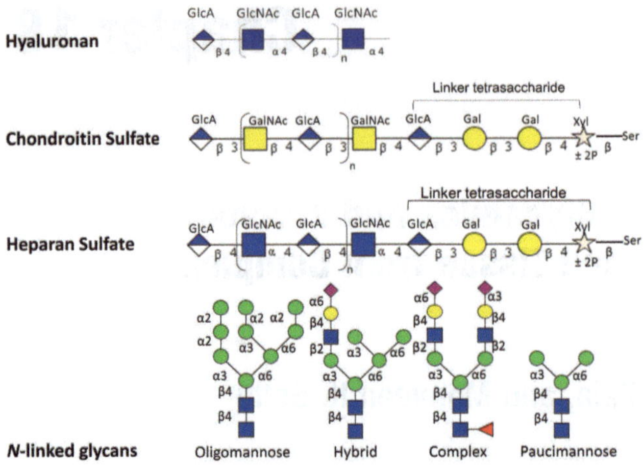

Lawrence code	Traditional nomenclature	*m/z* (z=1)
CS disaccharides		
D0a0	ΔHexA-GalNAc	378
D0a4	ΔHexA-GalNAc4S	458
D0a6	ΔHexA-GalNAc6S	458
D0a10	ΔHexA-GalNAc4S6S	538
HS disaccharides		
D0A0	ΔHexA-GlcNAc	378
D0A6	ΔHexA-GlcNAc(6S)	458
D2A0	ΔHexA(2S)-GlcNAc	458
D2A6	ΔHexA(2S)-GlcNAc(6S)	538
D0S0	ΔHexA-GlcNS	416
D0S6	ΔHexA-GlcNS(6S)	496
D2S0	ΔHexA(2S)-GlcNS	496
D2S6	ΔHexA(2S)-GlcNS(6S)	576

Fig. 1 Summary of key disaccharides (hyaluronan, chondroitin sulfate, and heparan sulfate) and *N*-linked glycans (oligomannose, complex, hybrid, and paucimannose) studied in this protocol

glycoproteins, consist of a core protein with covalently linked long linear chains of polysaccharides that are referred to as glycosaminoglycans (GAGs) [5, 6]. PGs including aggrecan, decorin, and syndecan are decorated with GAGs such as chondroitin sulfate (CS), heparan sulfate (HS), and hyaluronan (HA) [7, 8]. PGs are also known to contain *N*- and *O*-linked glycosylation (Fig. 1). The composition of the GAG chains and their interaction with the protein cores play an important role in determining the biological functions of PGs [9]. Therefore, the confident identification and quantification of protein glycosylation constitute a critical aspect in decoding the mechanistic processes related to health and disease [10, 11].

Whole tissues (fresh frozen, FF, or formalin-fixed paraffin-embedded, FFPE) provide starting material for glycomics and proteomics analysis on either matrix-assisted laser desorption/ionization (MALDI) or high-performance liquid chromatography (HPLC)-tandem mass spectrometry (MS/MS) platforms [12, 13]. However, the in-solution digestion step required for whole tissues is a time-consuming process wherein multiple sample workup and clean-up steps are required prior to mass spectrometric analysis. The variability in protocols used for sample handling and instrumentations for analysis across various research laboratories introduces a major heterogeneity in the identification and quantification of the different saccharides and peptides [14–20]. In addition, it is often not possible to capture and extract GAGs and *N*-linked glycans as well as intact (glyco)peptides from the precise region of interest. To overcome the complexity of handling whole wet tissue, the on-slide tissue digestion method was developed in the Zaia Lab to enzymatically digest and sequentially profile all the different biomolecules from a specific anatomical and targeted region of interest from the surface of tissue slides [21–23].

On-slide tissue digestion, compatible with both FF and FFPE tissues, is a simple and cost-effective approach wherein a target as small as 1 mm in diameter on a 10-μm-thick tissue slice affixed to a histological slide suffices to release (enzymatically and sequentially), extract, and desalt disaccharides, *N*-linked glycans, and intact (glyco)peptides. This method provides high coverage of the biomolecules without the need for additional derivatization or modification steps. Previously, the on-slide digestion method was adopted to profile GAGs from limited histological sites of the frozen brain cortex, brain stem, and cerebellum regions [22]. The positive outcomes of that study prompted the development of a comprehensive multi-enzyme workflow that simultaneously analyzed GAGs, *N*-linked glycans, and peptides from bovine cerebral cortex tissue surface areas with the use of microwave irradiation to reduce digestion times [24]. This workflow was particularly suitable for tissue microarrays and acted complementarily to MALDI imaging and laser capture microdissection array techniques, both of which need specific user training and instrumentation. This approach was further optimized by simultaneously digesting large *N*-linked glycans and small HS disaccharides and separating them using size-exclusion chromatography [21]. This protocol was then applied to investigate the glycomics and proteomics changes occurring in aging rat brains [25]. Finally, the on-slide digestion workflow was automated with the use of an inkjet chip printer for the addition of enzymes and a NanoMate robot for the extraction of digested molecules from clinically available glioblastoma tissue microarrays [26].

This protocol chapter provides a detailed and systematic step-by step guide to the multi-omics analysis of biomolecules from on-slide digested complex biological tissue using mass spectrometry. We provide in-depth experimental details for sample handling, extraction, clean-up, LC-MS/MS acquisition, and analysis of the biomolecules, as well as highlight key considerations and troubleshooting steps for the workflow (Fig. 2).

2 Materials

For on-slide tissue digestion and multi-omics analysis by mass spectrometry, we recommend a minimum of $n = 3$ biological or technical replicates to ensure consistent datasets. This protocol requires basic personal protective equipment including a laboratory coat, safety glasses, and gloves; any flammable solvents or hazardous chemicals should be handled carefully in a fume hood. In

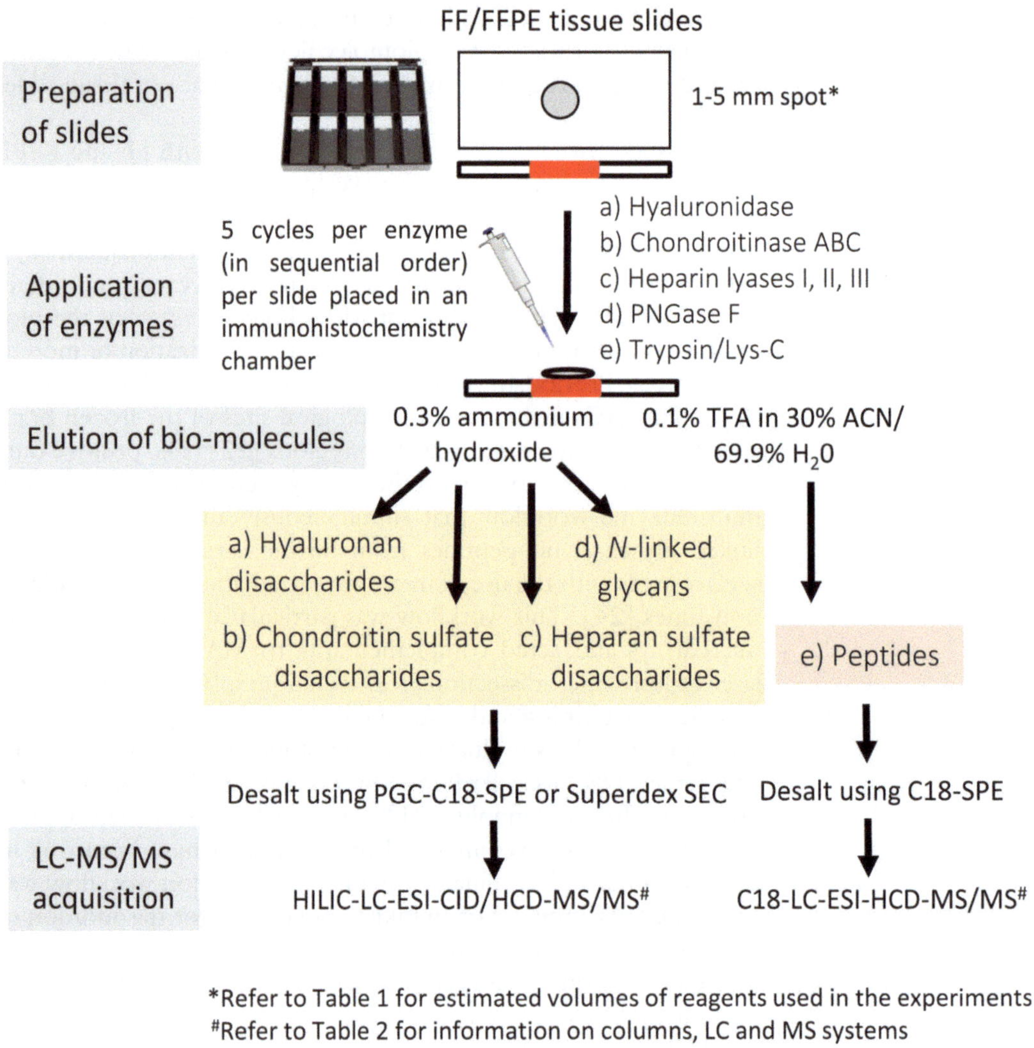

Fig. 2 On-slide tissue digestion workflow for sequential release and extraction of hyaluronan, chondroitin sulfate, and heparan sulfate disaccharides, *N*-linked glycans, and peptides

general, buffer and chemical solutions should be freshly prepared before each experiment, and the use of enzymes with repeated freeze-thaw cycles should be avoided. Other details for consumables (Tables 1 and 2) and equipment used in each individual operating step are mentioned below.

2.1 Tissue Sectioning

1. Kimtech Science Kimwipes tissues.
2. Ultrapure water.
3. 100% pure ethanol, acetone, and xylene (MS grade).
4. Cryostat.
5. Plastic Coplin staining jar.
6. Glass Coplin staining jar.

Table 1
Volume of reagents used to sequentially digest HA, CS, and HS disaccharides, *N*-linked glycans, and peptides from a single spot of on-slide biological tissue

Reagents	Preparation of solutions and storage conditions	Final concentrations used in the steps	Volumes used for 1–5 mm size spot
Ammonium acetate	5 mM in Milli-Q water, store solution for 1 month at room temperature	5 mM	–
Ammonium bicarbonate	25 mM in Milli-Q water, store solution for 1 month at room temperature	25 mM	–
Calcium chloride	5 mM in Milli-Q water, store solution for 1 month at room temperature	5 mM	–
Tris–HCl	20 mM in Milli-Q water, store solution for 3 months at room temperature	20 mM	–
Hyaluronidase enzyme	2 mU/μL in 5 mM ammonium acetate, freshly prepared enzyme solution	0.5 mU/μL	1–5 μL
Chondroitinase ABC enzyme	2 mU/μL in 20 mM Tris–HCl, 5 mM ammonium acetate, freshly prepared enzyme solution	0.5 mU/μL	1–5 μL
Heparin lyases I, II, and III enzymes	20 mU/μL in 5 mM ammonium acetate, freshly prepared enzyme solution	0.3 mU/μL	1–5 μL
Peptide: *N*-glycoside F enzyme	10 U/μL in the manufacturer's solution, store enzyme vial in 4 °C	0.5 U/μL	1–5 μL
Ammonium hydroxide	0.3% in Milli-Q water, store solution for 3 months at room temperature	0.3%	1–5 μL
Dithiothreitol	10 mM in 25 mM ammonium bicarbonate, store aliquoted solution for several months at −20 °C	10 mM	1–5 μL
Iodoacetamide	20 mM in 25 mM ammonium bicarbonate, freshly prepared enzyme solution, kept in the dark until use	20 mM	1–5 μL
Trypsin/Lys-C enzyme	0.1 μg/μL in 25 mM ammonium bicarbonate, store the aliquoted enzyme vial for 3 months at −20 °C	0.1 μg/μL	1–5 μL
ACN/TFA	0.1% TFA in 30% ACN, store solution for 1 month at room temperature	–	1–5 μL

Table 2
Columns and solvents used for desalting and LC-MS/MS acquisition of HA, CS, and HS disaccharides, *N*-linked glycans, and peptides

	Desalting		LC-MS/MS acquisition		
	Columns	Solvents	Columns and LC systems	Solvents	MS
HA, CS, and HS disaccharides	Porous graphitized carbon tips	Elute with 0.1% TFA in 20% ACN and 0.1% TFA in 40% ACN	GlycanPac AXH-1 HPLC column on Agilent 1200 HPLC system	Solvent A: 50 mM ammonium formate in 10% ACN (pH 4.4) Solvent B: 95% ACN	Agilent 6520 Q-TOF MS
N-linked glycans	Porous graphitized carbon tips	Elute with 25 mM ammonium acetate in 5% ACN (pH 4.4)	HILIC Amide-80 HPLC column on Nano Acquity UPLC system	Solvent A: 50 mM ammonium formate in 10% ACN (pH 4.4) Solvent B: 95% ACN	Q Exactive Plus MS
Peptides	Pierce™ C18-SPE spin columns or ZipTip C18-SPE tips	Elute with 0.1% TFA in 60% ACN	Acquity UPLC M-class symmetry C18 trap and analytical HPLC columns on Nano Acquity M-class system	Solvent A: 0.1% aqueous FA Solvent B: 0.1% FA in 99% ACN	Orbitrap Fusion Lumos Tribrid MS with TriVersa NanoMate electrospray ion source

7. Optimum cutting temperature (OCT) compound.
8. Superfrost Plus microscope slides.
9. 4% formaldehyde solution: Prepare 100 mL of 4% formaldehyde solution in Milli-Q water. The solution can be stored for 3 months at room temperature.
10. Paraffin wax.
11. Paraffin wax dispenser.
12. Tissue processor.
13. 10 mM Tris–hydrochloride (Tris–HCl) solution: Prepare 100 mL of 10 mM Tris–HCl solution in Milli-Q water (pH 8.0). The solution can be stored for 3 months at room temperature.
14. Incubators (37 °C and 95 °C).
15. Freezers (−20 °C and − 80 °C).

2.2 On-Slide Enzymatic Digestion and Extraction of Biomolecules

1. ImmPrint permanent marker resistant to acetone/ethanol/xylene.
2. Immunohistochemistry PAP pen.
3. Immunohistochemistry black staining boxes.
4. 5 mM ammonium acetate solution: Prepare 100 mL of 5 mM ammonium acetate solution in Milli-Q water. The solution can be stored for 1 month at room temperature.
5. 25 mM ammonium bicarbonate solution: Prepare 100 mL of 25 mM ammonium bicarbonate solution in Milli-Q water. The solution can be stored for 1 month at room temperature.
6. 0.3% ammonium hydroxide solution: Prepare 100 mL of 0.3% ammonium hydroxide solution in Milli-Q water. The solution can be stored for 3 months at room temperature.
7. 5 mM calcium chloride solution: Prepare 100 mL of 5 mM calcium chloride solution in Milli-Q water. The solution can be stored for 1 month at room temperature.
8. 20 mM Tris–HCl solution: Prepare 100 mL of 20 mM Tris–HCl solution in Milli-Q water (pH 8.0 and 7.5). The solutions can be stored for 3 months at room temperature.
9. 100% pure acetonitrile (ACN) (MS grade).
10. 100% pure trifluoroacetic acid (TFA) (MS grade).
11. 0.1% TFA in 30% ACN solution: Prepare 20 mL of 0.1% TFA in 30% ACN and top up the remaining solution with 69.9% Milli-Q water. The solution can be stored for 1 month at room temperature.
12. Ultrapure water.
13. 2 mU/μL hyaluronidase enzyme (MS grade derived from *Streptomyces hyalurolyticus*) stock solution: Prepare 2 mU/μL hyaluronidase enzyme stock in 5 mM ammonium acetate solution. The enzyme stock solution must be freshly prepared prior to use.
14. 2 mU/μL chondroitinase ABC enzyme (MS grade derived from *Proteus vulgaris*) stock solution: Prepare 2 mU/μL of chondroitinase ABC enzyme stock in 20 mM Tris–HCl (pH 8.0) and 5 mM ammonium acetate solutions. The enzyme stock solution must be freshly prepared prior to use.
15. 20 mU/μL heparin lyases I, II, III enzymes (MS grade from *Bacteroides*) stock solution: Prepare 20 mU/μL of heparin lyases enzyme stock in 20 mM Tris–HCl (pH 7.5), 5 mM ammonium acetate, and 5 mM calcium chloride solutions. The enzyme stock solution must be freshly prepared prior to use.

16. 10 U/μL peptide: *N*-glycosidase F enzyme (PNGaseF) (MS grade derived from *Elizabethkingia miricola*): Use the enzyme directly from the vial and store the vials at 4 °C.
17. Hyaluronan from bovine testes (Type I-S) (positive control).
18. Aggrecan from bovine articular cartilage (positive control for both chondroitin sulfate and peptides).
19. Heparan sulfate sodium salt from bovine kidney (positive control).
20. Alpha-1-acid glycoprotein from human plasma (positive control for both *N*-linked glycans and peptides).
21. 10 mM dithiothreitol (DTT) solution: Prepare 1 mL of 10 mM DTT solution in 25 mM ammonium bicarbonate solution. The solution can be aliquoted and stored at −20 °C for several months.
22. 20 mM iodoacetamide (IAA) solution: Prepare 1 mL of 20 mM IAA solution in 25 mM ammonium bicarbonate solution. The solution must be freshly prepared and kept in the dark prior to use.
23. 0.1 μg/μL trypsin/Lys-C (MS grade): Resuspend lyophilized enzyme in 25 mM ammonium bicarbonate solution and store the aliquoted enzyme vials for 3 months at −20 °C.
24. Safe-lock snap-cap microcentrifuge 1.5-mL tubes.
25. Vortex.
26. Microcentrifuge.
27. Vacuum centrifuge concentrators.
28. Incubators (37 °C and 55 °C).
29. Freezers (−20 °C and − 80 °C).

2.3 Desalting of Biomolecules

1. HA, CS, and HS disaccharides and *N*-linked glycans: Porous graphitized carbon (PGC) resin (particle/pore size of 30 μm/ 250 Å) in solid phase extraction (SPE) cartridges. The cartridges need to be cracked open to get the PGC resin.
2. Combined fraction of HS disaccharides and *N*-linked glycans: Cytiva Superdex 30 Increase Precision 3.2/300 GL Prepacked SEC Column on a System Gold HPLC setup with solvent module and detector.
3. Peptides:
   - Commercial Pierce C18-SPE spin columns (suitable for peptide clean-up of >10 μg starting amount, from ~5 mm spot).
   - Commercial ZipTip C18-SPE tips (suitable for preparing PGC-C18-SPE micro-columns for disaccharide and glycan clean-up of <10 μg starting amount, from ~1 mm spot).

4. 200 μL TipOne pipette tips.
5. 10 μL TipOne pipette tips.
6. Safe-lock snap-cap microcentrifuge 2-mL tubes.
7. Safe-lock snap-cap microcentrifuge 1.5-mL tubes.
8. Ultrapure water.
9. 100% methanol (MS grade).
10. 100% ACN (MS grade).
11. 0.1% aqueous TFA solution: Prepare 20 mL of 0.1% aqueous TFA solution in Milli-Q water. The solution can be stored for 1 month at room temperature.
12. 0.1% TFA in 2% ACN solution: Prepare 20 mL of 0.1% TFA in 2% ACN and top up the remaining solution with 97.9% Milli-Q water. The solution can be stored for 1 month at room temperature.
13. 0.1% TFA in 20% ACN solution: Prepare 20 mL of 0.1% TFA in 20% ACN and top up the remaining solution with 79.9% Milli-Q water. The solution can be stored for 1 month at room temperature.
14. 0.1% TFA in 40% ACN solution: Prepare 20 mL of 0.1% TFA in 40% ACN and top up the remaining solution with 59.9% Milli-Q water. The solution can be stored for 1 month at room temperature.
15. 0.1% TFA in 80% ACN solution: Prepare 20 mL of 0.1% TFA in 80% ACN and top up the remaining solution with 19.9% Milli-Q water. The solution can be stored for 1 month at room temperature.
16. 25 mM ammonium acetate solution: Prepare 100 mL of 25 mM ammonium acetate solution in 5% ACN (pH 4.4). The solution can be stored for 1 month at room temperature.
17. Vortex.
18. Microcentrifuge.
19. Vacuum centrifuge concentrators.
20. Freezers (−20 °C and − 80 °C).

2.4 Liquid Chromatography-Tandem Mass Spectrometry (LC-MS/MS) Data Acquisition

HA, CS, and HS disaccharides:

1. ΔUA-2S® GlcNCOEt-6S (I-P) internal standard disaccharide for samples.
2. Chondroitinase standards for MS.
3. Heparin standards for MS.
4. GlycanPac AXH-1 HPLC column, 175 Å (pore size), 1.9 μm (particle size), 300 μm (internal diameter) × 150 mm (length) (custom-made from Thermo Fisher Scientific).

5. Solvent A: 50 mM ammonium formate in 10% ACN (pH 4.4). Prepare 1 L of 50 mM ammonium formate stock solution in 10% ACN and top up the remaining volume with 90% Milli-Q water. Sonicate the solvent bottle in a water bath for at least 15 min to remove all air bubbles, purge the lines, and perform syringe washes before use. Solvent A must be freshly prepared before a new batch of runs and can be stored for up to 2 weeks at room temperature.
6. Solvent B: 95% ACN. Prepare 1 L of 95% ACN solution and top up the remaining volume with 5% Milli-Q water. Sonicate the solvent bottle in a water bath for at least 15 min to remove all air bubbles, purge the lines, and perform syringe washes before use. Solvent B must be freshly prepared before a new batch of runs and can be stored for up to 2 weeks at room temperature.
7. 1200 Series Gradient LC (Agilent Technologies).
8. 6520 Quadrupole Time-of-Flight (Q-TOF) (Agilent Technologies).

N-linked glycans:

9. AdvanceBio 2-AB labeled *N*-glycan internal standard.
10. HILIC Amide-80 HPLC column, 80 Å, 5 μm, 100 μm × 15 cm.
11. Solvent A: 50 mM ammonium formate in 10% ACN (pH 4.4). Prepare 1 L of 50 mM ammonium formate stock solution in 10% ACN and top up the remaining volume with 90% Milli-Q water. Sonicate the solvent bottle in a water bath for at least 15 min to remove all air bubbles, purge the lines, and perform syringe washes before use. Solvent A must be freshly prepared before a new batch of runs and can be stored for up to 2 weeks at room temperature.
12. Solvent B: 95% ACN. Prepare 1 L of 95% ACN solution and top up the remaining volume with 5% Milli-Q water. Sonicate the solvent bottle in a water bath for at least 15 min to remove all air bubbles, purge the lines, and perform syringe washes before use. Solvent B must be freshly prepared before a new batch of runs and can be stored for up to 2 weeks at room temperature.
13. Nano Acquity UPLC (Waters).
14. Q Exactive Plus MS (Thermo Fisher Scientific).

Peptides:

15. Pierce™ Peptide Retention Time Calibration Mixture.
16. Pierce™ HeLa Protein Digest Standard.
17. Pierce™ LTQ Velos ESI Positive Ion and Negative Ion Calibration Solution.

18. ACQUITY UPLC M-Class Symmetry C18 Trap Column, 100 Å, 5 μm, 180 μm × 20 mm.
19. ACQUITY UPLC M-Class Peptide BEH C18 Column, 130 Å, 1.7 μm, 150 μm × 100 mm.
20. Solvent A: 0.1% aqueous FA. Prepare 1 L of 0.1% FA solution in 99% Milli-Q water and top up the remaining volume with 0.9% ACN solution. Sonicate the solvent bottle in a water bath for at least 15 min to remove all air bubbles, purge the lines, and perform syringe washes before use. Solvent A must be freshly prepared before a new batch of runs and can be stored for up to 2 weeks at room temperature.
21. Solvent B: 0.1% FA in 99% ACN. Prepare 1 L of 0.1% FA solution in 99% ACN and top up the remaining volume with 0.9% Milli-Q water. Sonicate the solvent bottle in a water bath for at least 15 min to remove all air bubbles, purge the lines, and perform syringe washes before use. Solvent B must be freshly prepared before a new batch of runs and can be stored for up to 2 weeks at room temperature.
22. Nano Acquity M-Class (Waters).
23. Orbitrap Fusion Lumos Tribrid MS (Thermo Fisher Scientific) with TriVersa NanoMate electrospray ion source (Advion).
24. Polypropylene MS vials and caps.
25. Vortex.
26. Microcentrifuge.
27. Water bath sonicator.

2.5 Liquid Chromatography-Tandem Mass Spectrometry (LC-MS/MS) Data Analysis

1. MassHunter Qualitative Analysis vB.06 (Agilent Technologies).
2. Xcalibur Qual Browser v4.1.50 (Thermo Fisher Scientific).
3. msConvert v3.0.22041 (ProteoWizard) (https://proteowizard.sourceforge.io/).
4. GlycReSoft v0.4.22 (https://www.bumc.bu.edu/msr/glycresoft/).
5. PEAKS Studio X+ and Xpro (Bioinformatics Solutions Inc).
6. GlycoWorkBench v2.1 (https://glycoworkbench.software.informer.com/2.1/).
7. GAGfinder (https://github.com/jdhogan/gagfinder/).
8. GlycoMod (https://web.expasy.org/glycomod/).
9. GlyToucan (https://glytoucan.org/).
10. GlycoPost (https://glycopost.glycosmos.org/).
11. PRIDE (https://www.proteomexchange.org/).

3 Methods

3.1 Tissue Sectioning

3.1.1 Fresh Frozen (FF) Tissue

1. Section fresh frozen tissue in a cryostat. Sterilize the cryostat with ultra-violet rays for 30 min and then wipe the surfaces with 70% ethanol prior to use. Set the head and body temperature to −22 °C and − 20 °C, respectively. Mount a new blade on the cryotome (*see* **Note 1**).
2. Pre-chill 100% acetone to −20 °C in a plastic Coplin staining jar in the cryostat.
3. Mount the tissue on the chucks using minimal (~5–10 μL) OCT polymers and allow the tissue to harden for 3–5 min (*see* **Note 2**).
4. Use the cryotome to trim the tissue block at 10–20 μm thickness until the surface is even and then prepare desirable sections of 5–40 μm thickness (*see* **Note 3**).
5. Mount the tissue section on the Superfrost Plus charged slides by pressing the slide directly on the tissue (*see* **Notes 4** and **5**).
6. Immerse the slide in the chilled acetone jar for 1 min. Dry the slides for a few min in the cryostat.
7. Repeat **steps 3–6** for more tissue blocks and prepare at least 3–5 serial sections per tissue for technical replicates.
8. Store the slides at −20 °C for short-term storage or at −80 °C for long-term storage.
9. For rehydration of slides before enzymatic digestion, set up six plastic Coplin staining jars at room temperature with ethanol in varied concentrations: 100%, 90%, 70%, 50%, and 30%. Immerse the slides in the jars in decreasing concentration of ethanol for 2 min each, and then, rinse the slides in LC-MS grade water.

3.1.2 Formalin-Fixed Paraffin-Embedded Tissue (FFPE)

1. Fix fresh wet tissue in 4% formaldehyde solution and incubate overnight.
2. Embed the formalin-fixed tissue in the desired orientation in the paraffin block using a tissue processor (*see* **Note 6**). FFPE tissue blocks are stable at room temperature for several months.
3. Before sectioning, store the FFPE tissue blocks at −20 °C for at least 1–2 h (overnight is better).
4. Use a microtome to trim the FFPE tissue block until the surface is even and then prepare serial ribbon sections of 5–10 μm thickness.
5. Transfer the ribbons to water for a few min and mount them on the Superfrost Plus charged slides.

6. Allow the slides to dry and bake them at 37 °C overnight. FFPE slides are stable at room temperature for several months.
7. For deparaffinization and rehydration of FFPE slides, set up eight plastic Coplin staining jars at room temperature, two containing 100% xylene and the remaining with ethanol in varied concentrations: 100%, 90%, 70%, 50%, and 30%. Immerse the FFPE slides in each jar of xylene for 3 min each, followed by decreasing the concentration of ethanol for 2 min each. Rinse the slides in LC-MS grade water (*see* **Note** 7).
8. For antigen retrieval prior to enzymatic digestion, add 40 mL of 10 mM Tris–HCl (pH 8.0) in a glass Coplin staining jar placed in an oven at 95 °C to ensure uniform heating across the jar. Immerse the FFPE slides in the jar for 20 min, and then, rinse them in LC-MS grade water (*see* **Note 8**).

3.2 On-Slide Enzymatic Digestion and Extraction of Biomolecules

1. For both fresh frozen and FFPE tissue sections (*see* **Note 9**), target the region of interest (1–5 mm) for enzymatic digestion and mark on the back of the slide using a permanent marker or a PAP pen (*see* **Notes 10** and **11**).
2. Take a dark immunohistochemistry chamber with sufficient water at the bottom (to create a humidified chamber) and place the slides in a blinded and randomized order. This step ensures that humidity is maintained in the chamber, preventing enzymes from evaporating during the incubation period.
3. Spot well-known standards (such as aggrecan for chondroitin sulfate and alpha-1-acid glycoprotein for peptides) and relevant tissue (such as mouse brain or bovine cortex tissue) with similar thickness and spot size as positive controls for each of the biomolecules and tissue types and process them simultaneously with the rest of the samples to monitor workflow efficiency and reproducibility.
4. Hyaluronan (HA): For HA release and extraction, add 1–5 μL (depending on the target spot size of 1–5 mm; typically, 1 μL per 1 mm spot) of freshly prepared 2 mU/μL hyaluronidase enzyme solution (final concentration of 0.5 mU/μL) in five cycles—1 h incubation for four cycles and overnight incubation for the last cycle at 37 °C. On the next day, using the same volume as the enzyme, extract the HA disaccharides in 0.3% ammonium hydroxide solution, repeating five times (*see* **Notes 12–15**).
5. Chondroitin sulfate (CS) and heparan sulfate (HS): Repeat **step 3** for release and extraction of CS (using freshly prepared 2 mU/μL chondroitinase ABC enzyme solution, final concentration of 0.5 mU/μL) and HS (using freshly prepared 20 mU/μL heparin lyase I, II, III enzyme solution, final concentration of 0.3 mU/μL) disaccharides, respectively.

6. *N*-linked glycans: Repeat **step 3** for release and extraction of *N*-linked glycans using 10 U/μL PNGase F enzyme solution (final concentration of 0.5 U/μL). *Optional:* HS disaccharides and *N*-linked glycans can be simultaneously released and extracted by using both the enzymes (heparin lyase I, II, and III and PNGase F) together in the same step (*see* **Note 16**).
7. Peptides: Prior to peptide release and extraction, perform reduction and alkylation on the spots. For reduction, add 10 mM DTT and incubate at 55 °C for 20 min. For alkylation, add 20 mM IAA and incubate in the dark at room temperature for 20 min. Add 0.1 μg/μL trypsin/Lys-C enzyme solution for release and 0.1% TFA in 30% ACN for extraction of peptides.
8. Spin down the extracted biomolecules. Using vacuum centrifugation, dry all the extracted disaccharides, *N*-linked glycans, and (glyco)peptides prior to the desalting step. Dried glycans and peptides can be stored at −80 °C for several months.

3.3 Desalting of Biomolecules

3.3.1 HA, CS, and HS Disaccharides and N-linked Glycans

1. To prepare PGC-C18-SPE tips for desalting of disaccharides and glycans, first, make a PGC slurry by mixing 25 mg of PGC resin in 500 μL of 50% (*v/v*) methanol (as per the manufacturer's instructions) (*see* **Note 17**).
2. For each sample, deposit ~250 μL of PGC slurry (approximately ~10 mg of PGC resin in each tip as per the manufacturer's instructions) into commercially available or in-house C18-SPE 200 μL tips and place them in 2-mL tubes with microcentrifuge adapters. Spin at 1000 × *g* for 30 s or until no solution is left in the tip to form the PGC-C18-SPE micro-columns (*see* **Note 18**).
3. Equilibrate the PGC-C18-SPE micro-columns by sequentially washing them with
   - 100 μL of 0.1% TFA in 80% ACN, repeating three times
   - 100 μL of 0.1% TFA in 99.9% water, repeating three times

   The PGC-C18-SPE micro-columns are now ready for the desalting step (*see* **Note 19**).
4. Solubilize the dried disaccharides and glycans in 50 μL of water (*see* **Note 20**). Apply the resuspended samples to the equilibrated PGC-C18-SPE micro-columns in fresh 1.5-mL tubes. Spin at 1000 × *g* for 60 s or until no solution is left in the micro-column.
5. Wash the original sample tube with another 50 μL of water and apply them to the same PGC-C18-SPE micro-columns and spin. Re-load the samples (~100 μL) onto the micro-columns, repeating three times (*see* **Note 21**). This is the flow-through fraction (*see* **Note 22**).

6. Wash the PGC-C18-SPE micro-columns in fresh 2-mL tubes with 50 μL of water, repeating five times. Discard the washes.
7. Elute the disaccharides and glycans in fresh 1.5-mL tubes from the PGC-C18-SPE micro-columns by washing them with
   - 50 μL of 0.1% TFA in 20% ACN, repeating two times
   - 50 μL of 0.1% TFA in 40% ACN, repeating two times

   Pool all elution fractions and dry the desalted disaccharides and glycans in a vacuum centrifuge. Dried desalted disaccharides and glycans can be stored at −80 °C for several months until LC-MS/MS analysis.

3.3.2 Combined Fraction of HS Disaccharides and N-linked Glycans (Optional)

1. Using a size exclusion chromatography setup, HS disaccharides and *N*-linked glycans are simultaneously separated and desalted. Use the Superdex 3.2/300 peptide column with 25 mM ammonium acetate in 5% ACN (pH 4.4) as the mobile phase at an isocratic flow rate of 40 μL/min for 60 min at an operating pressure of 5–20 bar (*see* **Notes 23–25**).
2. Solubilize the disaccharide and glycan mixture in 20 μL of water and inject each sample one by one onto the column.
3. HS disaccharides will elute between 35 and 45 min (this is applicable to CS disaccharides also), and *N*-linked glycans will elute between 25 and 35 min, both detected using UV absorbance at 232 nm. Collect the fractions eluting between each time frame and dry them down in a vacuum centrifuge. Dried, desalted disaccharides and glycans can be stored at −80 °C for several months until LC-MS/MS analysis.

3.3.3 Peptides

1. Use commercially available C18-SPE spin columns (for 3–5 mm spots) or C18-SPE tips (for 1–2 mm spots) to desalt peptides. For C18-SPE tips, use smaller volumes in the range of 10–30 μL for equilibration, loading, wash, and elution steps. The volumes indicated below are for C18-SPE spin columns.
2. Place the C18-SPE spin columns in 2-mL tubes with microcentrifuge adapters. Equilibrate the C18-SPE spin columns by sequentially washing them with.
   - 200 μL of 50% ACN
   - 200 μL of 0.1% TFA in 2% ACN, repeating three times

   For every round, spin at 1000× *g* for 30 s or until the tip is devoid of any solution. The C18-SPE spin columns are now ready for the desalting step.
3. Solubilize the dried peptides in 100 μL of 0.1% TFA in 2% ACN. Apply the resuspended samples to the equilibrated C18-SPE spin columns in fresh 1.5-mL tubes. Spin at 1000× *g* for 60 s.

4. Wash the original sample tube with another 50 µL of 0.1% TFA in 2% ACN and apply them to the same C18-SPE spin columns and spin. Re-load the samples (~150 µL) onto the tips, repeating two times. This is the flow-through fraction.
5. Wash the C18-SPE tips in fresh 2-mL tubes with 200 µL of 0.1% TFA in 2% ACN, repeating 4 times. Discard the washes.
6. Elute the peptides from the C18-SPE tips by washing them with 30 µL of 0.1% TFA in 60% ACN, repeating three times. Pool all elution fractions and dry the desalted peptides in a vacuum centrifuge. Dried desalted peptides and glycopeptides can be stored at −80 °C for several months until LC-MS analysis (*see* **Note 26**).

3.4 Liquid Chromatography-Tandem Mass Spectrometry (LC-MS/MS) Data Acquisition

3.4.1 HILIC LC-MS/MS for HA, CS, and HS Disaccharides

1. Solubilize the dried disaccharides in solvent B, vortex, and spin them at 14,000× *g* for 5 min to pellet any remaining debris. Transfer the supernatant to MS sample vials leaving ~1 µL behind in the original tube. Add 800 femtomole of a known synthetic disaccharide to the resuspended HA, CS, and HS disaccharides as an internal standard.
2. MS and MS/MS standard curves for disaccharide data analysis: Run a range of HA, CS, and HS disaccharide standards with 800 femtomole of spiked-in synthetic disaccharide at different concentrations (for example, 500 femtomole, 1 picomole, 2 picomole, 5 picomole, and 10 picomole) and in triplicate using the same method parameters as described below to plot an MS standard curve required for data analysis. Acquire targeted MS/MS at different isomer ratios (for example, 0%, 20%, 40%, 60%, 80%, and 100%) using the same method parameters as described below to plot an MS/MS standard curve required for data analysis.
3. Mount the commercially available LC column onto the Agilent 1200 LC system (Agilent Technologies) coupled to Agilent 6520 Q-TOF (Agilent Technologies) and perform general instrument maintenance and calibration as per the manufacturer's instructions before injecting the samples.
4. For each sample, inject ~30% of the total amount and separate the HA, CS, and HS disaccharides using a 20-min isocratic method (85% solvent B) at a constant flow rate of 5 µL/min for HA and CS and 7 µL/min for HS.
5. Detect the HA, CS, and HS disaccharides using negative ion polarity mode and fixed normalized collision energy of 20 V. Perform targeted MS/MS analyses to differentiate between the disaccharides and their isomers: for example, for CS disaccharides, the *m/z* 458 (z = 1) isomers are *m/z* 282 (D0a6) and *m/z*

300 (D0a4); for HS disaccharides, the *m/z* 458 (z = 1) isomers are *m/z* 237 (D2A0) and *m/z* 357 (D0A6), and the *m/z* 496 (z = 1) isomers are *m/z* 338 (D0S6) and *m/z* 357 (D2S0) [27].

3.4.2 N-linked Glycans

1. Solubilize the *N*-linked glycans in solvent B, vortex, and centrifuge them at 14,000× *g* for 5 min to pellet any remaining debris. Transfer the supernatant to MS sample vials leaving ~1 μL behind in the original tube.
2. *Optional*: Add 400 femtomole of labeled *N*-glycan internal standard to the resuspended *N*-linked glycan samples (*see* **Note 27**).
3. Mount the commercially available HILIC column onto the Nano Acquity UPLC (Waters) coupled to the Q Exactive Plus MS and perform general instrument maintenance and calibration as per the manufacturer's instructions before injecting the samples.
4. For each sample, inject ~1 μg and separate the *N*-linked glycans using a 60-min linear gradient run at a constant flow rate of 0.6 μL/min from 92.5%–50%–20% (*v/v*) solvent B in 35 min.
5. Detect the *N*-linked glycans using negative ion polarity mode with a constant source voltage of 30 eV. Set the MS1 acquisition scan range to *m/z* 450 2000 with a precursor ion isolation window of *m/z* 3. Set the resolving power for MS1 and MS2 at 60,000 and 15,000 respectively with 1 microscan per spectrum for both. Set the automatic gain control and ion accumulation time for MS1 and MS2 at auto. Using data-dependent acquisition, select the 10 most abundant precursors in each MS1 full scan for fragmentation using stepped collision-induced dissociation at a normalized energy of 27 V and 35 V (*see* **Notes 28–30**).

3.4.3 C18 LC-MS/MS for Peptides

1. Solubilize the dried peptides and glycopeptides in solvent A, vortex, and spin them at 14,000× *g* for 5 min to pellet any remaining debris. Transfer the supernatant to MS sample vials leaving ~1 μL behind in the original tube.
2. *Optional:* Add 50 femtomole of peptide retention time calibration mixture to the resuspended peptide samples (*see* **Note 27**).
3. Mount the commercially available trap and analytical C18 columns onto the Nano Acquity M-Class system (Waters) coupled to Orbitrap Fusion Lumos Tribrid MS (Thermo Fisher Scientific) with an Advion NanoMate electrospray ion source and perform general instrument maintenance and calibration as per the manufacturer's instructions before injecting the samples.

4. For each sample, inject ~1 μg peptide mixture on the trap column with 100% solvent A for 4 min at a constant flow rate of 4 μL/min. Then, separate the peptides using a 120-min linear gradient run at a constant flow rate of 0.5 μL/min from 2%–40%–98% (*v/v*) solvent B in 100 min. The column temperature was maintained at 40 °C.
5. Detect the peptides using positive ion polarity mode with a constant source voltage of 1.7 kV, ion transfer tube at 275 °C, and radio frequency lens kept at 30%. Set the MS1 acquisition scan range to *m/z* 350–2000 with a resolving power of 120,000 and precursor ion isolation window of *m/z* 1.6. Set the automatic gain control for MS1 at 3×10^6 ions with a maximum ion accumulation time of 50 ms and a minimum intensity of 5×10^4. Acquire data in profile mode with 1 microscan per spectrum. Set MS2 resolution at 30,000 with automatic gain control of 1×10^5 ions and a maximum ion accumulation time of 60 ms. Using data-dependent acquisition, select the 20 most abundant multi-charged (z = 2–8) precursors in each MS1 full scan for fragmentation using stepped higher-energy collision-induced dissociation (HCD) energy of 27 and 35 V. Enable dynamic exclusion of 10 s after a single isolation and fragmentation of a given precursor ion. Acquire data in profile mode with 2 microscan per spectrum.
6. If electron transfer dissociation (ETD) is available, perform EThCD analyses by triggering tandem MS at a normalized HCD energy of 25 V when at least two diagnostic oxonium ions are observed from the list: *m/z* 138.0545, *m/z* 163.0606, *m/z* 204.08, and *m/z* 366.1396. Keep the rest of the MS settings the same as described above (*see* **Notes 28–30** and **33**).

3.5 Liquid Chromatography-Tandem Mass Spectrometry (LC-MS/MS) Data Analysis

3.5.1 Glycomics Data Analysis

1. Browse the LC-MS/MS raw data files using Agilent Mass Hunter software (for disaccharides) and Xcalibur Qual Browser software (for *N*-linked glycans) and confirm that the data is high quality and consistent across the runs based on (a) narrow LC peak width, (b) retention time, (c) isomeric separation, (d) mass accuracy and resolution, (e) signal-to-noise ratio, and (f) efficient fragmentation with the presence of diagnostic ions.
2. For disaccharide analysis, using a Microsoft Excel spreadsheet, first plot the standard curve for different ratios of disaccharides using triplicate data acquired for varied concentrations and internal standard of all the disaccharide standards. Next, calculate the area under the curve for each peak from the extracted ion chromatogram for each disaccharide. Normalize the data to the spiked internal control and then to the standard curve to obtain an absolute abundance of the disaccharides. Calculate

the relative abundances for each disaccharide per sample. For separated isomers, extract the peaks for the relevant diagnostic fragments as described in Subheading 3.4.1, **step 5**, calculate the area under the curve for each peak, and obtain the absolute and relative abundances.

3. For *N*-linked glycan analysis, search the data using in-house and publicly available GlycReSoft software [28, 29]. Build a combinatorial glycan database with possible glycan compositions using the following rules: HexNAc 2-9, Hex 1-10, Fuc 0-5, NeuAc/NeuGc 0-4, Phos/Sulf 0-2, HexNAc > Fuc, HexNAc - 1 > NeuAc. Depending on the sample type, the monosaccharide compositions can be altered which will result in a different number of total glycan compositions. The search will generate a report containing *m/z* of precursor ions with different charge states, retention time, and area-under-the-curve values for each peak and observed *N*-glycan classes (such as paucimannose, oligomannose, bi-antennary complex, tri-antennary complex, and hybrid). Use the area under the curve to determine the absolute and relative abundances of each *N*-linked glycan in the sample.
4. Use GAGfinder [30], GlycoMod [31], GlyTouCan [32], and GlycoWorkBench [33] tools to aid in the confident identifications of disaccharides and glycans [34, 35] (*see* **Note 34**).
5. Deposit all LC-MS/MS raw data files to a public repository such as GlycoPost [36].

3.5.2 Proteomics Data Analysis

1. Browse the LC-MS/MS raw data files using Xcalibur Qual Browser software and confirm that the data is high quality and consistent across the runs based on (a) total ion chromato gram signal, (b) high fragmentation efficiency, and (c) any diagnostic fragments (*see* **Notes 31** and **32**). Convert the LC-MS/MS raw data files into mzML format using ProteoWizard msConvert [37, 38].
2. Search the proteomics data on PEAKS Studio using PeaksDB and PeaksPTM [39, 40] features against the reviewed and validated UniProt/SwissProt databases (for *Homo sapiens*) and generate a list of (glyco)peptides with a 1% false discovery rate and 2 unique peptides. Set 10 ppm and 0.02 Da error tolerance for precursor (MS1) and product (MS2) ions, respectively. Specify trypsin/Lys-C enzyme (cleavage sites after arginine and lysine when not followed by proline) with 3 missed cleavages at the C-terminal end. Set carbamidomethylation as fixed modification and use PeaksPTM for a larger list of variable modifications including deamidation N, hexNAc ST, hex-hexNAc ST, hydroxylation K, hydroxyprolination P, oxidation M, phosphorylation STY, and ubiquitination K.

3. After generating a final protein list combining PeaksDB and PeaksPTM searches, use PeaksQ to perform label-free quantitation of proteins and peptides normalized to total ion chromatogram signal with mass error tolerance of 10 ppm and retention time shift tolerance of 2.0 min. Use the quantitation values to calculate the absolute and relative abundances of proteins and peptides in the sample.
4. For glycopeptide identification, appropriate modifications can be made to the above search strategy on PEAKS Studio using Glycan Finder [41] and assisted tools such as GlycReSoft [29] (*see* **Note 34**).
5. Deposit all LC-MS/MS raw data files to a public repository such as PRIDE [42].

4 Notes

1. Ensure that the blade mounted on the cryotome for sectioning is rust-free and sharp.
2. Minimal OCT should be added and ensure that it does not touch the tissue regions that will get sectioned as it can lead to contamination of the tissues.
3. Be careful of very thin tissue slices as they tend to peel off easily during the sample workup steps.
4. Tissue slices adhere well to positively charged slides; using other types of slides may lead to peeling off of tissue slices from the slides.
5. Avoid lifting and transferring the tissue slices from the cryotome to the slides with brushes or other tools to avoid tissue contamination.
6. For FFPE tissue blocks, you may consider dipping the block in ammonium hydroxide to avoid cracking of the tissue. However, ensure that the ammonium hydroxide–coated sections are discarded in the trimming process.
7. Deparaffinization with xylene is important to unmask latent epitopes and prepare the samples for enzyme access. The series of ethanol washes removes the xylene completely. Other solvents are available for deparaffinization steps, but a combination of xylene and ethanol solvents works effectively for this protocol.
8. The antigen retrieval step is only necessary for FFPE tissue slides and not needed for fresh frozen tissue slides. This step denatures chemical crosslinking made by formalin to allow for efficient enzymatic digestion. Tris–HCl is the preferred choice of solvent in this protocol as it has been shown to reveal a

higher number of epitopes post antigen retrieval step. In this step, 4 or 5 slides per jar can be processed at the same time. Make sure that the slides cool down before washing them with water to avoid any sudden cracks and perform sufficient washes with water to remove any residual Tris–HCl.

9. To avoid sectioning biases, process all sections for both fresh frozen or FFPE tissues, on the same day in random order to maintain consistency.
10. If using a marker instead of a PAP pen, ensure that it is resistant to ethanol/xylene washes. Take photographs of the slides with the marked region of interest as a reference profile.
11. To ensure that the tissue target chosen for enzymatic digestion is uniform across different slides, use information from a trained pathologist or use parallel H&E stained slides as references.
12. Optimal pH is a critical factor for efficient enzymatic digestion. Each enzyme should be prepared fresh before the experiment and kept on ice until use.
13. When adding the enzymes, make sure that it is retained specifically on the spot of interest and does not flow on other tissue regions to avoid biased results.
14. When adding enzymes or extracting the concerned molecules, make sure that the pipette tip does not touch the tissues.
15. When taking out the immunohistochemistry chamber from the 37 °C incubator, ensure that the condensation is wiped off with Kimwipes tissues so that the enzymes do not get washed off. Also, be careful that the water at the bottom of the chamber does not touch the tissue regions, to avoid any contamination.
16. Investigators interested in the analysis of all the biomolecules from the single tissue spot should follow the release and extraction process in the order of HA, CS, and HS disaccharides, *N*-linked glycans, and peptides. If you want to profile intact *N*-linked glycopeptides along with peptides, do not perform the *N*-linked glycan release and extraction step.
17. Preparation of the PGC slurry and the amount of PGC resin deposition into the tips may slightly differ based on the manufacturer's instructions as well as the starting material of the samples. Also, note the particle/pore size of the resin which may differ based on different vendors.
18. Ensure that all the PGC-C18-SPE micro-columns for a given batch of samples are of similar heights and consistency to avoid biases during desalting steps.

19. Equilibrated PGC-C18-SPE micro-columns can be prepared in advance and stored hydrated at room temperature for a few hours.
20. Resuspension of dried samples is dependent on the starting amount of the samples, but usually maintaining a 1 μg/μL concentration is a good starting point for all types of biomolecules.
21. The repeated re-loading of samples onto the micro-columns or tips is an important step to achieve maximum recovery of all the molecules with minimal loss.
22. Flow-through fractions usually can be discarded but may be retained for troubleshooting purposes in case all the molecules are not detected.
23. Size exclusion chromatography is an alternative to PGC-C18-SPE for desalting both individual and combined disaccharide and glycan samples, especially if your sample has highly sulfated HS disaccharides, which are harder to elute using PGC resin. While the former method will allow for the separation of HS disaccharide and *N*-linked glycans from a combined sample, it is a more time-consuming process as the desalting can be only done one sample at a time.
24. A blank run on the size exclusion chromatography system is important prior to actual samples to ensure that the column is clean and that the eluting fractions are real biomolecules and not just contaminants. CS and/or HS disaccharide standard mixtures (100 picomoles) should be run at the beginning and at the end of the sample queue to ensure that the elution time window is consistent across the entire batch.
25. The parameters for size exclusion chromatography presented here are optimized based on the column and sample types in use. These may differ depending on other column types, samples, flow rates, and run times.
26. For in-depth profiling of intact *N*-glycopeptides, after the desalting step, take a fraction of the desalted peptides and perform HILIC enrichment for glycopeptides prior to LC-MS/MS following the protocol described here [43].
27. Spiking in internal standards in *N*-linked glycans and peptides is not mandatory; however, it does control for run-to-run variability during data acquisition.
28. The LC-MS/MS parameters presented here are guided by the HPLC and MS used in this laboratory, and the settings may differ with other similar types of instruments (Q-TOF or Orbitrap MS) including appropriate tune files with fragile source conditions that will be essential to accurately estimate labile glycans.

29. Prior to LC-MS/MS acquisition, instrument calibration, maintenance and appropriate performance standards for disaccharides, glycans, and peptides should be run including blank washes to ensure stability and reproducibility of the instrument.
30. Sample runs should be randomized and run at least in triplicate to eliminate any sample or instrument bias that may arise from data acquisition.
31. For proteomics data, ensure that the signal for the total ion chromatogram is typically within twofold for the entire cohort in order to accurately estimate differential quantitation between the groups.
32. If there is a high total ion chromatogram intensity but a low number of proteins have been identified, either there is a polyethylene glycol contamination or there is a loss in MS2 sensitivity. For the former, carefully make sure all LC-MS grade solvents have been used for each experimental step, while for the latter, use the performance standard mixtures to tune the MS2 settings.
33. While electron transfer dissociation parameters can be applied for simple glycopeptide detection from peptide mixtures on the LC-MS/MS, more detailed settings can be added to the existing LC-MS/MS methods for detailed identification of glycopeptides as described here [44, 45].
34. Several other glycomics and proteomics data analysis software tools exist, both freely available and licensed. For accurate and confident identification of disaccharides, glycans, and peptides, it is essential to manually cross-check data while performing informed searches on the relevant software with constraints and biosynthetic pathway rules so that substantial false positive results can be eliminated [46–48].

Acknowledgements

The authors acknowledge funding by National Institute of Health (NIH) grant R01AI155975, BrightFocus Foundation Research Fellowship Award A2020687F, and Boston University Clinical Translational Science Institute Pilot Grant Award 1UL1TR001430.

Conflict of Interest The authors declare no conflict of interest.

References

1. Spiro RG (2002) Protein glycosylation: nature, distribution, enzymatic formation, and disease implications of glycopeptide bonds. Glycobiology 12(4):43r–56r
2. Varki A (2017) Biological roles of glycans. Glycobiology 27(1):3–49
3. Varki A, Kornfeld S (2015) Historical background and overview. In: Varki A et al (eds) Essentials of glycobiology. Cold Spring Harbor Laboratory Press Copyright 2015–2017 by The Consortium of Glycobiology Editors, La Jolla, California. All rights reserved, Cold Spring Harbor (NY), pp 1–18
4. Rudd P et al (2015) Glycomics and glycoproteomics. In: Varki A et al (eds) Essentials of glycobiology. Cold Spring Harbor Laboratory Press Copyright 2015–2017 by The Consortium of Glycobiology Editors, La Jolla, California. All rights reserved, Cold Spring Harbor (NY), pp 653–666
5. Iozzo RV, Schaefer L (2015) Proteoglycan form and function: a comprehensive nomenclature of proteoglycans. Matrix Biol 42:11–55
6. Klein JA, Meng L, Zaia J (2018) Deep sequencing of complex proteoglycans: a novel strategy for high coverage and site-specific identification of glycosaminoglycan-linked peptides. Mol Cell Proteomics 17(8): 1578–1590
7. Cummings RD (2009) The repertoire of glycan determinants in the human glycome. Mol BioSyst 5(10):1087–1104
8. Zappe A et al (2022) State-of-the-art glycosaminoglycan characterization. Mass Spectrom Rev 41(6):1040–1071
9. Merry CLR et al (2022) Proteoglycans and sulfated glycosaminoglycans. In: Varki A et al (eds) Essentials of glycobiology. Cold Spring Harbor Laboratory Press Copyright © 2022 The Consortium of Glycobiology Editors, La Jolla, California; published by Cold Spring Harbor Laboratory Press; All rights reserved, Cold Spring Harbor (NY), pp 217–232. https://doi.org/10.1101/glycobiology.4e.17
10. Reily C et al (2019) Glycosylation in health and disease. Nat Rev Nephrol 15(6):346–366
11. de Haan N, Falck D, Wuhrer M (2020) Monitoring of immunoglobulin N- and O-glycosylation in health and disease. Glycobiology 30(4):226–240
12. Grgic A, Krestensen KK, Heeren RMA (2023) Optimized protocol for MALDI MSI of N-glycans using an on-tissue digestion in fresh frozen tissue sections. Sci Rep 13(1): 2776
13. Coscia F et al (2020) A streamlined mass spectrometry-based proteomics workflow for large-scale FFPE tissue analysis. J Pathol 251(1):100–112
14. Casadonte R, Caprioli RM (2011) Proteomic analysis of formalin-fixed paraffin-embedded tissue by MALDI imaging mass spectrometry. Nat Protoc 6(11):1695–1709
15. Cintron-Diaz YL et al (2022) Spatially resolved neuropeptide characterization from neuropathological formalin-fixed, paraffin-embedded tissue sections by a combination of imaging MALDI FT-ICR mass spectrometry Histochemistry and liquid extraction surface analysis-trapped ion mobility spectrometry-tandem mass spectrometry. J Am Soc Mass Spectrom 33(4):681–687
16. Hinneburg H et al (2017) Unlocking cancer Glycomes from histopathological formalin-fixed and paraffin-embedded (FFPE) tissue microdissections. Mol Cell Proteomics 16(4): 524–536
17. O'Rourke MB, Padula MP (2016) Analysis of formalin-fixed, paraffin-embedded (FFPE) tissue via proteomic techniques and misconceptions of antigen retrieval. BioTechniques 60(5):229–238
18. Zhu Y et al (2019) High-throughput proteomic analysis of FFPE tissue samples facilitates tumor stratification. Mol Oncol 13(11): 2305–2328
19. Sethi MK, Downs M, Zaia J (2020) Serial in-solution digestion protocol for mass spectrometry-based glycomics and proteomics analysis. Mol Omics 16(4):364–376
20. Gill VL et al (2013) Disaccharide analysis of Glycosaminoglycans using hydrophilic interaction chromatography and mass spectrometry. Anal Chem 85(2):1138–1145
21. Raghunathan R, Sethi MK, Zaia J (2019) On-slide tissue digestion for mass spectrometry based glycomic and proteomic profiling. MethodsX 6:2329–2347
22. Shao C et al (2013) Mass spectral profiling of glycosaminoglycans from histological tissue surfaces. Anal Chem 85(22):10984–10991
23. Raghunathan R et al (2019) Proteomics, glycomics, and glycoproteomics of matrisome molecules. Mol Cell Proteomics 18(11): 2138–2148
24. Turiak L et al (2014) Workflow for combined proteomics and glycomics profiling from histological tissues. Anal Chem 86(19):9670–9678
25. Raghunathan R et al (2018) Glycomic and proteomic changes in aging brain nigrostriatal

pathway. Mol Cell Proteomics 17(9): 1778–1787
26. Sethi MK et al (2022) In-depth matrisome and glycoproteomic analysis of human brain glioblastoma versus control tissue. Mol Cell Proteomics 21(4):100216
27. Zaia J (2013) Glycosaminoglycan glycomics using mass spectrometry. Mol Cell Proteomics 12(4):885–892
28. Maxwell E et al (2012) GlycReSoft: a software package for automated recognition of glycans from LC/MS data. PLoS One 7(9):e45474
29. Wang M et al (2021) Glycoproteomic sample processing, LC-MS, and data analysis using GlycReSoft. Curr Protoc 1(3):e84
30. Hogan JD et al (2018) Software for peak finding and elemental composition assignment for glycosaminoglycan tandem mass spectra. Mol Cell Proteomics 17(7):1448–1456
31. Cooper CA, Gasteiger E, Packer NH (2001) GlycoMod—a software tool for determining glycosylation compositions from mass spectrometric data. Proteomics 1(2):340–349
32. Fujita A et al (2021) The international glycan repository GlyTouCan version 3.0. Nucleic Acids Res 49(D1):D1529–d1533
33. Ceroni A et al (2008) GlycoWorkbench: a tool for the computer-assisted annotation of mass spectra of glycans. J Proteome Res 7(4): 1650–1659
34. Klein J, Zaia J (2019) Glypy: an open source glycoinformatics library. J Proteome Res 18(9): 3532–3537
35. Klein J, Carvalho L, Zaia J (2018) Application of network smoothing to glycan LC-MS profiling. Bioinformatics 34(20):3511–3518
36. Watanabe Y et al (2021) GlycoPOST realizes FAIR principles for glycomics mass spectrometry data. Nucleic Acids Res 49(D1):D1523–d1528
37. Klein J, Zaia J (2019) Psims – a declarative writer for mzML and mzIdentML for python. Mol Cell Proteomics 18(3):571–575
38. Holman JD, Tabb DL, Mallick P (2014) Employing ProteoWizard to convert raw mass spectrometry data. Curr Protoc Bioinformatics 46:13.24.1–13.24.9
39. Zhang J et al (2012) PEAKS DB: de novo sequencing assisted database search for sensitive and accurate peptide identification. Mol Cell Proteomics 11(4):M111.010587
40. McCafferty J et al (2012) Peak studio: a tool for the visualization and analysis of fragment analysis files. Environ Microbiol Rep 4(5): 556–561
41. Sun W et al (2023) Glycopeptide database search and de novo sequencing with PEAKS GlycanFinder enable highly sensitive glycoproteomics. Nat Commun 14(1):4046
42. Perez-Riverol Y et al (2019) The PRIDE database and related tools and resources in 2019: improving support for quantification data. Nucleic Acids Res 47(D1):D442–d450
43. Chau TH et al (2023) Glycomics-assisted Glycoproteomics enables deep and unbiased N-Glycoproteome profiling of complex biological specimens. Methods Mol Biol 2628:235–263
44. Chau TH et al (2023) Critical considerations in N-glycoproteomics. Curr Opin Chem Biol 73: 102272
45. Hu H et al (2016) A review of methods for interpretation of glycopeptide tandem mass spectral data. Glycoconj J 33(3):285–296
46. Khatri K, Klein JA, Zaia J (2017) Use of an informed search space maximizes confidence of site-specific assignment of glycoprotein glycosylation. Anal Bioanal Chem 409(2):607–618
47. Klein J, Zaia J (2020) Relative retention time estimation improves N-glycopeptide identifications by LC-MS/MS. J Proteome Res 19(5): 2113–2121
48. Klein JA, Zaia J (2020) A perspective on the confident comparison of glycoprotein site-specific glycosylation in sample cohorts. Biochemistry 59(34):3089–3097

Chapter 19

Multiplexed and Multiomic Mass Spectrometry Imaging of MALDI-IHC Photocleavable Mass Tag Tissue Probes, *N*-Glycomics, and the Extracellular Matrisome from FFPE Tissue Sections

Harrison B. Taylor, Jaclyn B. Dunne, Mark J. Lim, Gargey Yagnik, Kenneth J. Rothschild, Anand Mehta, Richard R. Drake, and Peggi M. Angel

Abstract

Recent work in single-cell imaging has allowed unprecedented insight into single-cell interactions that control disease progression. However, approaches to understanding the combined extracellular and cellular microenvironment are limited. In the current protocol, we describe an approach that allows single-cell type imaging using matrix-assisted laser desorption/ionization immunohistochemistry (MALDI-IHC) of UV (ultraviolet) photocleavable mass tags combined with *N*-glycomic and ECM-targeted proteomic imaging from the same formalin-fixed paraffin-embedded tissue section. These approaches use the same imaging mass spectrometry platform to provide a comprehensive view of both the cellular and extracellular tissue microenvironment.

Key words Imaging mass spectrometry, Photocleavable mass tags, MALDI-IHC, Antibody, Single-cell type, Extracellular matrix, *N*-Glycan, *N*-Glycosylation, Fibrosis, Proteomics, Tissue imaging, Collagen peptide imaging, Collagen, Histology

1 Introduction

Current research in spatial biology is revealing the interactive landscape of single-cell types in tissue by transcriptomics and immunodetection of protein markers [1–5]. However, there is a significant gap in understanding how the extracellular microenvironment contributes to cell interaction and response during disease processes. We have previously reported approaches to the multiplexed analysis of the extracellular microenvironment using the standard pathology stain hematoxylin and eosin followed by *N*-glycan and extracellular matrix (ECM)-targeted mass spectrometry imaging (MSI)

Taufika Islam Williams (ed.), *Tissue Proteomics: Methods and Protocols*, Methods in Molecular Biology, vol. 2884, https://doi.org/10.1007/978-1-0716-4298-6_19,

[6, 7]. According to these MSI reports, targeting the ECM, including fibrillar collagens and their post-translational modifications (PTMs), an assortment of 40–60, other ECM proteins are involved in the regulation of collagen fibers. We have further shown that the ECM proteome accessed by collagenase type III has on average 4.3 sites per protein that may be differentially *N*-glycosylated [8]. *N*-Glycosylation is a major PTM of the ECM proteome, influencing protein folding, secretion, and organization. Removing the *N*-glycan structure not only allows spatial multiplexing of the glycome with the ECM-targeted proteome but also improves sequence coverage and detection of the extracellular proteome [6]. Each of these approaches leverages the use of enzymes to provide targeted specificity to the analytes. For *N*-glycans, the enzyme peptide *N*-glycosidase F (PNGase F) is used to cleave all asparagine-linked (*N*-linked) glycans from the amino acid asparagine (N) regardless of their structural complexity within the consensus sequence N-X-S/T where X ≠ P. Imaging of the glycans released by PNGase F results in a broad overview of all *N*-glycan structures within a tissue microenvironment, including significant *N*-glycan structure complexity [9]. The enzyme collagenase type III (*Clostridium histolyticum*) is used to target the ECM from formalin-fixed paraffin-embedded (FFPE) tissue. This enzyme has a preference for the sequences G-P-X and G-X-P, where the proline can also be hydroxylated [10]. Collagenase thus uniquely reports on fibrillar collagen triple helical structures that have high glycine and proline content. Together, PNGase F and collagenase provide significant coverage of the extracellular microenvironment.

Recent work [7] extends this concept of MSI-based spatial reporting of both the cellular and extracellular microenvironment by using novel photocleavable mass tags (PCMTs) linked to antibody probes (Miralys™ antibody probes), in a process termed MALDI-immunohistochemistry (MALDI-IHC)[11, 12]. PCMTs are short (7–15 amino acids) peptide reporters attached to an antibody by a novel UV-cleavable linker. Once released by UV prior to matrix application, the PCMTs remain localized in place. In addition, the photo-released PCMTs have a positively charged amino group on the C-terminus, thereby increasing sensitivity in the positive ion mode [11, 12]. PCMTs are detected using the same instrumentation and types of chemical matrices as *N*-glycan and ECM-targeted workflows. We have previously reported that the MALDI-IHC results depended on whether it was performed before or after MSI of *N*-glycans and ECM peptides [7]. In the current protocol, we describe a highly multiplexed and multimodel workflow, which combines MALDI-IHC with *N*-glycan and ECM-targeted imaging to obtain spatial information on both the cellular and extracellular microenvironments (Fig. 1). The protocol is reported for FFPE tissue sections, allowing access to clinically archived tissue. The multiplexed MSI technique provides a

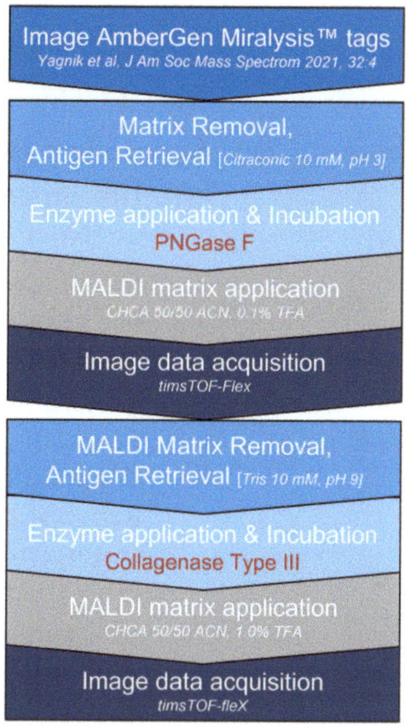

Fig. 1 Complete workflow combining Miralys™ photocleavable mass tag tissue imaging probes with *N*-glycomic and collagen-targeted imaging mass spectrometry

complete view of the tissue microenvironment, reporting on both cellular and extracellular changes toward enhancing basic research, translational research, and therapeutic studies.

2 Materials

All safety and disposal regulations should be followed according to BSL-2 guidelines. Suppliers listed by name are those that have been identified as suitable for the optimization of this method.

1. HPLC-Grade H_2O (*see* **Note 1**)
2. Solutions for Use with MALDI-IHC
   (a) Deparaffinization and Rehydration Solutions
      (i) Xylenes, semiconductor grade/reagent grade (Thermo Fisher).
      (ii) 100% reagent grade 200 proof ethanol.
      (iii) Xylene:ethanol (1:1) solution (*see* **Note 2**).
      (iv) 95% ethanol solution (*see* **Note 3**).
      (v) 70% ethanol solution.
      (vi) 50% ethanol solution.

(vii) 10× Tris-buffered saline (TBS): 500 mM Tris (pH 7.5), 2 M NaCl (*see* **Note 4**).

(viii) 1× TBS buffer (*see* **Note 5**).

(b) Antigen Retrieval Solutions

(i) 10× citrate retrieval buffer, pH 6.0 (Sigma-Aldrich, Ca. No. C9999).

(ii) 1× citrate retrieval buffer (pH 6.0): 1:10 solution of 10× citrate retrieval buffer in HPLC-grade H2O (*see* **Note 6**).

(iii) Concentrated alkaline retrieval buffer: Choose one from 10× EDTA Buffer, pH 8.5, antigen retriever (Sigma-Aldrich, Ca. No. E1161), VisUCyte Antigen Retrieval Reagent-Basic VCTS021 (R&D Systems), or antigen retrieval buffer (100× Tris-EDTA Buffer, pH 9.0) (ABCAM, Ca. No. ab93684).

(iv) 1× alkaline retrieval buffer: dilute in HPLC-grade H_2O based on the concentration of the chosen alkaline retrieval buffer (Subheading 2, **item 2(a)(iii)**) (*see* **Note 6**).

(c) Tissue Blocking and Probe Solutions

(i) Stock normal rabbit and mouse sera (Jackson ImmunoResearch Labs, Inc.) (*see* **Note** 7).

(ii) 25% bovine serum albumin (BSA) in 1× PBS (*see* **Note 8**).

(iii) TBS-OBG: 0.05% octyl ß-D-glucopyranoside (OBG) detergent in 1× TBS (Subheading 2, **item 2 (a)(viii)**) (*see* **Note 9**).

(iv) Tissue blocking buffer: 2% (v/v) normal rabbit serum, 2% (v/v) normal mouse serum, and 5% (v/v) BSA prepared in TBS-OBG (Subheading 2, **item 2(c)(iii)**) (*see* **Note 10**).

(v) Miralys™ Probes (AmberGen) (*see* **Note 11**).

(vi) Working probe mixture: 1–2.5 μg/mL probe in 1× Tissue Blocking Buffer (Subheading 2, **item 2(a) (viii)**) (*see* **Note 12**).

(d) Bulk Washing Solutions

(i) Ammonium bicarbonate (ABC) buffer: 50 mM ammonium bicarbonate in HPLC-grade H_2O (*see* **Note 13**).

(e) Matrix Application Solutions

(i) 0.1% trifluoroacetic acid (TFA).

(ii) Angiotensin II internal standard ([M+H] 2+ = 1046.54): 100 μM ProteoMass™ Angiotensin II MALDI-MS Standard (Sigma-Aldrich, Ca. No. A8846-5X1VL), 0.1% TFA (*see* **Note 14**).

(iii) Ammonium phosphate stock solution: 34 mM ammonium phosphate monobasic in HPLC-grade H_2O.

(iv) Matrix buffer: 70% acetonitrile (ACN), 30% HPLC-grade H_2O, 0.1% TFA, 10 mM ammonium phosphate) (*see* **Note 15**).

(v) α-Cyano-4-hydroxycinnamic acid (CHCA) matrix solution with angiotensin II standard: 10 mg/mL CHCA, 2 nM angiotensin II standard (Subheading 2, **item 2(e)(ii)**) (*see* **Note 16**).

(f) Matrix Recrystallization Solution

(i) 5% LC-MS grade isopropyl alcohol (IPA) (*see* **Note 17**).

3. Solutions for *N*-Linked Glycomic Mass Spectrometry Imaging

(a) PNGase F antigen retrieval solution (pH ~3): 10 mM citraconic anhydrous buffer (≥98%), 0.5 mM HCl.

(b) PNGase F enzyme solution: 0.1 μg/μL PNGase F (*see* **Note 18**).

(c) Ammonium Hydroxide (5% v/v) cleaning solution (*see* **Note 19**).

(d) TFA (1% v/v) cleaning solution (*see* **Note 20**).

(e) PNGase F matrix solvent: 50% ACN, 50% HPLC H2O, 0.1% TFA (*see* **Note 20**).

(f) Methanol (50% v/v) cleaning solution (*see* **Note 20**).

(g) N-Glycan MALDI matrix: 7 mg/mL CHCA (*see* **Note 21**).

4. Solutions for Collagenase Mass Spectrometry Imaging

(a) High-pH tissue clearing solution/collagenase antigen retrieval solution: 10 mM Trizma base (Millipore Sigma) (pH 9) (*see* **Note 22**).

(b) CaCl2 stock: 500 mM CaCl2 (*see* **Note 23**).

(c) Collagenase type III buffer: 10 mM ammonium bicarbonate, 1 mM CaCl2, pH 7.35 (*see* **Note 24**).

(d) Collagenase type III enzyme solution: 0.1 μg/μL collagenase type III in collagenase type III buffer (*see* **Note 25**).

(e) Collagenase type III matrix solvent: 50% ACN, 50% HPLC-grade H2O, 1% TFA (*see* **Note 26**).

(f) Glu-Fib internal standard stock solution: 1 mg/mL Glu1-fibrinopeptide B (Millipore Sigma) (*see* **Note 27**).

(g) Collagen MALDI matrix solution: 7 mg/mL CHCA, 1.67 μg/mL Glu-Fib internal standard (*see* **Note 28**).

(h) Ammonium phosphate (AP) monobasic stock solution: 500 mM AP monobasic (Fisher Scientific) (*see* **Note 29**).

(i) AP dip solution: 5 mM AP stock in HPLC-grade H2O (*see* **Note 30**).

5. Equipment

(a) Mini incubator or lab oven (Fisher Biotech or VWR®, digital mini incubator).

(b) Glass Coplin staining jars (Fisher Scientific).

(c) Polypropylene Coplin staining jars (Fisher Scientific).

(d) Water bath capable of 95 °C (Fisher Scientific).

(e) Hydrophobic (PAP) barrier pen (Sigma-Aldrich).

(f) Humidification chamber/staining tray (Fisher Scientific).

(g) Vacuum desiccator (Bel-Art Space Saver Vacuum Desiccators White PP Bottom; 230 mm diameter: desiccators, Fisher Scientific).

(h) Near-UV light box (AmberGen).

(i) 100 mm × 20-mm glass Petri dish bottoms (Pyrex).

(j) 150-mm plastic Petri dish lids (Fisher Scientific).

(k) 6 × 2-mm disc neodymium magnets (Amazon.com: FINDMAG 120 pcs. 6 × 2 mm).

(l) Standard microcentrifuge.

(m) M5 TM-Sprayer™ (HTX Imaging).

(n) Isocratic pump (BlueShadow Pump 20P, Knauer).

(o) Syringe pump (NE-1000, Pump Systems Inc.).

(p) M5 TM-Sprayer™ (HTX Imaging).

(q) Disposable, Luer-lock 16-ga probe needle (Grainger).

(r) 3-mL Luer-lock disposable syringes (Becton Dickinson).

(s) 5-mL SGE™ gas-tight Luer-lock glass syringe (Millipore Sigma).

(t) 13 mm PTFE hydrophilic HPLC-grade Millex-LG 0.2 μm filters (Millipore Sigma).

(u) timsTOF flex (Bruker Daltonics).

(v) Vegetable steamer (Rival Model CKRVSTLM21).

(w) Decloaking chamber (NxGen Model DC2012).

(x) Tabletop thermomixer (Eppendorf Model 5350).

3 Methods

3.1 Tissue Heating and Dewaxing

The AmberGen Miralys™ workflow is the initial set of methods implemented in the overall protocol for formalin-fixed paraffin-embedded (FFPE) tissues (Fig. 2). First, the paraffin must be removed to expose the protein structure for antibody and enzymatic access. Paraffin is melted by heating at 60 °C for 2 h, which also serves to denature protein structure, exposing epitopes for antibody binding, and helps adhere the tissue to the slide. Paraffin

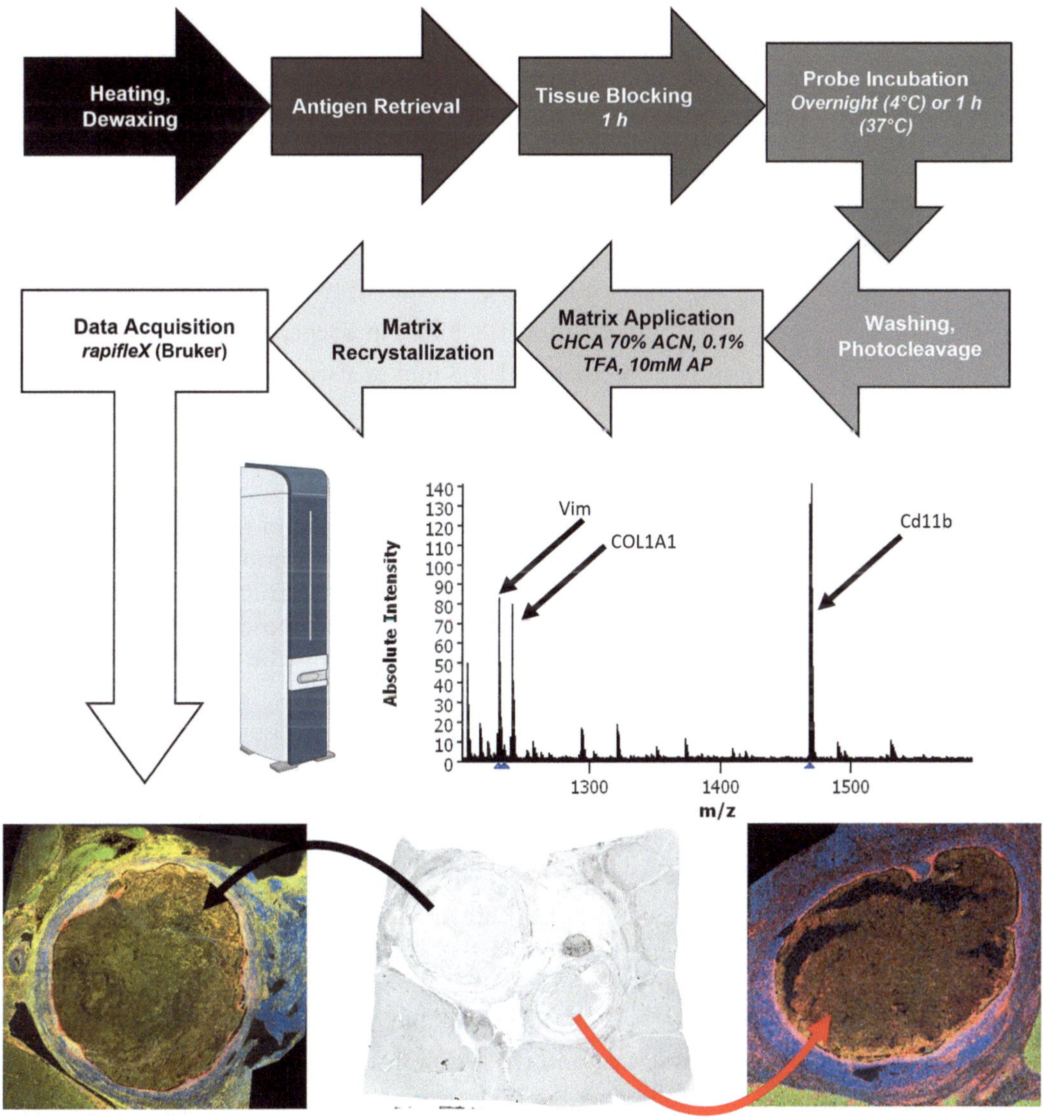

Fig. 2 Workflow outlining the steps for imaging via Miralysis™ photocleavable mass tags. Example workflow steps and results from Miralys™ imaging on hepatocellular carcinoma (human). The image shows differential tumor detection of UV-released tags from vimentin (VIM), collagen α-1 (1) (Col1a1), and CD11b

is then removed through washing steps that include xylenes to solubilize the paraffin. This also serves to remove most of the remaining chemical noise due to lipids and metabolites. Removal of paraffins and chemical noise is required for high-quality detection of PCMTs and analytes produced by enzymatic digestion. The tissues must then be washed in sequentially decreasing concentrations of alcohol for rehydration.

1. To promote tissue adhesion, incubate the slides for 2 h at 60 °C oriented flat and tissue-side up, with the back of the slide in direct contact with the heated surface. Then, dry slides for 5 min in a vacuum desiccator.
2. Deparaffinize the slides in glass Coplin jars with solutions poured to the fill line in the following sequence:
   (a) Xylene, 3 × 5 min in three separate jars.
   (b) Xylene:ethanol 1:1, 1 × 3 min in a separate jar.
3. Re-hydrate the tissue sections in Coplin jars using the following steps:
   (a) 100% ethanol, 2 × 2 min in two separate jars.
   (b) 95% ethanol, 1 × 3 min in a separate jar.
   (c) 70% ethanol, 1 × 3 min in a separate jar.
   (d) 50% ethanol, 1 × 3 min in a separate staining jar.
4. Wash slides in 1× TBS for 10 min in a staining jar.

3.2 Antigen Retrieval (Perform Either Step 1 or Step 2)

Antigen retrieval is used to remove some of the formalin-induced protein cross-links and further expose epitopes to allow for antibody interaction. Enzymes act on specific structures and amino acid compositions. The buffers required to expose these sites are optimized for the target enzymes. Care must be taken to limit tissue disruption or loss of tissue from the slide.

1. For brain tissues and amyloid plaques: In a capped polypropylene staining jar, immerse slides in ~60 mL of 1× citrate retrieval buffer (Subheading 2, **item 2(b)(ii)**) and incubate for 1 h at 95 °C. Allow to cool back to room temperature for 30 min (*see* **Note 31**).
2. For most other tissues: In a capped polypropylene staining jar, immerse slides in ~60 mL of 1× alkaline retrieval buffer (Subheading 2, **item 2(b)(iv)**) and incubate for 30 min at 95 °C. Allow to cool to room temperature for 30 min.
3. After either antigen retrieval step, wash slides once in TBS for 10 min.

3.3 Tissue Blocking and Probe Incubation

After antigen retrieval, treatment of the tissues with a blocking buffer is used to reduce background interference by blocking non-specific antibody-binding sites.

1. Treat the slides by soaking them in a Coplin jar containing ~60 mL tissue blocking buffer (Subheading 2, **item 2(c)(iv)**) for 1 h.
2. Prepare the working probe mixture (Subheading 2, **item 2(c)(vi)**), and then, filter the solution through a filter unit (EMD Millipore, Ca. No. UFC30HV00) (*see* **Note 32**) at room temperate for 1 min at 15,000 revolutions per minute (rpm).
3. Remove the slides from the tissue blocking buffer and prop up the slides at a high angle against the staining jar to allow the liquid to run off via gravity for no longer than 2 min to avoid complete drying of the tissue (*see* **Note 33**). Dry the back of the slide and the areas around the tissue. Avoid direct contact with the tissues on the front of the slides.
4. Circle the area around the tissue sections on the **front** side of the slide without touching/damaging the tissue sections using a hydrophobic peroxidase-antiperoxidase (PAP) barrier pen to retain the antibody solution which is subsequently added directly to the tissue sections as detailed below.
5. Incubate the tissue sections with 100 μL of working probe mixture for each cm^2 of tissue section using *one* of the following two methods:
   (a) Overnight incubation: Incubate the slides overnight in the absence of light at +4 °C in the humidification chamber with enough water in the bottom of the chamber for humidification.
   (b) 1-h incubation: Incubate each slide face up in a clean enclosed, preheated Petri dish with wetted filter paper at 37 °C for 1 h in a preheated oven to encourage humidification (*see* **Note 34**).

3.4 Washing and Photocleavage

Washing allows for the removal of the working probe mixture which contains excess unbound PCMT antibodies and buffer constituents (e.g., salts and detergents), which, if not removed, will substantially interfere with ionization during MALDI-MSI. Following washing and drying of the slides, the PCMTs on the antibodies must be photocleaved. Photocleavage by UV light is necessary to release the PCMTs before matrix application, as well as to allow for efficient PCMT co-crystallization with the subsequently applied matrix and to facilitate desorption/ionization during MSI.

1. Remove the working probe mixture by tilting the slide over a waste container and allowing the mixture to run off. Gently

shake the slide to remove excess fluid but do not fully dry the slide. Special care should be taken if slides contain multiple tissue sections (*see* **Note 35**).

2. Bulk wash the slides for 5 min in 35–40 mL of 1× TBS in a 10-cm diameter Petri dish tissue-side up on an orbital shaker at 40–50 rpm.
3. Remove the slide with clean metal forceps and shake off the excess liquid three times while holding the slide edges with a nitrile-gloved hand. Pour out the wash fluid into the waste container. Repeat **steps 2** and **3** for a total of three times, gently pouring the wash solution into the Petri dish away from the tissue.
4. Pour out the final TBS wash, replace the slide, and rinse in 35–40 mL ABC buffer (Subheading 2, **item 2(d)(i)**) for 10 s. Remove the slide and gently shake off the liquid as indicated in **step 3**, and then, pour off the buffer from the dish and place the slide back in it.
5. Wash in 35 mL of 50 mM ABC buffer for 2 min, and then, remove the slide and shake off excess water as indicated previously. Pour off the buffer from the dish and replace. Repeat a total of three times.
6. After shaking off the bulk liquid for the final time, place the slides in a vacuum desiccator to dry for 1.5 h at room temperature.
7. Slides containing stained tissues should be allowed to equilibrate to room temperature and then can be vacuum sealed in a pouch or capped in a polypropylene slide holder sealed within a second container containing desiccant pouches. These should be placed in the dark at 4 °C. Before use, these tissues must be allowed to warm back to room temperature to avoid condensation.
8. The dried slides may then be illuminated in a 365 nm lightbox (AmberGen, Inc.) for 10 min to allow for photocleavage of the PCMTs from the probes.

3.5 Matrix Application with CHCA

A matrix of CHCA must be applied to the tissues to absorb the MALDI laser energy and facilitate ionization/desorption of the analytes. CHCA has previously been shown to be the optimal matrix for detecting peptides via MALDI-MSI [13]. A standard (angiotensin II) is added to the matrix as a reference point to normalize for topological ionization bias and potential uneven matrix application/recrystallization.

1. Turn on TM-Sprayer™ and then the controlling computer and software.

2. Clean TM-Sprayer™ by running the following solutions through the pump for 5 min each in the following sequence: 5% ammonium hydroxide (Subheading 2, **item 3(iii)**), 1% TFA (Subheading 2, **item 3(iv)**), 0.1% TFA in 50% ACN (Subheading 2, **item 3(v)**), 50% methanol; turn the valve switch to "Load" when injecting the solution from the S.G.E. glass syringe, and then, turn it to "Spray" to allow the solution to flow through the machine. Run cleaning solutions at a flow rate of 150 μL/min.
3. Open the nitrogen gas tank valve, setting the regulator to 5 psi.
4. Set the TM-Sprayer™ temperature to 60 °C.
5. Program the *x-y* coordinates of the TM-Sprayer™ to cover the appropriate area for the number of slides being run, with an additional 5 mm edge to account for sprayer head turnaround.
6. After loading samples on the TM-Sprayer™ stage and fastening them with laboratory tape, fill the glass syringe (attached to the line going to the six-port valve) with prepared CHCA matrix with internal standard (Subheading 2, **item 2(e)(v)**).
7. With the valve switch in the "Load" position, inject the CHCA matrix with angiotensin II internal standard into the 5-mL loop.
8. Increase gas pressure to 10 psi and change the pump flow rate to 100 μL/min. Move the valve switch to the "Spray" position.
9. Program the TM-Sprayer™ method for CHCA to use eight passes, a crisscross pattern, a velocity of 1350 mm/min, a 3 mm track spacing, and a dry time of 10 s. The tip of the spray head should be 40 mm from the surface of the slide.
10. Ensure that the spray temperature has reached the appropriate temperature and start the syringe pump, waiting 1–3 min for the matrix to start spraying consistently.
11. After 1–3 min, press "Start" in the TM-Sprayer™ software, and the MALDI matrix solution will spray in a thin layer onto slides.
12. When finished, matrix-coated slides may be imaged immediately by MALDI-MSI or stored in a vacuum desiccator until ready to image, typically 4–6 h.
13. Turn the temperature back down to 30 °C and clean the sprayer using the same sequence of solutions seen in **step 2**, allowing each solution to flow through for 5 min. Once the sprayer is cleaned, turn the gas off, shut down the software, and turn off the instrument.

3.6 Matrix Recrystallization

Matrix recrystallization serves to increase extraction for improved signal sensitivity.

1. For each slide, place circular filter paper at the bottom of a clean Petri dish and add 1 mL of matrix recrystallization solution (Subheading 2, **item 2(f)(i)**).
2. On the inside of the Petri dish lid, attach the slide inverted (facing down over the filter paper) with the four small magnets. Close the Petri dish.
3. Place the dish in a preheated oven for 1–2 min at 55 °C (*see* **Note 36**). Remove the slides; they are now ready for matrix-assisted laser desorption/ionization-mass spectrometry imaging (MALDI-MSI) analysis.

3.7 Matrix Removal and Antigen Retrieval for PNGase F Application

Once the slides have been imaged after mass tag imaging for antibody localization, the *N*-glycan workflow is followed to map the glycome across the tissue (Fig. 3). The matrix and probes are removed from the slide to regain access to the tissues for enzyme digestion and MALDI imaging. Antigen retrieval is performed at a lower pH for optimal PNGase F access to the asparagine-linked glycosylation sites and release of the *N*-glycans. For *N*-glycomics, temperatures higher than 95 °C can result in the loss of certain structures including sialic acids and fucose. A vegetable steamer is preferred for *N*-glycomics, as limiting heat and pressure typically applied by decloaking devices limits the loss of glycan structural features. The Miralys™ workflow temperatures, which are also at 95 °C, show negligible disruption to the *N*-glycan structure.

1. Wash slides in 100% ethanol for 1 min with gentle agitation to remove the matrix.
2. Fill the vegetable steamer to the marked water level in the bottom chamber of the steamer and preheat for 5 min prior to antigen retrieval.
3. Using a plastic five-slide mailer with a side opening, prefill the mailer with ~ 10 mL of PNGase F antigen retrieval solution (Subheading 2, **item 3(ii)**).
4. Place a maximum of three slides into each five-slide mailer with a side opening. Slides in positions 1 and 5 are placed with tissue facing inward toward the solution, away from the walls. Position 3 may face either way. This allows good solvent access to the tissue.
5. Fill the rest of the slide mailer with the PNGase F antigen retrieval solution so that all tissue is completely covered. Any tissue not covered by the solution will show a variable signal.
6. Allow for the release of steam from the slide mailer during antigen retrieval by closing only one corner of the mailer.

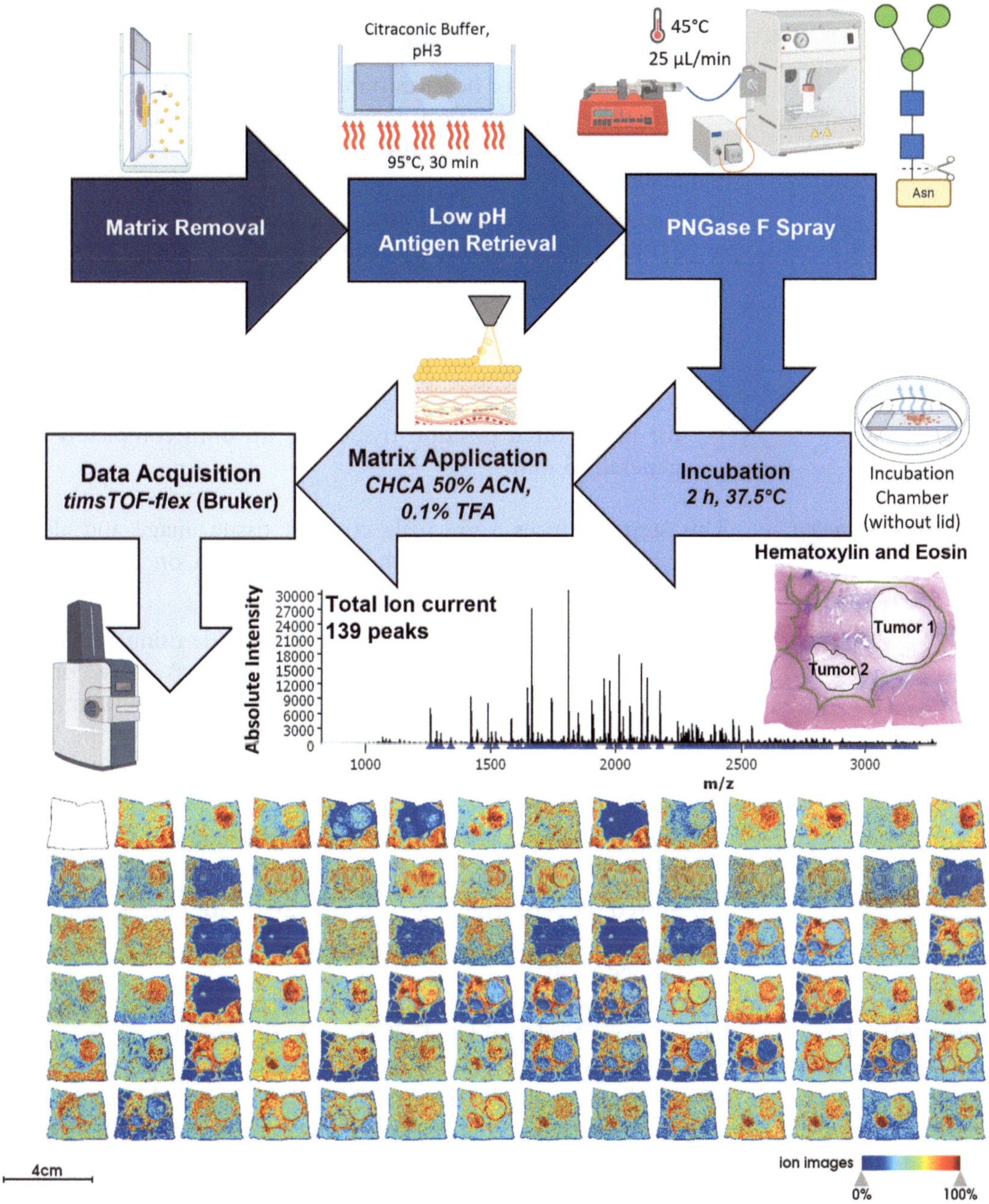

Fig. 3 Workflow displaying the steps to complete *N*-glycan imaging. Example workflow and *N*-glycan results from *N*-glycomic imaging after application of Miralys™ photocleavable mass tags. Total ion current shows the detection of 139 peaks from PNGase digestion on tissue. Images show strong localization to tissue features of tumor, stroma, and hepatocytes

7. Place the mailer in the center of the vegetable steamer and heat for 30 min. The temperature should reach 95 °C for a minimum of 20 min and a maximum of 30 min (*see* **Note 37**).

8. Remove the slide. Bubbles should be observed within the solution, indicating that the solution was near boiling.
9. Using hot pads or heat-resistant gloves, place the mailer with a hot solution in a room-temperature water bath for 5 min. Ensure that the water only comes midway up the side of the mailer.
10. Remove half the buffer from the mailer and replace it with HPLC water. Cool on the countertop for 5 min.
11. Repeat Subheading 3.7, **step 10** for a total of three times.
12. Remove the final volume, refill the mailer with HPLC water, and dump it out.
13. Remove the slides from the mailer. Dry the slides in a desiccator in a slanted position tissue-side up until completely dry, generally 5 min.

3.8 Slide Scanning

This step produces a grayscale contrast tissue image and allows reference points designating the tissue location on the slide for mass spectrometry imaging acquisition.

1. Draw small circles on each corner of the slide using a reflective metallic marker, and then, create a hash mark, cross, or dot pattern using black Sharpie ink within the metallic marked circle. The reflective marker allows visualization of the black mark in the foreground to be used as a reference point for image acquisition.
2. Scan the slide at a minimum of 3200 dpi and save the image as a JPEG.
3. Stored slides overnight at 4 °C or −20 °C for longer term (1–4 wk) storage. It is preferable to proceed with the next step immediately.

3.9 PNGase F Application for N-glycan Imaging

The enzyme *N*-glycosidase F (PNGase F) is sprayed onto the tissue, cleaving between the core *N*-acetylglucosamine and the asparagine within the consensus sequence N-X-S/T, where X can be any amino acid except proline. Highly localized digestion releases *N*-linked glycans from the protein structure within the tissue microenvironment, allowing for their visualization and relative quantification per histopathological feature by mass spectrometry imaging.

1. Turn on the M5 TM-Sprayer™ and open the software on the computer.
2. Clean TM-Sprayer™ line used for enzymes by loading 3 mL of the following solvents sequentially, running and collecting each solvent for 5 min each at 50 μL/min: 10 mM Tris buffer pH 9, 0.1% TFA, HPLC-grade water. Use a single 3-mL disposable syringe for these solutions. Check the pH of each solution being dispensed to ensure a pH change has occurred.

3. The syringe pump is screw-driven, and backpressure due to normal pump degradation or resistance in the lines is impossible to determine without measuring the system outflow. Perform a system suitability check on the syringe pump by catching the flow-through for each cleaning step from the pump in an empty tube (previously weighed out) and calculate the average volume of solvent flow-through for each tube by weighing out the filled tube and subtracting the weight of the empty tube from it. Average the three values and calculate the error from the expected volume (250 μL). Use this volume to recalculate the syringe pump flow rate to achieve the adjusted 25 μL. This allows the correct amount of enzyme deposited onto the tissue and minimizes batch variation.
4. Set the nozzle temperature to 45 °C.
5. Open the nitrogen gas tank valve and adjust the regulator to read 5 psi on the TM-Sprayer gauge.
6. Program the *x-y* coordinates of the TM-Sprayer™ to cover the appropriate area for the number of slides being run, with an additional 5 mm edge to account for sprayer turnaround.
7. After loading samples on the TM-Sprayer™ stage and fastening them with laboratory tape, fill a new 1-mL Luer-lock syringe with prepared PNGase F solution. Ensure there are no bubbles in the syringe before loading the solution into the pump (*see* **Note 38**).
8. Increase the pressure via the regulator to 10 psi and change the pump flow rate to the flow rate calculated in Subheading 3.9, **step 3**. Do not start pumping at this time.
9. Program the TM-Sprayer™ method for PNGase F to use 15 passes, a crisscross pattern, a velocity of 1200 mm/min, a 3.0 mm track spacing, and a dry time of zero. The tip of the spray head should be 40 mm from the surface of the slide.
10. Start the syringe pump, waiting 1–3 min for PNGase F to start spraying. Evaluate the spray by holding a blank slide under the nozzle. The spray area should show a circular pattern due to spraying without observation of sputtering.
11. Press "Start" in the TM-Sprayer™ software.
12. While waiting for the spraying protocol to complete, continue with incubation preparation for on-tissue digestion below.

3.10 PNGase F Incubation for On-Tissue Digestion

Incubation of the PNGase F-sprayed tissues at 37.5 °C will allow for the enzyme to thoroughly react within the tissues, cleaving *N*-glycans from asparagine residues within the tissues.

1. Prior to spraying, prepare one incubation chamber per slide for on-tissue digestion using a plastic 100 × 15-mm cell culture

dish. Place a single layer of paper towel (Wypall 60×) cut to fit the bottom of the chamber. Roll two 4 × 6-inch Kimwipes into short, stacked cylinders and place them on either side of the dish. Saturate the Kimwipes and the bottom of the dish with water (approximately 5 mL of water for a 4 × 6-inch Kimwipe)

2. Preheat the incubation chambers at 37.5 °C for 15–30 min (approximately the duration of the spraying protocol). A thin layer of condensation should be visible on the lid of the culture dish prior to loading in the sample. If humidity is not high enough to detect condensate on the lid, samples will not digest uniformly. A small hygrometer purchased from a pet store may be placed in the chamber when determining the humidity. Humidity should be ≥90%.
3. Once the TM-Sprayer™ protocol is complete, remove slides and place them tissue-side up in their own incubation chamber. Ensure the slide is pushed down on the saturated Kimwipes enough so that the lid of the incubation chamber does not come into contact with the tissue.
4. For PNGase F digestions, incubate tissues for 2 h at 38 °C. It is preferable to not stack the incubation chambers on top of one another but to have them all equally in contact with the surface for uniform heating.
5. After incubation, carefully remove the slide and wipe excess condensation off the back of the slide with a dry Kimwipe. Ensure no excess moisture drops onto tissue; this will delocalize analytes produced by digestion.
6. The slides are ready to proceed to the MALDI matrix application. If unable to spray the MALDI matrix immediately, let them dry in a desiccator for 1–3 min to prevent condensate from delocalizing analytes and then store at −20 °C. When removing after storage, move from freezer to desiccator and dry under vacuum. *N*-Glycans are extremely hydrophilic and may delocalize due to condensation from temperature changes.

3.11 MALDI Matrix Application for N-Glycan Imaging

The principles of application of the matrix are outlined above (Subheading 3.5).

1. Turn on the M5 TM-Sprayer™ and the controlling computer and software.
2. Clean the TM-Sprayer™ by running the following solutions through the pump for 5 min each in the following sequence: 5% ammonium hydroxide, 1% TFA, 0.1% TFA in 50% ACN, 50% methanol; turn the valve switch to "Load" when injecting the solution from the S.G.E. glass syringe, and then, turn it to "Spray" to allow the solution to flow through the machine. Run cleaning solutions at a flow rate of 150 μL/min.

3. Open the nitrogen gas (ultra-high purity) tank valve, setting the regulator to 5 psi.
4. Set the TM-Sprayer™ temperature to 79 °C.
5. Program the *x-y* coordinates of the TM-Sprayer™ to cover the appropriate area for the number of slides being run, with an additional 5 mm edge to account for sprayer head turnaround.
6. After loading samples on the TM-Sprayer™ stage and fastening them with laboratory tape, fill the glass syringe (attached to a line going to the six-port valve) with prepared MALDI matrix solution using the CHCA matrix for *N*-glycan imaging.
7. With the valve switch in the "Load" position, inject the MALDI matrix solution into the 5-mL loop.
8. Increase gas pressure to 10 psi and change the pump flow rate to 70 μL/min. Move the valve switch to the "Spray" position.
9. Program the TM-Sprayer™ method for CHCA to use 14 passes, a crisscross pattern, a velocity of 1300 mm/min, a 2.5 mm track spacing, and a dry time of zero. The tip of the spray head should be 40 mm from the surface of the slide.
10. Ensure that the spray temperature has reached the appropriate temperature and start the syringe pump, waiting 1–3 min for the matrix to start spraying consistently. Evaluate that the matrix is spraying by holding a blank glass slide under the nozzle. A white residue should be observed from the spray, and no sputtering should be observed. Sputtering indicates crystal buildup within the nozzle tip and may require additional cleaning.
11. After 1–3 min, press "Start" in the TM-Sprayer™ software, and the MALDI matrix solution will spray in a thin layer onto the slides.
12. When finished, matrix-coated slides may be imaged immediately or stored in a vacuum desiccator until ready to image. Once the matrix is added, tissues will maintain integrity in a vacuum.

3.12 Tissue Clearing of Matrix Between Enzymes

The matrix must be removed from the tissues to regain access for collagenase imaging preparation (Fig. 4). Additionally, *N*-glycans and the enzyme PNGase F must be removed prior to collagenase type III digestion as the strong *N*-glycan signal will limit the detection of collagenase type III peptides. The enzyme PNGase F must be removed to limit ion suppression.

1. Soak the slides in 100% ethanol for 1 min to remove the matrix from the slide.
2. Immerse the slide in 95% ethanol for 1 min, followed by 1 min in HPLC-grade H_2O.

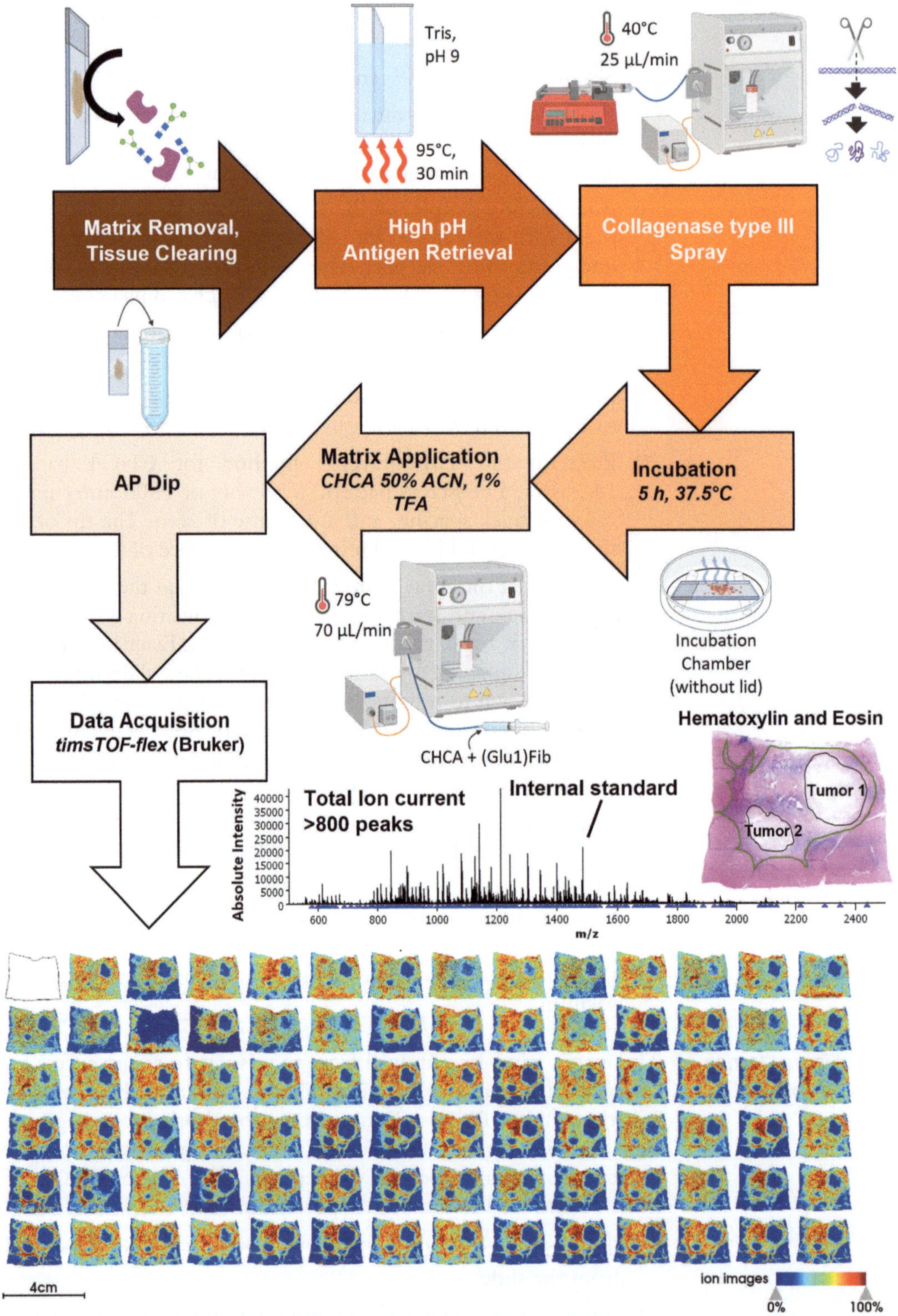

Fig. 4 Workflow displaying the steps involved in collagen imaging. Example workflow from collagenase imaging after application of Miralys™ photocleavable mass tags and *N*-glycan results. Total ion current shows the detection of >800 peaks from collagenase digestion on tissue. An internal standard (Glu1-

3. Immerse the slide in 70% ethanol for 1 min.
4. Remove cleaved *N*-glycans by incubating the slide in HPLC-grade H_2O for 1 min, high-pH tissue clearing solution for 1 min, HPLC-grade H_2O for 1 min, and low-pH tissue clearing solution for 1 min. At the end of each step, gently agitate the slides three to five times, and then, let them dry in a desiccator for 5 min.

3.13 Antigen Retrieval for Collagenase Type III Application

Similar to Subheading 3.7, incubating the tissues in a high pH buffer (pH 9) at temperatures close to boiling further releases formalin bonds and exposes optimal sites for collagenase type III activity. Antigen retrieval for collagenase enzyme activity may be done in either a decloaker or a vegetable steamer (*see* **Note 39**). This protocol discusses the use of the vegetable steamer.

1. Fill vegetable steamer or decloaker to marked water level and preheat prior to antigen retrieval (*see* **Note 40**).
2. Add approximately 10 mL collagenase antigen retrieval solution (Subheading 2, **item 2(d)(i)**) to a side-opening plastic slide mailer (*see* **Note 41**).
3. Place a maximum of three slides into each five-slide mailer, one slot apart from each other. Make sure tissues at the end slots are facing away from the walls. The middle position may face either way. This positioning allows good solvent access to the tissue.
4. Fill the rest of the mailer with a collagenase antigen retrieval solution so that all the tissues are completely covered.
5. Place the mailer in the center of the vegetable steamer and heat for 30 min with one end of the lid capped. The temperature should reach 95 °C for a minimum of 20 min (*see* **Note 37**).
6. Place the hot mailer in a container of cool or room-temperature water using hot pads or heat-resistant gloves. The water should come midway up the side of the mailer and should not get inside with the buffer. Let the mailer cool for 5 min.
7. Remove half the buffer from the mailer and replace it with HPLC-grade water. Cool on the countertop for 5 min.
8. Repeat Subheading 3.13, **step 7** for a total of three times.
9. Complete by filling the mailer with HPLC-grade water and dumping it out. Let the slides dry slanted, tissue-side up in a desiccator for at least 5 min or until dry.

Enzymes allow targeting of specific components of the tissue in imaging mass spectrometry. The spraying of collagenase onto the

Fig. 4 (continued) fibrinopeptide B peptide) is detected at 100 fmol/μL. Images show strong localization to stroma features surrounding the tumors

tissues at the appropriate temperature will allow for the targeted breakdown of fibrillar and other collagen types into peptides that can then be imaged by mass spectrometry imaging.

3.14 Collagenase Type III Application for Collagen Imaging

1. Turn on the M5 TM-Sprayer™ and then the controlling computer and software.
2. Clean the M5 TM-Sprayer™ line used for enzymes by loading 1.5 mL of the following solvents sequentially and running for 5 min each at 50 μL/min: Tris buffer (Subheading 2, **item 4 (i)**), 1% TFA (Subheading 2, **item 3(iv)**), and collagenase type III buffer (Subheading 2, **item 4(iii)**). Use a single 3-mL disposable syringe for these solutions. Check the pH of each solution being dispensed to ensure a pH change has occurred (*see* **Note 42**).
3. Calculate the adjusted flow rate by performing a system suitability check, indicated in Subheading 3.9, **step 3**.
4. Open the nitrogen gas tank valve, set the regulator to 5 psi, and set the temperature to 40 °C.
5. Program the *x-y* coordinates of the TM-Sprayer™ to cover the appropriate area for the number of slides being run, with an additional 5 mm edge to account for sprayer head turnaround.
6. After loading samples on the TM-Sprayer™ stage and fastening them with laboratory tape, fill a new 1-mL Luer-lock syringe with prepared collagenase type III solution (Subheading 2, **item 4(iv)**), ensuring that there are no bubbles in the syringe (*see* **Note 38**).
7. Change the pump flow rate to the adjusted flow rate calculated in Subheading 3.14, **step 3**. Do not start pumping at this time.
8. Program the TM-Sprayer™ method for collagenase III to use 15 passes, a crisscross pattern, a velocity of 1200 mm/min, a 3.0 mm track spacing, and a dry time of zero. The tip of the spray head should be 40 mm from the slide's surface.
9. Start the syringe pump, turn the gas up to 10 psi, and wait 1–3 min for collagenase type III to start spraying continuously.
10. After 1–3 min, press "Start" in the TM-Sprayer™ software.
11. While waiting for spraying to complete, continue with incubation preparation for on-tissue digestion below.

3.15 Collagenase Type III Incubation for On-Tissue Digestion

As with Subheading 3.10, incubation of the tissues at high humidity allows for localized digestion of the tissues by collagenase type III. This will maximize collagen peptide release for detection by mass spectrometry imaging.

1. Prepare one incubation chamber per slide for on-tissue digestion using a plastic 100 × 15-mm cell culture dish. Place a single layer of paper towel (Wypall 60×) cut to fit the bottom

of the chamber. Form two rolled cylinders using 4 × 6-inch Kimwipes and place them on each of the two sides of the dish. Saturate the Kimwipes and the bottom of the dish with water (approximately 5 mL of water for a 4 × 6-inch Kimwipe).

2. Preheat the incubation chambers at 38 °C for 20–30 min (approximately the duration of the spraying protocol). A thin layer of condensation should be visible on the lid of the culture dish prior to loading in the sample. If humidity is not high enough to detect condensate, samples will not digest uniformly.
3. Once the TM-Sprayer™ protocol is complete, remove slides and place them tissue-side up in their own incubation chamber. Ensure the slide is pushed down on the saturated Kimwipes so that the lid of the incubation chamber does not come into contact with the tissue.
4. For collagenase type III, incubate tissues for 5 h at 38 °C, relative humidity ≥90%.
5. After incubation, carefully remove the slide and wipe excess condensation off the slide backing. Ensure no excess moisture drops onto tissue; this will delocalize the analytes cleaved during digestion.
6. Proceed to MALDI matrix application or let dry in a desiccator for 1–3 min and then store at −20 °C until matrix application. Matrix application should be performed no longer than 3–5 days after application.

3.16 MALDI Matrix Application for Collagen Imaging

Similar to Subheadings 3.5 and 3.11, a matrix of CHCA must be applied. This matrix has a higher concentration of trifluoracetic acid (1%), which we have found improves peptide ionization. A Glu-Fib peptide (*see* Subheading 2, **item 4(vi)** and **(Vii)**) is added to the matrix as an internal standard and for use as a system suitability measure of the instrument method.

1. Turn on the M5 TM-Sprayer™ and then the controlling computer and software.
2. Clean the M5 TM-Sprayer™ by running the following solutions through the pump for 5 min each in the following sequence: 5% ammonium hydroxide, 1% TFA (TFA), 0.1% TFA in 50% ACN, 50% methanol; turn the valve switch to "Load" when injecting the solution from the S.G.E. glass syringe, and turn it to "Spray" to allow the solution to flow through the machine. Run cleaning solutions at a flow rate of 150 μL/min.
3. Open the nitrogen gas tank valve, setting the regulator to 5 psi and the temperature to 79 °C.

4. Program the *x*-*y* coordinates of the TM-Sprayer™ to cover the appropriate area for the number of slides being run, with an additional 5 mm edge to account for sprayer head turnaround.
5. Load the slide samples on the TM-Sprayer™ stage and fasten them with laboratory tape. Prepare the collagen MALDI matrix solution (Subheading 2, **item 4(vii)**) and fill the glass syringe (attached to a line going to the six-port valve), ensuring the absence of air bubbles in the syringe.
6. With the valve switch in the "Load" position, inject the MALDI matrix solution into the 5-mL loop.
7. Change the pump flow rate to 70 μL/min. Move the valve switch to the "Spray" position.
8. Program the TM-Sprayer™ method for CHCA to use 14 passes, a crisscross pattern, a velocity of 1300 mm/min, a 2.5 mm track spacing, and a dry time of zero. The tip of the spray head should be 40 mm from the surface of the slide.
9. Start the syringe pump, turn up the nitrogen gas to 10 psi, and wait 1–3 min for the matrix to start spraying continuously. Evaluate the spray by holding a clean glass slide under the nozzle. The spray should form an opaque circle with no observation of sputtering.
10. Ensure that the spray temperature has reached 79 °C and press "Start" in the TM-Sprayer™ software.
11. MALDI matrix solution will be sprayed in a thin layer onto the target tissues.

3.17 Ammonium Phosphate (AP), Monobasic Dip

Dipping a matrix-coated slide in cold ammonium phosphate solution allows for the reduction of matrix clusters and aids in boosting peptide signals.

1. Perform an AP dip of each slide with the solutions and steps indicated above. Solutions should be used at 4 °C for reproducible effects and minimal loss of tissue or matrix.
2. Add the solution to a clean Coplin jar or clean upright slide mailer.
3. Rapidly immerse the matrix-coated slides into the AP solution (Subheading 2, **item 4(ix)**), removing completely within 1 s.
4. Dry the slides vertically in a desiccator, ensuring that the AP solution does not pool on the tissue. Pooling will differentially change the peptide signal, appearing as an artifact in the image data.
5. Store slides under vacuum in a desiccator prior to MSI acquisition for up to 5 days. Achieve long-term storage by vacuum sealing the slides and storing them in a cool dark place for a month without loss of signal.

4 Notes

1. HPLC-grade H_2O should be used for solution preparation for all uses within this protocol.
2. Mix 500 mL of xylenes with 500 mL 100% ethanol to create the 1:1 xylene:ethanol solution, and then, pour to the fill line of a Coplin jar.
3. Upon preparation of ethanol solutions (95%, 75%, and 50%), gently shake the solution container to mix.
4. The 10× TBS solution should be stored at 4 °C when not in use.
5. Prepare a 500 mL solution of 1× TBS by adding 50 mL 10× TBS (Subheading 2, **item 2(a)(vii)**) to 450 mL HPLC-grade H_2O. Store at 4 °C for up to 1 week.
6. Prepare only enough 1× citrate retrieval buffer (Subheading 2, **item 2(b)(ii)**) and/or 1× alkaline retrieval buffer (Subheading 2, **item 2(b)(iv)**) for the number of slides being prepared. Store at 4 °C for up to 1 week.
7. Reconstitute each serum in HPLC-grade H_2O according to the manufacturer's instructions.
8. Mix gently and ensure complete dissolution of BSA in PBS, as it may be slow to dissolve. We typically prepare 100 mL at a time. Store at 4 °C.
9. Prepare 100 mL by adding 100 μL of 50% (w/v) OBG to 100 mL 1× TBS and mixing gently. Store for up to 1 week at 4 °C.
10. For 100 mL of blocking buffer, add 2 mL normal rabbit serum, 2 mL normal mouse serum, and 20 mL of stock (25%) BSA to 76 mL TBS-OBG. Mix gently. Prepare fresh before use, as activity decreases over time. The volume needed should be adjusted based on the number of slides used.
11. Miralys™ antibody probes are labeled with PCMTs (photocleavable mass tags) and in some cases fluorophores. Store in the dark at −20 °C but no colder. Dilute based on lot-specific concentrations in each stock. Probes do not need to be aliquoted prior to storage, as each probe species is shipped as a separate stock solution.
12. Between 1 and 2.5 μg/mL of each probe should be used as a recommended starting concentration, but the optimum concentration can be up to 10 μg/mL. Prepare fresh and do not store. Concentrations of each probe should be prepared based on tissue type and thickness, the matrix used, the antibody, and other factors. Multiple probes can be combined into one working mixture.

13. We recommend ammonium bicarbonate from Sigma-Aldrich (Ca. No. 5330050050). Store this solution for up to 1 week at 4°C.
14. Add 100 μL 0.1% to the 10 nmol vial of angiotensin II internal standard. Prepare 5 μL aliquots of 100 μM and store at −20 °C until use.
15. Prepare matrix buffer using ammonium phosphate stock (Subheading 2, **item 2(e)(iii)**). Prepare fresh and do not store. We recommend a volume preparation of 10 mL.
16. Prepare by adding 100 mg CHCA matrix (Sigma-Aldrich, Ca. No. 70990-1G-F) in 10 mL of matrix buffer (Subheading 2, **item 2(e)(iv)**), mix gently, and then, filter using specified 13 mm 0.2 μm PTFE hydrophilic HPLC-grade syringe filter (Millipore Sigma) to remove larger particles of undissolved CHCA that could potentially clog the sprayer. These filters have been verified in our lab to not leech polymers that could override the mass spectrometry signal. Add 495 μL HPLC H_2O to an angiotensin 5 μL stock vial for a 1:100 dilution and mix thoroughly, and then, add 20 μL to the filtered 10 mg/mL CHCA matrix solution.
17. Each recrystallization step requires 1 mL. Prepare fresh in HPLC-grade H_2O. Prepare based on the amount needed given the number of slides per reaction.
18. Add 1 mL of HPLC-grade H_2O to 100 μg of PNGase F and tap gently to mix. Do not vortex. This is enough volume to spray approximately four slides. Prepare fresh, as activity decreases over time.
19. We recommend preparing 1 L of 5% ammonium hydroxide (Subheading 2, **item 3(iii)**) for cleaning. Store at room temperature for 2–3 months.
20. Prepare under a fume hood. We recommended preparing 1 L. Store at room temperature for 2–3 months.
21. Prepare by adding 42 mg CHCA matrix (Sigma-Aldrich, Ca. No. 70990-1G-F) in 6 mL of PNGase F matrix solvent (Subheading 2, **item 3(vii)**). Vortex briefly, then sonicate for 5 min. Filter using a specified 13 mm 0.2 μm PTFE hydrophilic HPLC-grade syringe filter (Millipore Sigma) to remove larger particles of undissolved CHCA that could potentially clog the sprayer. These filters have been verified in our lab to not leech polymers that could override the mass spectrometry signal.
22. To a clean glass bottle containing 500 mL HPLC-grade H_2O, mix in 1.21 + 0.12 g Trizma base (Millipore Sigma). Shake to dissolve, and then, fill to 900 mL with HPLC-grade H_2O. Allow to mix overnight for complete dissolution, and then, adjust the pH to 9.0 + 0.03 using a calibrated pH meter

to ensure accuracy. Adjust volume to 1000 mL with HPLC-grade H_2O. Store at room temperature for 1–2 months.

23. We recommend preparing a volume of 50 mL by dissolving 2.775 g of $CaCl_2$ in 25 mL HPLC-grade H_2O and vortexing to mix. Fill to 50 mL with HPLC-grade H_2O and Store at room temperature.
24. To a 50-mL tube, add 39.5 mg ABC to 40 mL HPLC-grade H_2O and mix to dissolve. Then, add 100 μL of 500 mL $CaCl_2$ stock (Subheading 2, step 2(d)(ii)) to enhance enzyme activity. Fill to 50 mL with HPLC-grade H_2O and adjust the pH to 7.35 + 0.05. Store at 4 °C for up to 2 weeks.
25. Prepare collagenase type III enzyme solution immediately before use, as activity decreases rapidly over time. A volume of 1 mL is enough to cover three to four slides, depending on tissue size.
26. Prepare within a fume hood. Shake gently to mix the solution, and then, store at room temperature for 2–3 months.
27. Add 1 mL of PNGase F Matrix solvent (Subheading 2, **item 3 (v)**) directly to 1 mg vial of Glu1-fibrinopeptide B, and then, prepare 10 μL aliquots. Store at 4 °C for 4–5 months.
28. Prepare using the same methods outlined in **Note 21**. Add a 10 μL aliquot of Glu-Fib internal standard to the solution by transferring ~1 mL of the filtered matrix to the aliquot, pipetting up and down to mix, and then adding the entire mixture back to the rest of the filtered matrix and mix. Prepare immediately before use.
29. We recommend preparing a volume of 50 mL in HPLC-grade H_2O. Store at room temperature.
30. Add 500 μL stock AP solution (Subheading 2, **item 4(viii)**) to 49.5 mL cold (4 °C) HPLC-grade H_2O in a 50-mL conical tube. Invert three to five times to mix. Each AP dip solution accommodates two slides, so prepare as many solutions as needed.
31. As higher pH antigen retrievals may damage and distort certain tissue morphologies, certain tissues respond better to low pH antigen retrievals.
32. We recommend this particular filter unit, as it has proven to optimize results with our method.
33. Complete drying of the tissue can interfere with both the blocking process as well as probe attachment.
34. Water and/or wetted paper are not needed within the Petri dish for humidification as in the case of overnight incubation, which is humidified only from the working probe mixture on the slide. Additionally, the 1 h method typically proves higher

signal-to-background than overnight incubations. Antigens that are more difficult to detect may need overnight incubation, however.

35. If there are different tissue sections on the same slide, each with different probing conditions, each solution within a PAP pen-marked area needs to be separately vacuum aspirated from each tissue to prevent cross-contamination of the probe mixtures.
36. The time duration required for optimal recrystallization may be dependent on the type of tissue, MALDI-MS instrument, or spatial resolution required.
37. Certain tissues with high fat content or those with open areas such as normal skin, breast, and lung may detach during lengthy antigen retrieval. To limit the loss of tissue from the slide, reduce antigen retrieval time to 20 min.
38. When doing separate workflows, ensure that the same syringe is not being used for PNGase F and collagenase type III loading to prevent cross-contamination of enzymes.
39. If a lab-grade decloaker is not available, we have found that a vegetable steamer is sufficient for antigen retrieval.
40. If using the decloaker, the bottom should be filled with ~500 mL of HPLC-grade H_2O, and each of the three smaller chambers should be filled with 200–250 mL of water prior to preheating. The use of top-loading slide mailers in decloakers placed within the smaller chambers is a way to limit potential contamination.
41. Side-opening mailers should be used for the vegetable steamer, and top-opening mailers should be used for the decloaker. Top-opening mailers should remain partially open during antigen retrieval.
42. The pH can be checked with pH strips. Upon flowing through the hose, the pH should be ~9 for Tris buffer, 2–3 for 1% TFA, and 7.35 for collagenase type III buffer. Ensuring that the pH of the flowthrough has changed will ensure that the hose line has been cleaned thoroughly.

Acknowledgements

PMA and HBT are supported by NIH/NIAMS R21AR079765, NIH/NCI R21CA263464, and NIH/NCI R01CA253460 and in part by the Hollings Cancer Center's Cancer Center Support Grant P30 CA138313 and the Digestive Disease Research Center P30DK123704 at the Medical University of South Carolina. Additional support was provided by the South Carolina Centers of

Economic Excellence SmartState program to RRD and ASM. The MUSC Mass Spectrometry Facility is supported by the Office of the Provost and the South Carolina COBRE in Oxidants, Redox Balance and Stress Signaling (NIH/NIGMS P30tGM140964) and NIH/NIGMS/OD 1S10OD030212 (PMA, PI). This work was supported in part by grants to AmberGen, Inc. from the Nation Institutes of Health, including the following: R44CA236097, R44AG078097, and R44MH132196.

Conflict of Interest Authors MJL, GY, and KJR were employed by the company.

AmberGen, Inc. – The company has awarded and pending patents on some aspects of this work.

References

1. Bergholtz H, Carter JM, Cesano A, Cheang MCU, Church SE et al (2021) Best practices for spatial profiling for breast cancer research with the GeoMx® digital spatial profiler. Cancers 13:4456. https://doi.org/10.3390/cancers13174456
2. Park J, Kim J, Lewy T, Rice CM, Elemento O, Rendiero AF et al (2022) Spatial omics technologies at multimodal and single cell/subcellular level. Genome Biol 23:256. https://doi.org/10.1186/s13059-022-02824-6
3. Zhuang X (2021) Spatially resolved single-cell genomics and transcriptomics by imaging. Nat Methods 18:18–22. https://doi.org/10.1038/s41592-020-01037-8
4. Chang Q, Ornatsky OI, Siddiqui I, Loboda A, Baranov VI, Hedley DW (2017) Imaging mass cytometry. Cytometry Part A J Quant Cell Sci 91:160–169. https://doi.org/10.1002/cyto.a.23053
5. Keren L, Bosse M, Thompson S, Risom T, Vijayaragavan K, McCaffrey E et al (2019) MIBI-TOF: a multiplexed imaging platform relates cellular phenotypes and tissue structure. Sci Adv 5:eaax5851. https://doi.org/10.1126/sciadv.aax5851
6. Clift C, Mehta AS, Drake RR, Angel PM (2021) Multiplexed imaging mass spectrometry of histological staining, N-glycan and extracellular matrix from one tissue section: a tool for fibrosis research. Methods Mol Biol 2350: 313–329. https://doi.org/10.1021/jasms.1c00189
7. Dunne J, Griner J, Romeo M, Macdonald J, Krieg C, Lim M, Yagnik G, Rothschild KJ, Drake RR, Mehta AS, Angel PM (2023) Evaluation of antibody-based single cell type imaging techniques coupled to multiplexed imaging of N-glycans and collagen peptides by matrix-assisted laser desorption/ionization mass spectrometry imaging. Anal Bioanal Chem 415: 7011–7024. https://doi.org/10.1007/s00216-023-04983-2
8. Angel P, Comte-Walters S, Ball LE, Talbot K, Mehta AS, Brockbank KGM, Drake RR (2018) Mapping extracellular matrix proteins in formalin-fixed, paraffin-embedded tissues by MALDI imaging mass spectrometry. J Proteome Res 17(1):635–646. https://doi.org/10.1021/acs.jproteome.7b00713
9. Wallace E, West CA, McDowell CT, Lu X, Bruner E, Mehta AS, Aoki-Kinoshita KF, Angel PM, Drake RR (2024) An N-glycome tissue atlasx of 15 human normal and cancer tissue types determined by MALDI-imaging mass spectrometry. Sci Rep 14:489. https://doi.org/10.1038/s41598-023-50957-w
10. Eckhard U, Huesgen PF, Brandstetter H, Overall CM (2014) Proteomic protease specificity profiling of clostridial collagenases reveals their intrinsic nature as dedicated degraders of collagen. J Proteomics 100:102–114. https://doi.org/10.1016/j.jprot.2013.10.004
11. Lim M, Yagnik G, Henkel C, Frost SF, Bien T, Rothschild KJ (2023) MALDI HiPLEX-IHC: multiomic and multimodal imaging of targeted intact proteins in tissues. Front Chem 11:

1182404. https://doi.org/10.3389/fchem.2023.1182404

12. Yagnik G, Liu Z, Rothschild KJ, Lim MJ (2021) Highly multiplexed immunohistochemical MALDI-MS imaging of biomarkers in tissues. J Am Soc Mass Spectrom 32:977–988. https://doi.org/10.1021/jasms.0c00473
13. Hioki YKH, Hamana C, Takeyama K, Tanaka K (2013) An improved sample preparation method for the sensitive detection of peptides by MALDI-MS. J Mass Spectrom 48. https://doi.org/10.1002/jms.3283

Chapter 20

Proteomics-Based Analysis of Laser-Capture Micro-dissected, Formalin-Fixed Paraffin-Embedded Tissue Samples

Teresa Mendes Maia, Delphi Van Haver, Sara Dufour, Malaïka Van der Linden, Amélie Dendooven, Francis Impens, and Simon Devos

Abstract

The ability to bring spatial resolution to omics studies enables a deeper understanding of cell populations and interactions in biological tissues. In the case of proteomics, single-cell and spatial approaches have been particularly challenging, due to limitations in sensitivity and throughput relative to other omics fields. Recent developments at the level of sample handling, chromatography, and mass spectrometry have set the stage for proteomics to be established in these new disciplines.

Key words Spatial proteomics, LCM, FFPE, DIA, Amyloidosis

In this chapter, we describe a robust and sensitive workflow for spatial proteomics that combines laser microdissection and mass spectrometry. With a focus on the analysis of archival laser capture micro-dissected tissue, this workflow builds up on a previously optimized formalin-fixed paraffin-embedded sample preparation protocol, now adapted for lower amounts of input material. The combination of an Evosep One liquid chromatography system and a trapped ion mobility [1] time-of-flight (timsTOF) mass spectrometer, operated in diaPASEF [2] mode, offers the right setup for low-input/single-cell analysis, ensuring extremely high sensitivity and speed of operation. We illustrate the applicability of this approach to study clinical amyloidosis samples, with deep proteomics profiling of amyloid deposits, proving its potential for unambiguous disease subtyping.

Taufika Islam Williams (ed.), *Tissue Proteomics: Methods and Protocols*, Methods in Molecular Biology, vol. 2884,
https://doi.org/10.1007/978-1-0716-4298-6_20,

1 Introduction

Spatial omics has recently emerged as a new discipline that brings in situ resolution to high-throughput analysis of biological specimens, therefore allowing a better understanding of tissue physiology [3–5]. It shares very similar challenges with the single-cell omics field: it requires an analytical platform that is both sensitive and fast, and it involves handling small tissue regions or even single cells (with the exception of imaging-based platforms) and the robust processing thereof. One additional factor is that, besides the case of in situ molecular labeling methodologies, in spatial technologies, each sample has to be precisely excised from a microscope slide preparation and this position needs to be stored. Despite being a fairly challenging technology, spatial omics is increasingly used in research projects on developmental, tumor, or infection biology [6–10].

In the case of spatial proteomics, until now, published work relied mostly on the use of antibody-based technologies, which have since reached the "hyper-plex" level [3, 5], meaning they can profile hundreds of target proteins per sample [11]. Typically, panels of 100–200 protein markers can now be assayed by imaging tissue regions with fluorescent or metal-labeled antibodies, for instance, to identify new cell states or aberrant gene expression in disease that can potentially guide the design of new therapeutic strategies [12, 13]. However, a deeper and unbiased assessment of the spatial proteome is often crucial to understanding biology, and then, liquid chromatography-tandem mass spectrometry (LC-MS/MS) technology comes into play. An excellent example is the recent work by Mund et al. [14], presenting a novel methodology termed deep visual proteomics (DVP). This approach consists of the proteomics analysis of restricted tissue regions with known phenotype, based on histology, to show in-depth molecular profiling that can discern nuclei from whole cells, normal tissue from cancerous tissue, and that can reveal different cell states within a tumor [14]. In another proof-of-concept study, Piehowski et al. rightfully profiled two cell types from mouse uterine tissue by LC-MS/MS-based proteomics [15]. Specifically, protein expression levels from minute tissue samples that represented each of the cell types correlated with the corresponding immunohistochemical measurements and reflected the expected cell type–related functional annotations.

This type of study became possible due to tremendous progress recently made in the different steps of the LC-MS/MS proteomics workflow to increase both sensitivity and throughput [16, 17]. The sensitivity that is needed to analyze low-input samples or even single cells was made possible through the development of miniaturized and automated sample preparation methods on the one hand (for example nanoPOTs [18], SCoPE-MS [19, 20], and proteoCHIP [21]) and through the introduction of ultra-sensitive mass spectrometers like the timsTOF SCP or Ultra (Bruker Daltonics [17, 22]) and the Orbitrap Astral (Thermo Scientific [23]). In proteomics, advances in throughput can be achieved through sample multiplexing by isobaric labeling, but also by increasing the analysis speed. The Evosep One from Evosep is a nano-LC system that allows very fast and robust sample analysis by using offline loading and desalting steps and generating a pre-formed solvent gradient with the sample, which then gets very efficiently separated in the analytical column. This LC system has been shown to perform very well in low-input workflows when combined with data-independent acquisition (DIA) methods, an MS modality that guarantees better data completeness and reproducibility between runs, in comparison to classic data-dependent acquisition (DDA) methods [14, 24, 25]. Importantly, data processing software tools have been keeping pace with the developments described above [2, 24–26].

In this chapter, we describe a robust, sensitive, and high-throughput LC-MS method to analyze laser capture micro-dissected (LCM), formalin-fixed, paraffin-embedded (FFPE) tissue samples for spatial proteomics. As a test case, we applied this on LCM-captured amyloidosis deposits from kidney tissue. Amyloidosis is a rare disease characterized by the deposition of protein aggregates, termed fibrils or amyloid substances in the extracellular space, eventually leading to tissue or organ failure [27]. Fibrils originate from one single protein that is either present in excess (e.g., monoclonal, auto-aggregating immunoglobulin light chains κ or λ) or that loses conformation due to mutation (e.g., transthyretin (TTR), fibrinogen A-α chain) or abnormal proteolytic cleavage [e.g., serum amyloid A, transthyretin (TTR)] [28]. Because patient treatment differs according to the disease-causing amyloidogenic protein, it is fundamental to get a precise diagnosis. As such, analysis of tissue biopsies or fat aspirates starts by screening with Congo red dye for the presence of amyloid deposits and is then followed by amyloid subtyping. Although this can be done by immunohistochemistry, limitations related to epitope sensitivity and specificity [29–31] led laboratory experts in amyloidosis diagnosis toward MS-based proteomics as an alternative method [28, 32–34]. Indeed, shotgun bottom-up proteomic analysis of Congo red positive tissue sections allows for subtyping of a majority of the ca. 10% of cases that formerly remained inconclusive when analyzed by other methodologies [32, 33, 35]. Specifically, given its unbiased nature and depth of analysis, LC-MS performed on presumptive amyloidosis patient samples can reveal both (1) the three main signature components of amyloid protein deposits (the blood circulating proteins serum amyloid P component (SAP), apolipoprotein E (ApoE), and apolipoprotein A-IV (ApoA-IV)) [33] and

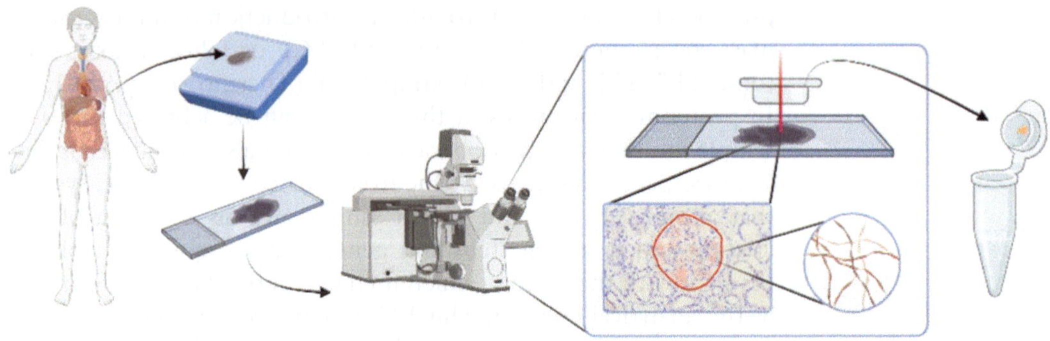

Fig. 1 Laser capture microdissection (LCM) of patient's formalin-fixed paraffin-embedded archived tissue. A tissue section from a patient's kidney biopsy is submitted to LCM. In the specific case of an amyloidosis sample, Congo red staining reveals amyloid fibril areas, which then get excised and collected in the cap of microfuge tubes. (Created with BioRender.com)

(2) the identity of over 30 amyloidogenic proteins that allow for amyloidosis subtyping [36].

Taking advantage of our previous experience in the proteomic analysis of FFPE amyloidosis samples [37], we now implemented a high-sensitivity and high-resolution spatial proteomics workflow (Figs. 1 and 2). We demonstrate the application of this workflow for the diagnosis and typing of amyloidosis in kidney tissue from two patients. Samples were gathered through LCM and consisted of cut-out Congo red–positive glomeruli and blood vessels (sample 1) or glomeruli exclusively (sample 2), allowing for directly assaying proteins present in amyloid deposits. Experiments were carried out on a timsTOF SCP instrument, preceded by an Evosep One LC system, a setup that was tailored for low-input/single-cell analysis, as it offers extremely high sensitivity and speed of operation [22].

A total of 7339 and 5920 peptides and 1733 and 1348 proteins were identified in samples 1 and 2, respectively. In Table 1, we list the 20 most abundant proteins in each sample, as determined with intensity-based absolute quantification (iBAQ), a quantification unit that corrects the raw protein intensity values to account for theoretical detectability [38]. The presence of IGKC protein in sample 1 allowed for its classification as an AL case (i.e., immunoglobulin light chain amyloidosis), while SAA1 protein in sample 2 led to an AA (i.e., systemic amyloidosis) diagnosis. Here, immunohistochemistry imaging was able to conclusively type the cases as well and confirmed the validity of the methodology (Fig. 3). The three disease signature proteins could also be detected in the top 20 most abundant proteins for sample 1. For sample 2, ApoE and SAP were found among the top 20, while ApoA-IV fell outside the top-ranking protein IDs but was identified nevertheless with an estimated relative abundance of 0.04%. Evidence from the literature shows it is common to completely miss the detection of ApoA-IV in

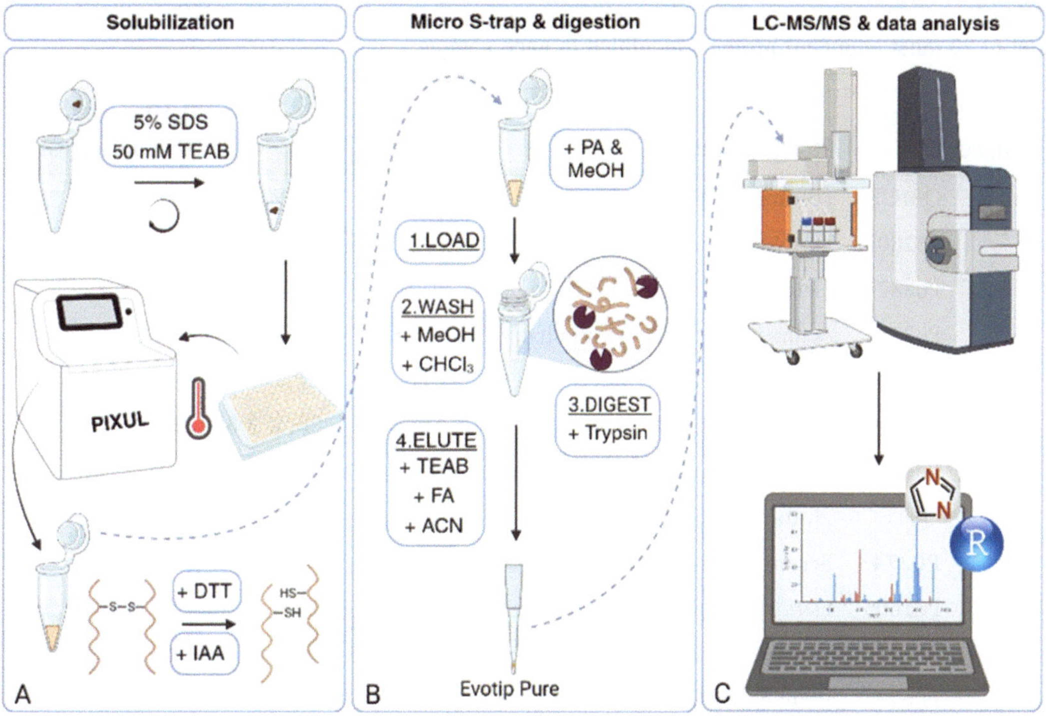

Fig. 2 Schematic outline of the laser capture micro-dissected (LCM) formalin-fixed paraffin-embedded (FFPE) sample preparation protocol for liquid chromatography (LC) and mass spectrometry (MS) analysis. (a) LCM-FFPE samples are resuspended in lysis buffer, transferred to a PIXUL plate, and submitted to mechanical friction, heating, and sonication, in order to get proteins extracted and break protein cross-links. Cysteine residues are then reduced with dithiothreitol (DTT) and alkylated with iodoacetamide (IAA). (b) Phosphoric acid (PA) and methanol (MeOH) are added, and the protein extract is loaded on an S-Trap micro column. After several washes with MeOH and chloroform ($CHCl_3$), trypsin is added to digest proteins, and finally, triethylammonium bicarbonate (TEAB), formic acid (FA), and acetonitrile (ACN) are used for the elution of peptides from the Evotips. (c) LC-MS/MS and data analyses are carried out using DIA-NN and R/RStudio. (Created with BioRender.com)

AA, and so being able to find it at low intensity confirms the high sensitivity of our assay.

2 Materials

All solutions should be prepared using ultrapure water (with a maximum resistance of 18.2 MΩ-cm at 25 °C, purified from deionized water) and high-quality LC-MS grade reagents. Stock solutions should be stored at room temperature unless stated otherwise. When disposing of waste material, meticulously follow waste disposal regulations.

Table 1
List of 20 most abundant proteins identified in each sample

Sample	Protein ID	Protein name	Gene name	iBAQ	Relative abundance (%)
Sample 1	Q8IUE6	Histone H2A type 2-B	H2AC21	295,928.57	14.9
	P62805	Histone H4	H4C16	153,280.17	7.7
	P68871	Hemoglobin subunit beta	HBB	125,766.36	6.3
	P63267	Actin, gamma-enteric smooth muscle	ACTG2	103,353.04	5.2
	P04004	Vitronectin	VTN	75,153.85	3.8
	P02649	**Apolipoprotein E**	**APOE**	**55,473.57**	**2.8**
	P69905	Hemoglobin subunit alpha	HBA2	43,679.78	2.2
	Q5TEC6	Histone H3–7	H3–7	37,355.64	1.9
	P02743	**Serum amyloid P-component**	**APCS**	**26,518.55**	**1.3**
	P81605	Dermcidin	DCD	26,375.38	1.3
	P61626	Lysozyme C	LYZ	25,490.33	1.3
	P06703	Protein S100-A6	S100A6	22,555.00	1.1
	P08493	Matrix Gla protein	MGP	19,099.54	1.0
	Q969L2	Protein MAL2	MAL2	16,049.60	0.8
	P08670	Vimentin	VIM	15,282.95	0.8
	P31151	Protein S100-A7	S100A7	14,985.29	0.8
	P06727	**Apolipoprotein A-IV**	**APOA4**	**14,121.10**	**0.7**
	P10909	Clusterin	CLU	13,720.04	0.7
	P05109	Protein S100-A8	S100A8	13,438.33	0.7
	P01834	***Immunoglobulin kappa constant***	***IGKC***	***11,050.92***	***0.6***
Sample 2	P62805	Histone H4	H4C16	254,935.00	9.7
	P68871	Hemoglobin subunit beta	HBB	244,548.18	9.3
	P69905	Hemoglobin subunit alpha	HBA2	197,447.78	7.5
	P02743	**Serum amyloid P-component**	**APCS**	**128,771.82**	**4.9**
	P04004	Vitronectin	VTN	81,335.77	3.1
	Q9H9F9	Actin-related protein 5	ACTR5	62,119.38	2.4
	P02768	Albumin	ALB	61,244.26	2.3
	P02649	**Apolipoprotein E**	**APOE**	**61,063.21**	**2.3**
	Q5TEC6	Histone H3–7	H3–7	58,938.91	2.2
	Q5T750	Skin-specific protein 32	XP32	38,167.42	1.5
	P81605	Dermcidin	DCD	33,896.25	1.3

(continued)

Table 1 (continued)

Sample	Protein ID	Protein name	Gene name	iBAQ	Relative abundance (%)
	Q13103	Secreted phosphoprotein 24	SPP2	31,587.40	1.2
	Q969L2	Protein MAL2	MAL2	30,619.60	1.2
	P0DJI8	***Serum amyloid A-1 protein***	*SAA1*	*29,402.20*	*1.1*
	P61626	Lysozyme C	LYZ	29,041.75	1.1
	P10599	Thioredoxin	TXN	26,360.63	1.0
	P08670	Vimentin	VIM	25,639.25	1.0
	P04406	Glyceraldehyde-3-phosphate dehydrogenase	GAPDH	23,661.82	0.9
	P06703	Protein S100-A6	S100A6	22,567.86	0.9
	Q15517	Corneodesmosin	CDSN	18,789.31	0.7

2.1 Tissue Sectioning and Staining

1. FFPE tissue block.
2. MembraneSlide 1.0 polyethylene naphthalene (PEN) (Zeiss), which facilitates laser cutting.
3. Poly-L-lysine solution, 0.01%.
4. Microtome.
5. Disposable microtome blades.
6. Cool and hot plate.
7. Filter paper.
8. Oven.
9. Congo red staining kit (ready-to-use, for automated staining) + accompanying auxiliary reagents (deparaffinization solution, wash solution, liquid coverslips).
10. Absolute alcohol.
11. Microscope slide box.

2.2 Laser Capture-Mmicrodissection

1. PALM MicroBeam device (Zeiss) with pulsed Uva laser for sample collection.
2. AdhesiveCap opaque collection tube (Zeiss) and corresponding tube collector.
3. PALM RoboSoftware.

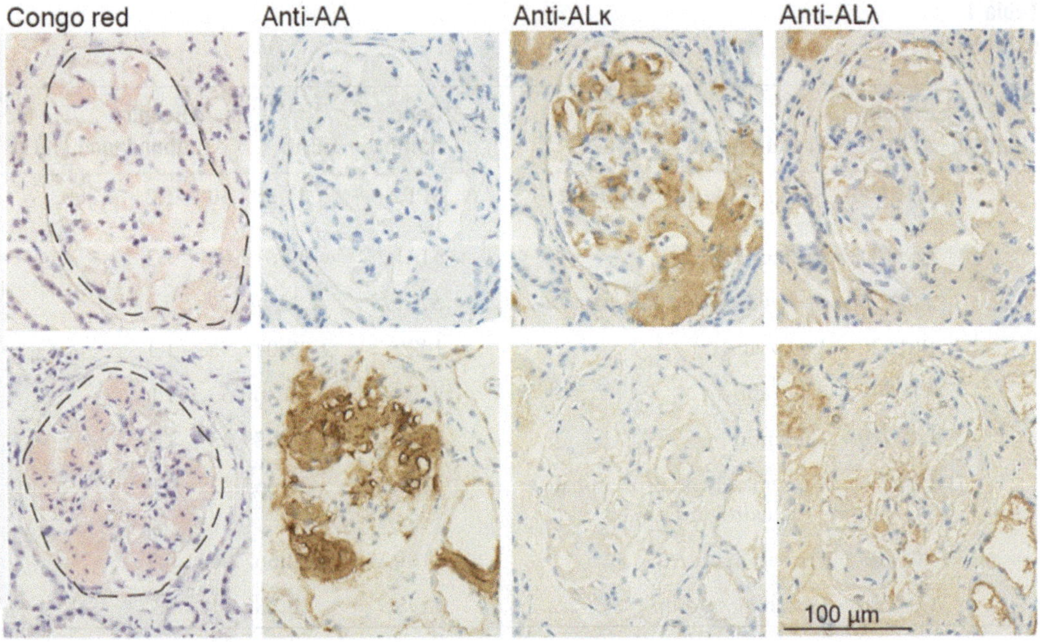

Fig. 3 Light microscopy images of kidney formalin-fixed paraffin-embedded tissue from two amyloidosis patients. (Top) Histological stainings from serial tissue sections of clinical case 1. (Bottom) Equivalent stainings for clinical case 2. Stainings are seen in the following order: Congo red, immunostaining against human amyloid A, kappa light chains, and lambda light chains. While orange-pink, Congo red–positive staining appears for both patient cases, brown immunostaining is visible only on the kappa light chain panel for clinical case 1 and on the amyloid A panel for clinical case 2. Black dashed lines represent sections of kidney glomeruli (and of a blood vessel for case 1) subjected to microdissection. Antibodies are from Agilent. (Created with BioRender.com)

2.3 Protein Extraction and Solubilization from LCM-FFPE Samples

1. LCM-FFPE sample.
2. Lysis buffer: 5% sodium dodecyl sulfate (SDS), 50 mM triethylammonium bicarbonate (TEAB) in water, pH 8.5 (*see* **Note 1**).
3. PIXUL 96-well plate with sealer (Active Motif).
4. PIXUL coupling liquid (Active Motif).
5. Polypropylene 96-well plate or microcentrifuge tubes.
6. PIXUL sonicator (Active Motif).
7. Benchtop centrifuge with adaptors for microcentrifuge tubes and 96-well plates.
8. Incubator.

2.4 Reduction/Alkylation of Proteins

1. 0.5 M dithiothreitol (DTT) in water. Store at −20 °C (*see* **Note 2**).
2. 0.5 M iodoacetamide (IAA) in water (*see* **Note 3**).

3. pH indicator strip,
4. Incubator.

2.5 S-Trap, Digestion of Proteins, and Peptide Purification

1. S-Trap micro column (ProtiFi).
2. 2.0-mL microcentrifuge tube.
3. 1.5-mL Protein LoBind microcentrifuge tube.
4. 27.5% phosphoric acid in water (*see* **Note 4**).
5. pH indicator strip.
6. Binding buffer: 100 mM TEAB in 90% methanol, pH 7.55 (*see* **Note 5**).
7. Wash buffer: 50% chloroform, 50% methanol (*see* **Note 6**).
8. Digestion buffer: 50 ng/μL sequencing grade modified trypsin in 50 mM TEAB (*see* **Note 7**).
9. Elution buffer 1: 50 mM TEAB in water (*see* **Note 8**).
10. Elution buffer 2: 0.2% formic acid (FA) in water (*see* **Note 9**).
11. Elution buffer 3: 0.2% FA in water/acetonitrile (ACN, 50:50, v/v) (*see* **Note 10**).
12. Incubator.
13. Benchtop centrifuge with adaptors for microcentrifuge tubes.
14. Vacuum concentrator.

2.6 Evotip Loading

1. Dried peptides.
2. Loading solvent: 0.1% trifluoroacetic acid (TFA) in water/acetonitrile (ACN, 98:2, v/v) (*see* **Note 11**).
3. Solvent A: 0.1% FA in water (*see* **Note 12**).
4. Solvent B: 0.1% FA in ACN (*see* **Note 13**).
5. Isopropanol.
6. Polypropylene multichannel reservoirs.
7. Polypropylene single-well plate.
8. Multichannel pipette 20–100 μL.
9. Evotip Pure (Evosep).
10. Evotip box (Evosep).
11. Evotip adaptor rack (Evosep).
12. Benchtop centrifuge with adaptors for deep well plates (*see* **Note 14**) or pipette tips.

2.7 LC-MS/MS Analysis

1. MS solvent A: 0.1% FA in water.
2. MS solvent B: 0.1% FA in ACN.

3. Aurora™ Elite Generation 3 CSI column (15 cm × 75 μm I.D., 1.7 μm beads, nanoZero®; IonOpticks) or any other high-resolution and high-throughput nano-UPLC column.
4. LC-MS/MS system, for example, Evosep One LC system (Evosep) in-line connected to a timsTOF SCP (Bruker), or any other LC-MS/MS system with comparable sensitivity and analysis speed.

2.8 Data Analysis

1. DIA-NN [25, 39] (freeware, version 1.8.1) or other software package dedicated to DIA data analysis.

3 Methods

3.1 Tissue Sectioning and Staining

We use Congo red–stained tissue sections to determine which areas should be micro-dissected. Tissue sections are prepared following the manufacturer's protocol for protein handling for LC-MS/MS (poly-L-lysin coating, overnight incubation at 56 °C) on the one hand and Congo red staining (10-μm-thick sections) on the other hand.

1. Coat polyethylene naphthalene (PEN) membrane slides with 0.01% poly-L-lysine for better adhering. Incubate the slides for 5 min, rinse with deionized water, and air-dry.
2. Cool paraffin blocks, using a cool plate, prior to sectioning. Distribute a drop of deionized water on top of the coated PEN membrane slide. Mount 10 μM tick FFPE tissue sections (see **Note 15**), number depending on the amount of material required, onto the slide and let sections stretch on a plate at 40 °C. Press the slide on a wet filter paper and leave it on the hotplate for another 30 min. Additionally, dry slides overnight in an oven at 56 °C, to improve adhesion of the sections to the membrane.
3. Deparaffinize, rehydrate, and stain sections in the conventional manner:
   (a) Deparaffinization: 12 min.
   (b) Congo red: 32 min at 37 °C.
   (c) Hematoxylin: 12 min.
4. Dehydrate sections in three changes absolute alcohol (10 dips each time).
5. Do not cover the slip.
6. Store slides in a microscope slide box at room temperature for LCM.

3.2 Laser Capture Microdissection

Laser capture microdissection is performed using the PALM MicroBeam device with a pulsed UVa laser and RoboSoftware.

1. Switch on the system before you start the software.
2. Check that automatic focusing is turned on (*see* **Note 16**).
3. Open the windows "Element List," "Navigator," and "RoboMover/CapMover" for ease of use.
4. Move the stage to the load position and insert the slide making sure that the section side is facing up. Return the stage to its original position.
5. After inserting the collector and collection tube, check and eventually adjust the distance between the cap of the tube and the tissue so that the material is captured properly (via the "RoboMover/CapMover" window).
6. Optionally, scan your sample first (via the "Navigator" window). Use the 5× objective and autofocus (*see* **Note 17**).
7. Choose the 20× objective for LCM and focus the image.
8. Check the laser beam path as described in the PALM MicroBeam user manual. Start cutting and adjust the laser power and focus if necessary. The perfect laser setting is achieved when using a minimum of laser energy produces a fine cutting line. The settings for laser focus and energy have to be adjusted to the objective used. Usually, the following parameters work well:
   (a) Cutting speed: 15.
   (b) Laser energy: 60.
   (c) Laser focus: 69.
9. For LCM, choose the laser function "RoboLPC" (cuts and subsequently catapults the tissue into the cap) and define the areas you want to dissect with the graphic tool (*see* **Note 18**). You can indicate several areas one after another. The size of the areas is measured automatically but is an estimated value. For sample 1, approximately 385.571 μm^2 tissue (or nine Congo red positive glomeruli) was collected from two sections. For sample 2, approximately 207,485 μm^2 tissue (or eight Congo red positive vessels) was collected from two sections. As soon as you start the laser, the stage automatically moves along the defined areas (*see* **Notes 19** and **20**).
10. Inspect the cap (via the "CapCheck" function in the "RoboMover/CapMover" window) to see the catapulted material (*see* **Note 21**). Use the 10× objective (and optionally auto live for less light) and focus. Move the stage back.

11. Remove the tube (via the "RoboMover/CapMover" window) and move the stage to the load position to remove the slide.
12. Exit the software before you switch off the system.

3.3 Protein Extraction and Solubilization from LCM-FFPE Samples

As a result of the FFPE process, proteins are cross-linked, and samples are embedded in water-insoluble paraffin. To extract and dissolve proteins in an aqueous buffer, samples are homogenized by mechanical friction, heating, and sonication in a buffer containing SDS. Additionally, DNA present in the samples is sheared during the sonication step, as full-length DNA can clog S-Trap columns.

1. Resuspend the LCM-FFPE sample in 50 μL lysis buffer and transfer it to a PIXUL 96-well plate. Seal the plate well.
2. Incubate the sample overnight at 50 °C.
3. Check if the plate is still sealed completely. If necessary, fix the seal better or replace the seal.
4. Ultrasonicate the sample with a PIXUL sonicator for 5 min or until pieces have been dissolved completely with the following settings: 50 N pulse, 1 kHz, 20 Hz burst rate.
5. Spin down shortly (*see* **Note 22**) and place the sample in an incubator at 80 °C for 1 h.
6. Repeat ultrasonication of **step 4** for 6 min.
7. Centrifuge 96-well plate containing the homogenized lysate at 16,000× *g* for 15 min (*see* **Note 22**).
8. Transfer the soluble fraction to a new 96-well plate or microcentrifuge tube.

3.4 Reduction/Alkylation of Proteins

1. Reduce the protein extract by adding 1.5 μL 0.5 M DTT for a final concentration (f.c.) of 15 mM and incubate in a shaking incubator for 30 min at 55 °C.
2. Cool the sample to room temperature (until it feels neither warm nor cold) by putting it on ice.
3. Pipette 1 μL of a sample on a pH strip to check that the pH is approximately 8. If the pH is too low, add TEAB to increase the pH.
4. Add 3 μL 0.5 M IAA (30 mM f.c.) and incubate in a shaking incubator in the dark for 15 min at room temperature (*see* **Note 23**).

3.5 S-Trap, Digestion of Proteins, and Peptide Purification

After completely unfolding all proteins with SDS, DTT, and IAA, these reagents need to be removed to allow efficient digestion of proteins into peptides for LC-MS/MS analysis. In the following part, we describe an adapted version of the S-Trap micro and the HYPERsol FFPE protocol [40, 41].

1. Add 5 μL 27.5% aqueous phosphoric acid to the reduced and alkylated lysate (2.75% f.c.) and mix well.
2. Check the pH of a sample by spotting 1 μL on a pH strip and add more 27.5% phosphoric acid if the pH is higher than 1.
3. Add 330 μL binding buffer to the acidified lysate and mix well to form the colloidal protein particulate (*see* **Note 24**).
4. Place the S-Trap micro column on top of a microcentrifuge tube.
5. Transfer the colloidal suspension to the S-Trap micro column in parts of 200 μL, close the S-Trap micro column, and centrifuge for 30 s at 4000× *g* or until all solution has passed (*see* **Note 25**). Discard the flow-through.
6. Turn the S-Trap 180° and wash the S-Trap column containing the protein particulates once with 150 μL binding buffer and centrifuge for 30 s at 4000× *g*. Visually check that all buffers have passed through. The flow-through should be discarded.
7. Repeat **step 6** three times with wash buffer (*see* **Note 26**) and twice with binding buffer to remove all wax. Ensure that all buffer has passed through at each wash step. Discard the flow-through.
8. Centrifuge the S-Trap column for an additional minute at 4000× *g*.
9. Transfer the S-Trap column to a newly labeled Protein LoBind microcentrifuge tube.
10. Pipette 20 μL digestion buffer containing 1 μg trypsin (*see* **Note 27**) directly on top of the protein capture matrix. Ensure the protease solution covers the complete matrix (*see* **Note 28**).
11. Cap the S-Trap loosely and incubate overnight at 37 °C in a non-shaking incubator (*see* **Note 29**).
12. Elute the peptides by adding 40 μL elution buffer 1 to the S-Trap column and centrifuge for 1 min at 4000× *g*. Do not centrifuge before adding elution buffer 1.
13. Repeat elution **step 12** once with 40 μL elution buffer 2 and once with 40 μL elution buffer 3.
14. Dry the combined eluate with a vacuum concentrator.
15. Store dried peptides at −20 °C.

3.6 Evotip Loading

To remove any residual contaminants and minimize carry-over between samples, peptides are loaded on Evotips before injection on the Evosep One LC system (*see* **Note 30**). The protocol below is based on the original instructions published by Evosep [42].

1. Dissolve dried peptides in 20 μL loading solvent.
2. Dilute the sample 10 times by adding 2.5 μL sample to 22.5 μL solvent A.
3. Fill 1 multichannel reservoir with solvent A and 1 reservoir with solvent B. Fill the single-well plate with isopropanol.
4. Using a multichannel pipette, transfer 20 μL solvent B to all Evotips (*see* **Note 31**).
5. Load the Evotips in the centrifuge with the appropriate counterweight and centrifuge for 1 min at 800× *g*.
6. Ensure all liquid passes through the Evotips and remove all liquid from the Evotip box. Repeat the centrifugation step if any liquid remains on the Evotips.
7. Soak the Evotips for a minimum of 10 s in isopropanol (IPA) by placing the Evotips (in the yellow rack) on the Evotip adaptor rack stacked on the IPA-containing single-well plate.
8. Visually inspect all Evotips to ensure the peptide-binding material is pale white before continuing with the next step (*see* **Note 32**).
9. Using a multichannel pipette, transfer 20 μL solvent A to all Evotips. Transfer the Evotips (in the yellow rack) back to the Evotip box.
10. Repeat **steps 5** and **6**.
11. Transfer 20 μL of the 10× diluted sample to the Evotip.
12. Repeat **steps 5** and **6**.
13. Using a multichannel pipette, transfer 20 μL solvent A to all Evotips.
14. Repeat **steps 5** and **6**.
15. Using a multichannel pipette, transfer 100 μL solvent A to all Evotips.
16. Load the Evotips in the centrifuge with the appropriate counterweight and centrifuge for 10 s at 800× *g*.
17. Fill the Evotip box containing the Evotips with solvent A until the bottom of the Evotips is submerged. Close the Evotip box, seal it with parafilm, and store it at +4 °C.

3.7 LC-MS/MS Analysis

In theory, any LC-MS/MS system can be used to measure FFPE samples. We advise using a clinical application with a sensitive approach for low-input samples (e.g., Evosep One LC system connected to a timsTOF SCP mass spectrometer). Below, we provide a description of our setup and how the LC-MS/MS analysis of the amyloidosis samples is performed in our laboratory.

1. Remove the parafilm and lid from the Evotip box.
2. Place the Evotip box on the source tray from the Evosep One LC system.
3. Samples are run in data-independent acquisition—parallel accumulation serial fragmentation (diaPASEF) mode [17] on an Evosep One LC system in-line connected to a timsTOF SCP equipped with a CaptiveSpray ion source. Peptides were analyzed with the 20 SPD whisper method using the Aurora Gen3 Elite™ CSI column, heated to 50 °C. Peptides were eluted from the column through the predefined 20 SPD whisper gradient. Eluting peptides were measured in positive polarity with a full-scan range of 100–1700 *m/z*. The TIMS was operated at a fixed duty cycle close to 100%, with a ramp and accumulation time of 100 ms, ranging from $1/K_0 = 0.64$ Vscm^2 to $1/K_0 = 1.50$ Vscm^2. Collision energy was linearly ramped as a function of the inverse mobility from 20 eV at $1/K_0 = 0.60$ Vscm^2 to 59 eV at $1/K_0 = 1.60$ Vscm^2. A diaPASEF mass range of 400 Da to 1000 Da was used in a mobility range of $1/K_0 = 0.64$ Vscm^2 to $1/K_0 = 1.37$ Vscm^2 using a window size of 25 Da according to Table 2, resulting in a cycle time of 0.96 s. The source voltage was set to 1.5 kV.

3.8 Data Analysis

Different search engines and data analysis packages can be used to analyze the LC-MS/MS spectral data. The data analysis of the LC-MS/MS runs reported here was performed with the DIA-NN algorithm, in library-free mode (*see* **Note 33**). LC-MS/MS runs of the amyloidosis samples were searched separately, and the specific search settings are listed below.

1. Go to the "Input" pane. Load the diaPASEF .d files and add the correct database(s) in FASTA file format. We used the human UniProt Reference Proteome (database release version of January 2023, downloaded from http://www.uniprot.org) containing 20,594 protein sequences.
2. Check the "FASTA digest for library-free search/library generation" option in the Precursor ion generation tab. The "Deep learning-based spectra, RTs and IMs prediction" option is automatically enabled.
3. Choose "Trypsin/P" (*see* **Note 34**) as protease and set the maximum number of missed cleavages at 2 (*see* **Note 35**). Enable the following modifications: "Ox(M)," "Ac (N-term)," and "C Carbamidomethylation" as a fixed modification (*see* **Note 36**) and set the precursor *m/z* range to "400–1000."
4. Set MS2 and MS1 mass accuracy to 20 and 15 ppm, respectively, in the algorithm pane. Disable heuristic protein inference.

Table 2
diaPASEF window scheme

Cycle ID	Start IM [1/K_0]	End IM [1/K_0]	Start mass [*m/z*]	End mass [*m/z*]
1	0.64	0.83	400	425
2	0.64	0.85	425	450
3	0.64	0.87	450	475
4	0.64	0.9	475	500
5	0.64	0.92	500	525
6	0.64	0.94	525	550
7	0.64	0.97	550	575
8	0.64	0.99	575	600
1	0.83	1.01	600	625
2	0.85	1.04	625	650
3	0.87	1.06	650	675
4	0.9	1.09	675	700
5	0.92	1.11	700	725
6	0.94	1.13	725	750
7	0.97	1.16	750	775
8	0.99	1.18	775	800
1	1.01	1.37	800	825
2	1.04	1.37	825	850
3	1.06	1.37	850	875
4	1.09	1.37	875	900
5	1.11	1.37	900	925
6	1.13	1.37	925	950
7	1.16	1.37	950	975
8	1.18	1.37	975	1000

5. Specify the main output folders, enable the “Use existing . quant files when available” option, set the number of threads to a number that does not exceed the number of physical cores available on the used computer, and start the database search.

Downstream analysis can be performed by different analysis platforms. Our downstream analysis was performed in R (version 4.2.2).

6. When the database search is finished, load the DIA-NN main report output table.
7. Filter the table on precursor and protein library q-value cut-off of 1% and filter out only proteotypic hits.
8. Pivot the table into a wide format.
9. Calculate the iBAQ intensity values with the DIAgui's R package "get_iBAQ" function.
10. Remove possible contaminants from the protein list. Contaminant list downloaded from [43].

4 Notes

1. For 100 mL lysis buffer, weigh 5 g SDS in a weighing dish and transfer to a 100-mL graduated cylinder containing a magnetic stir bar. Add 5 mL 1 M TEAB solution (pH 8.5) and add water to a total volume of 100 mL. Mix on a magnetic stirrer until all SDS is dissolved. To avoid foam production, limit the stirring speed when mixing the lysis buffer. The lysis buffer can be stored at room temperature for up to 1 year. Never store the lysis buffer at 4 °C, as SDS precipitates at this temperature. If precipitation of SDS would occur during sample preparation, dissolve the precipitate by mixing and heating at 37 °C.
2. Weigh 77 mg DTT in a microcentrifuge tube and dissolve in 1 mL water. Transfer 50 μL aliquots to new (PCR) tubes and store at −20 °C for up to 1 year.
3. Weigh 92 mg IAA in a microcentrifuge tube in a fume hood, wrap it with tinfoil, or put it in a light protection bag and keep it at room temperature. Only add 1 mL water immediately before use. Always prepare fresh and protect from light.
4. Add 324 μL 85% phosphoric acid stock solution to 676 μL water to obtain a 27.5% solution that can be stored for up to 1 year at room temperature when protected from air.
5. Due to the high percentage of organic solvent, adjusting the pH is not possible after mixing TEAB and methanol. To circumvent this problem, first adjust the pH of 1 M TEAB with phosphoric acid to pH 7.55. Freeze 10 mL aliquots at −20 °C for future use. For 100 mL binding buffer, add 10 mL 1 M TEAB pH 7.55 to 90 mL methanol in the fume hood. Store for up to 1 year at 4 °C.
6. Add 50 mL methanol to 50 mL chloroform in the fume hood.
7. Dissolve 100 μg lyophilized trypsin in 2 mL 50 mM TEAB for 50 ng/μL final concentration. Store aliquots at −20 °C.

8. For 100 mL, mix 5 mL 1 M TEAB with 95 mL water. Store for up to 1 year at 4 °C.
9. For 100 mL 0.2% FA, add 200 μL FA to 99.8 mL water under the fume hood. Store for up to 1 year at 4 °C.
10. For 100 mL 50% ACN and 0.2% FA, mix 50 mL ACN, 200 μL FA, and 49.8 mL water in the fume hood. Store for up to 1 year at 4 °C.
11. For 100 mL 2% ACN and 0.1% TFA, mix 2 mL ACN, 100 μL FA, and 97.9 mL water in the fume hood. Store for up to 1 year at room temperature.
12. For 100 mL 0.1% FA in MQ, mix 100 μL FA and 99.9 mL water in the fume hood. Store for up to 1 year at room temperature.
13. For 100 mL 0.1% FA in ACN, mix 100 μL FA and 99.9 mL ACN in the fume hood. Store for up to 1 year at room temperature.
14. Make sure the centrifuge is equipped with adaptors for deep well plates with a height of at least 56 mm. If the maximum height is lower than 50 mm, the S-Trap combined with a deep well plate will get stuck, and buffers will not flow through correctly.
15. When the paraffin sections are too thin (<5 μM), likely not sufficient amyloid would be available to take up Congo red, yielding false-negative results. For this reason, paraffin sections of 10 μM are preferred for staining.
16. Each area indicated for microdissection, the so-called element, gets coordinates describing the position (*x* and *y* coordinates) and the focus value of the microscope (*z* coordinate). When starting to cut a set of elements, the focus is automatically set to the appropriate value as you approach an element. When an element is out of focus, cutting and catapulting might not be successful.
17. Scanning your sample first makes navigation easier.
18. The AxioVision image processing program enables you to localize specific areas of interest with an automated search function. If not already done, scan your sample first and transfer it to the program for analysis. Set the parameters via the "Life Image" window.
19. If the tissue is not catapulted, give an extra pulse in the middle of the area via the drawing tool "Dot."
20. Microscope images can be saved automatically before and/or after the laser function (via the "Setup" menu).

21. It can be possible that tissue is not being captured into the cap and still is on the slide. In such a case, check the slide for the "missed" tissue and collect the tissue via an extra pulse.
22. After sonication, some coupling liquid might stick to the bottom of the PIXUL 96-well Plate. To avoid pollution of the centrifuge, remove this liquid before centrifugation. Tapping the plate on a tissue and removing the liquid with a vacuum aspiration pump work best.
23. Efficient alkylation of sulfhydryl groups by IAA requires pH 8 and a dark environment. Wrapping the plate or rack with microcentrifuge tubes in tinfoil works fine. Longer incubation times must be avoided to prevent over-alkylation of other chemical groups [44].
24. Do not centrifuge samples at this point!
25. Transfer the complete sample, including any insoluble parts, to the S-Trap column. Pipetting up and down may be necessary to enable good transfer. If the column gets clogged during sample loading, increase the centrifuge speed to a maximum of 15,000× *g* (a higher speed will damage the column). Alternatively, negative pressure (vacuum manifold) or positive pressure can be applied to the column.
26. As chloroform is highly toxic, perform all steps involving the wash buffer in the fume hood.
27. Do not use less than 1 μg trypsin as this might result in incomplete digestion. Other proteases (Tryp-N, Lys-C, Arg-C, ...) can be used to digest proteins to peptides but might need different digestion buffers and/or temperatures. MS-compatible detergents like RapiGest can also be used to aid digestion.
28. Do not damage the matrix with the pipette tip when transferring protease solution to the S-Trap columns.
29. Do not seal the S-Trap column airtight as pressure will build up, and this might cause the solution to flow through prematurely.
30. The Evosep One LC system was preferred here, but any other nano-LC system can be used as well.
31. Evotips are seated in the Evotip box for all following steps unless stated otherwise.
32. The Evotip packing material needs to be properly activated before loading the sample, which is characterized by a pale white color.
33. DIA-NN performs optimally when run in library-free mode (in silico predicted library) but can also be run using a spectral library.

34. The "Trypsin/P" setting allows the detection of peptides resulting from cleavage C-terminally from Arg or Lys residues, even when followed by a Pro residue.
35. Change this setting according to the protease that was used for the digestion of the proteins. The number of missed cleavages can be increased to three or more, but this will expand the search space and significantly lengthen the database search time.
36. The software automatically sets Ox(M) and Ac(N-term) as variable modification and C carbamidomethylation as fixed modification. Other modifications can be added to the additional options pane using the —var-mod [name],[mass],[sites] and —fixed-mod [name],[mass],[sites] command.

Acknowledgements

The authors would like to thank Ran Rumes and Irem Kaya for their technical contribution to the histopathology of amyloidosis samples. We would also like to thank Prof. Ward De Spiegelaere and the Light Microscopy CORE of Ghent University (GLiM) for providing access to their LCM equipment and for guidance. Support from the VIB Grand Challenges Be.Amycon grant (Belgium's Amyloidosis expertise consortium) is gratefully acknowledged.

References

1. Michelmann K, Silveira JA, Ridgeway ME, Park MA (2015) Fundamentals of trapped ion mobility spectrometry. J Am Soc Mass Spectrom 26:14–24. https://doi.org/10.1007/s13361-014-0999-4
2. Meier F, Brunner A-D, Frank M, Ha A, Bludau I, Voytik E, Kaspar-Schoenefeld S, Lubeck M, Raether O, Bache N et al (2020) diaPASEF: parallel accumulation-serial fragmentation combined with data-independent acquisition. Nat Methods 17:1229–1236. https://doi.org/10.1038/s41592-020-00998-0
3. Bressan D, Battistoni G, Hannon GJ (2023) The dawn of spatial omics. Science 381: eabq4964. https://doi.org/10.1126/science.abq4964
4. Marx V (2021) Method of the year: spatially resolved transcriptomics. Nat Methods 18:9–14. https://doi.org/10.1038/s41592-020-01033-y
5. Park J, Kim J, Lewy T, Rice CM, Elemento O, Rendeiro AF, Mason CE (2022) Spatial omics technologies at multimodal and single cell/subcellular level. Genome Biol 23:256. https://doi.org/10.1186/s13059-022-02824-6
6. Asp M, Giacomello S, Larsson L, Wu C, Fürth D, Qian X, Wärdell E, Custodio J, Reimegård J, Salmén F et al (2019) A spatiotemporal organ-wide gene expression and cell atlas of the developing human heart. Cell 179: 1647–1660.e19. https://doi.org/10.1016/j.cell.2019.11.025
7. Grauel AL, Nguyen B, Ruddy D, Laszewski T, Schwartz S, Chang J, Chen J, Piquet M, Pelletier M, Yan Z et al (2020) TGFβ-blockade uncovers stromal plasticity in tumors by revealing the existence of a subset of interferon-licensed fibroblasts. Nat Commun 11:6315. https://doi.org/10.1038/s41467-020-19920-5
8. Guilliams M, Bonnardel J, Haest B, Vanderborght B, Wagner C, Remmerie A, Bujko A, Martens L, Thoné T, Browaeys R et al (2022) Spatial proteogenomics reveals

distinct and evolutionarily conserved hepatic macrophage niches. Cell 185:379–396.e38. https://doi.org/10.1016/j.cell.2021.12.018

9. Quintana JF, Chandrasegaran P, Sinton MC, Briggs EM, Otto TD, Heslop R, Bentley-Abbot C, Loney C, de Lecea L, Mabbott NA et al (2022) Single cell and spatial transcriptomic analyses reveal microglia-plasma cell crosstalk in the brain during Trypanosoma brucei infection. Nat Commun 13:5752. https://doi.org/10.1038/s41467-022-33542-z
10. Winkler EA, Kim CN, Ross JM, Garcia JH, Gil E, Oh I, Chen LQ, Wu D, Catapano JS, Raygor K et al (2022) A single-cell atlas of the normal and malformed human brain vasculature. Science 375:eabi7377. https://doi.org/10.1126/science.abi7377
11. Jackson HW, Fischer JR, Zanotelli VRT, Ali HR, Mechera R, Soysal SD, Moch H, Muenst S, Varga Z, Weber WP et al (2020) The single-cell pathology landscape of breast cancer. Nature 578:615–620. https://doi.org/10.1038/s41586-019-1876-x
12. Kinkhabwala A, Herbel C, Pankratz J, Yushchenko DA, Rüberg S, Praveen P, Reiß S, Rodriguez FC, Schäfer D, Kollet J et al (2022) MACSima imaging cyclic staining (MICS) technology reveals combinatorial target pairs for CAR T cell treatment of solid tumors. Sci Rep 12:1911. https://doi.org/10.1038/s41598-022-05841-4
13. Winkler EA, Kim CN, Ross JM, Garcia JH, Gil E, Oh I, Chen LQ, Wu D, Catapano JS, Raygor K et al (2022) Single cell atlas of the normal and malformed human brain vasculature. Science 375:eabi7377. https://doi.org/10.1126/science.abi7377
14. Mund A, Coscia F, Kriston A, Hollandi R, Kovács F, Brunner A-D, Migh E, Schweizer L, Santos A, Bzorek M et al (2022) Deep visual proteomics defines single-cell identity and heterogeneity. Nat Biotechnol 40:1231–1240. https://doi.org/10.1038/s41587-022-01302-5
15. Piehowski PD, Zhu Y, Bramer LM, Stratton KG, Zhao R, Orton DJ, Moore RJ, Yuan J, Mitchell HD, Gao Y et al (2020) Automated mass spectrometry imaging of over 2000 proteins from tissue sections at 100-μm spatial resolution. Nat Commun 11:8. https://doi.org/10.1038/s41467-019-13858-z
16. Bruker Corporation, BC (2016) Bruker introduces innovative timsTOF mass spectrometer. Analytical Science News. https://analyticalscience.wiley.com/do/10.1002/sepspec.15526f331b8/
17. Meier F, Beck S, Grassl N, Lubeck M, Park MA, Raether O, Mann M (2015) Parallel Accumulation–Serial Fragmentation (PASEF): multiplying sequencing speed and sensitivity by synchronized scans in a trapped ion mobility device. J Proteome Res 14:5378–5387. https://doi.org/10.1021/acs.jproteome.5b00932
18. Zhu Y, Piehowski PD, Zhao R, Chen J, Shen Y, Moore RJ, Shukla AK, Petyuk VA, Campbell-Thompson M, Mathews CE et al (2018) Nanodroplet processing platform for deep and quantitative proteome profiling of 10–100 mammalian cells. Nat Commun 9:882. https://doi.org/10.1038/s41467-018-03367-w
19. Budnik B, Levy E, Harmange G, Slavov N (2018) SCoPE-MS: mass spectrometry of single mammalian cells quantifies proteome heterogeneity during cell differentiation. Genome Biol 19:161. https://doi.org/10.1186/s13059-018-1547-5
20. Specht H, Emmott E, Petelski AA, Huffman RG, Perlman DH, Serra M, Kharchenko P, Koller A, Slavov N (2021) Single-cell proteomic and transcriptomic analysis of macrophage heterogeneity using SCoPE2. Genome Biol 22:50. https://doi.org/10.1186/s13059-021-02267-5
21. Hartlmayr D, Ctortecka C, Seth A, Mendjan S, Tourniaire G, Mechtler K (2021) An automated workflow for label-free and multiplexed single cell proteomics sample preparation at unprecedented sensitivity. Preprint at bioRxiv. https://doi.org/10.1101/2021.04.14.439828
22. Meier F, Brunner A-D, Koch S, Koch H, Lubeck M, Krause M, Goedecke N, Decker J, Kosinski T, Park MA et al (2018) Online Parallel Accumulation-Serial Fragmentation (PASEF) with a novel trapped ion mobility mass spectrometer. Mol Cell Proteomics 17:2534–2545. https://doi.org/10.1074/mcp.TIR118.000900
23. Stewart HI, Grinfeld D, Giannakopulos A, Petzoldt J, Shanley T, Garland M, Denisov E, Peterson AC, Damoc E, Zeller M et al (2023) Parallelized acquisition of orbitrap and astral analyzers enables high-throughput quantitative analysis. Anal Chem 95:15656–15664. https://doi.org/10.1021/acs.analchem.3c02856
24. Brunner A-D, Thielert M, Vasilopoulou C, Ammar C, Coscia F, Mund A, Hoerning OB, Bache N, Apalategui A, Lubeck M et al (2022) Ultra high sensitivity mass spectrometry

quantifies single-cell proteome changes upon perturbation. Mol Syst Biol 18:10.15252/msb.202110798

25. Demichev V, Szyrwiel L, Yu F, Teo GC, Rosenberger G, Niewienda A, Ludwig D, Decker J, Kaspar-Schoenefeld S, Lilley KS et al (2022) dia-PASEF data analysis using FragPipe and DIA-NN for deep proteomics of low sample amounts. Nat Commun 13:3944. https://doi.org/10.1038/s41467-022-31492-0
26. Staes A, Maia T, Dufour S, Bouwmeester R, Gabriels R, Martens L, Impens F, Devos S (2023) Benchmarking DIA data analysis workflows. Preprint at bioRxiv. https://doi.org/10.1101/2023.06.02.543441
27. Bustamante JG, Zaidi SRH (2023) Amyloidosis. In: StatPearls. StatPearls Publishing
28. Dasari S, Theis JD, Vrana JA, Rech KL, Dao LN, Howard MT, Dispenzieri A, Gertz MA, Hasadsri L, Highsmith WE et al (2020) Amyloid typing by mass spectrometry in clinical practice: a comprehensive review of 16,175 samples. Mayo Clin Proc 95:1852–1864. https://doi.org/10.1016/j.mayocp.2020.06.029
29. Picken MM, Herrera GA (2007) The burden of "sticky" amyloid: typing challenges. Arch Pathol Lab Med 131:850–851. https://doi.org/10.5858/2007-131-850-TBOSAT
30. Satoskar AA, Burdge K, Cowden DJ, Nadasdy GM, Hebert LA, Nadasdy T (2007) Typing of amyloidosis in renal biopsies: diagnostic pitfalls. Arch Pathol Lab Med 131:917–922. https://doi.org/10.5858/2007-131-917-TOAIRB
31. Solomon A, Murphy CL, Westermark P (2008) Unreliability of immunohistochemistry for typing amyloid deposits. Arch Pathol Lab Med 132:14; author reply 14-15. https://doi.org/10.5858/2008-132-14b-IR
32. Canetti D, Brambilla F, Rendell NB, Nocerino P, Gilbertson JA, Di Silvestre D, Bergamaschi A, Lavatelli F, Merlini G, Gillmore JD et al (2021) Clinical amyloid typing by proteomics: performance evaluation and data sharing between two centres. Molecules 26:1913. https://doi.org/10.3390/molecules26071913
33. Vrana JA, Gamez JD, Madden BJ, Theis JD, Bergen HR, Dogan A (2009) Classification of amyloidosis by laser microdissection and mass spectrometry-based proteomic analysis in clinical biopsy specimens. Blood 114:4957–4959. https://doi.org/10.1182/blood-2009-07-230722
34. Winter M, Tholey A, Kristen A, Röcken C (2017) MALDI mass spectrometry imaging: a novel tool for the identification and classification of amyloidosis. Proteomics 17:1700236. https://doi.org/10.1002/pmic.201700236
35. Canetti D, Rendell NB, Gilbertson JA, Botcher N, Nocerino P, Blanco A, Vagno LD, Rowczenio D, Verona G, Mangione PP et al (2020) Diagnostic amyloid proteomics: experience of the UK National Amyloidosis Centre. Clin Chem Lab Med 58:948–957. https://doi.org/10.1515/cclm-2019-1007
36. Benson MD, Buxbaum JN, Eisenberg DS, Merlini G, Saraiva MJM, Sekijima Y, Sipe JD, Westermark P (2018) Amyloid nomenclature 2018: recommendations by the International Society of Amyloidosis (ISA) nomenclature committee. Amyloid 25:215–219. https://doi.org/10.1080/13506129.2018.1549825
37. Van Haver D, Dendooven A, Impens F (2023) Proteomics-based analysis and diagnosis of formalin-fixed paraffin-embedded amyloidosis samples. In: Gevaert K (ed) Mass spectrometry-based proteomics methods in molecular biology. Springer US, pp 213–233. https://doi.org/10.1007/978-1-0716-3457-8_12
38. Schwanhäusser B, Busse D, Li N, Dittmar G, Schuchhardt J, Wolf J, Chen W, Selbach M (2011) Global quantification of mammalian gene expression control. Nature 473:337–342. https://doi.org/10.1038/nature10098
39. Demichev V, Messner CB, Vernardis SI, Lilley KS, Ralser M (2020) DIA-NN: neural networks and interference correction enable deep proteome coverage in high throughput. Nat Methods 17:41–44. https://doi.org/10.1038/s41592-019-0638-x
40. Marchione DM, Ilieva I, Devins K, Sharpe D, Pappin DJ, Garcia BA, Wilson JP, Wojcik JB (2020) HYPERsol: high-quality data from archival FFPE tissue for clinical proteomics. J Proteome Res 19:973–983. https://doi.org/10.1021/acs.jproteome.9b00686
41. S-Trap™ micro spin column digestion protocol (203AD)
42. Loading protocol for Evotip Pure (2022)
43. Frankenfield AM, Ni J, Ahmed M, Hao L (2022) Protein contaminants matter: building universal protein contaminant libraries for DDA and DIA proteomics. J Proteome Res 21:2104–2113. https://doi.org/10.1021/acs.jproteome.2c00145
44. Boja ES, Fales HM (2001) Overalkylation of a protein digest with Iodoacetamide. Anal Chem 73:3576–3582. https://doi.org/10.1021/ac0103423

Chapter 21

Decision Tree for Protein Biomarker Selection for Clinical Applications

Katharina Waury

Abstract

Discovery of novel protein biomarkers for clinical applications is an active research field across a manifold of diseases. Despite some successes and progress, the biomarker development pipeline still frequently ends in failure. Biomarker candidates that are discovered by appropriate technologies such as unbiased mass spectrometry cannot be validated or translated to immunoassays in many cases. Selection of strong disease biomarker candidates that further constitute suitable targets for antibody binding in immunoassays is thus important to allow routine clinical use. This essential selection step can be supported and rationalized using bioinformatics tools such as protein databases. Here, we present a workflow in the form of decision trees to computationally investigate biomarker candidates and their available affinity reagents in depth. This analysis can identify the most promising biomarker candidates for assay development, while minimal time and effort are required.

Key words Protein biomarker, Bioinformatics, Online databases, Immunoassays, Antibodies

1 Introduction

Biomarkers are an important cornerstone of clinical research and practice as they support patient diagnosis, disease monitoring, and the implementation of clinical drug trials [1]. Especially protein biomarkers have received increased interest, and the use of fluid protein markers in clinical practice is now strongly established for multiple diseases [2, 3]. Nevertheless, many needs in the medical field are still unmet, and the development of protein biomarkers is an ongoing and active research field [4, 5]. Novel candidates are constantly reported, often based on the results of high-throughput proteomics studies of body fluids that compare the protein concentrations between diseased and healthy groups to identify proteins with meaningful differences in concentration [5]. Despite the plethora of candidate proteins, rarely do any biomarkers reach the clinical practice, as the biomarkers pipeline is challenging and

Taufika Islam Williams (ed.), *Tissue Proteomics: Methods and Protocols*, Methods in Molecular Biology, vol. 2884, https://doi.org/10.1007/978-1-0716-4298-6_21,

failure-prone after the discovery phase [6]. One especially prevalent bottleneck is the technology-translation gap between the technologies used for discovery and the ones used for validation and clinical implementation [4, 7]. Unbiased mass spectrometry is ideally suited for the discovery phase. However, equipment costs, tedious sample preparation, and required staff expertise limit its use in a clinical setting [8]. Note that continuous innovations of this technology show promise to translate mass spectrometry systems from research to clinical applications in the future [8]. While not suitable for large-scale biomarker discovery, immunoassays are currently still essential for using biomarkers in routine clinical practice [9]. Thus, to facilitate clinical biomarkers, it is important to establish that a protein is 1) a reliable marker for the disease or biological process of interest and 2) a feasible target molecule in an immunoassay. Examination and confirmation of these two properties using laboratory experiments are time-consuming and costly. Considering that biomarker discovery proteomics studies often lead to a long list of candidates, it is not feasible to test all proteins thoroughly, and a selection of the most promising candidates is required [7].

Bioinformatics resources have been suggested to be a suitable approach to support this selection process [7, 10]. Importantly, this step requires minimal effort while potentially saving a vast amount of energy and costs otherwise spent on unsuccessful assay development. The inclusion of information provided in online databases and other resources would allow for corroboration of two main attributes for an investigated protein:

1. The protein is a strong biomarker candidate for a disease of interest based on available information including its known disease associations, function, and interactions.
2. The protein is a suitable target for antibody binding to facilitate its use in clinical immunoassays considering the available antibody reagents and the antibody-binding region within the protein.

This chapter introduces biomarker researchers to the available resources and illustrates how to utilize the provided information for biomarker candidate selection. First, all proposed online resources for biomarker candidate screening are summarized in the following *Resources* section. Next, in the *Methods* section, a workflow in the form of two decision trees is presented. Each decision tree is suited to confirm one of the two aforementioned, desired attributes of clinical biomarkers using computational resources.

2 Resources

The suggested resources for a biomarker suitability survey are described shortly below. All included tools are free to use, are thoroughly maintained and updated, and are meant to be easily usable by non-bioinformatics researchers. All tools are summarized in Table 1 which also provides the current web address and the year

Table 1
Overview of bioinformatics databases for biomarker candidate selection

Database	Information	URL	Ref.	Last update
UniProt.	Various	https://www.uniprot.org/	[11]	Cont.
MarkerDB	Biomarkers-disease associations	https://markerdb.ca/	[12]	2020
DisGeNet	Gene-disease associations, biomarker-disease associations	https://www.disgenet.org/; https://www.disgenet.org/biomarkers/	[13] [14],	2020, 2022
STRING	Protein-protein interactions	https://string-db.org/	[15]	2023
HBFP	Body fluid proteins	https://bmbl.bmi.osumc.edu/HBFP/	[16]	2021
HPA	Various	https://www.proteinatlas.org/	[17]	2023
ProteomicsDB	Various	https://www.proteomicsdb.org/	[18]	Cont.
PaxDb	Protein abundance	https://www.pax-db.org/	[19]	2023
SheddomeDB	Ectodomain shedding proteins	https://bal.lab.nycu.edu.tw/sheddomedb/	[20]	2023
Vesiclepedia	Extracellular vesicle cargo proteins	http://microvesicles.org/	[22]	2023
Antibodypedia	Antibody validation	https://www.antibodypedia.com/	[23]	Cont.
CiteAb	Antibody validation	https://www.citeab.com/	[24]	Cont.
iPTMnet	PTMs	https://research.bioinformatics.udel.edu/iptmnet/	[25]	2024
InterPro	Domains	https://www.ebi.ac.uk/interpro/	[27]	Cont.
AlphaFold database	Predicted protein structures	https://alphafold.ebi.ac.uk/	[29]	2022
MobiDB	Disorder, interacting residues	https://mobidb.bio.unipd.it/	[31]	2022

Cont. Continuously updated, *HBFP* Human Body Fluid Proteome, *HPA* Human Protein Atlas, *PTM* Post-translational modification

of their last major update (*see* **Note 1**). The suggested use of these tools in the context of biomarker candidate discovery and selection is explained in the following section.

(a) The UniProt Knowledge Database [11] is the most comprehensive data resource for proteins currently available. Especially for human proteins, a wide range of high-quality, manually curated annotations is collected and presented in a consistent manner. The database is constantly updated, and new cross-references to other major bioinformatics tools are included. Important annotations for each protein include gene ontology terms, subcellular location, and post-translational modifications (PTMs), including cross-links and disulfide bridges. UniProt provides the canonical sequence and potential isoforms of each protein.

(b) The MarkerDB database [12] collects all known clinical and pre-clinical biomarkers in one resource. It currently contains 142 protein biomarkers for which information is provided on the associated condition and the biomarker type. It also provides the currently established biomarker concentration cutoffs in biological fluids used in clinical practice.

(c) Gene-disease associations are compiled in DisGeNet [13] based on scientific literature, curation by experts, and high-throughput studies, e.g., GWAS. Disease associations at the gene level can indicate a role of the linked protein, and thus, this information can be suitable to identify protein biomarkers. The collection of relevant publications provides an estimate of how long and how well a specific disease association has already been studied by the scientific community. Recently, the DisGeNet database was expanded to include the Clinical Biomarker App [14]. Here, currently, over 3000 proteins are presented regarding the diseases and clinical trials they are utilized for as a biomarker.

(d) STRING [15] is a protein-protein interaction database that derives its information from curations, scientific literature by text mining, and computational predictions (*see* **Note 2**). As a large number of organisms are included, it is important to select the human proteome when searching for interactions of a protein of interest. The interactions are visualized as a graph network which also includes the interplay between a protein's interaction partners. Every interaction is annotated with a confidence score based on the available evidence.

(e) The Human Body Fluid Proteome database [16] currently contains over 11,000 proteins detected in 17 body fluids. For many less well-researched fluids, this might be the single resource available for collecting the results of multiple proteomics studies. Every protein entry provides a confidence score per body fluid, and the references of the studies are included in the database.

(f) The Human Protein Atlas [17] is a constantly expanding project that aims to map the entire human proteome regarding various aspects. Presently, it contains sections on tissue and cell type expression, subcellular location, and several specific atlases, e.g., of the brain proteome and the blood proteome. Importantly, the atlas provides isoform-specific sequence annotations. Different technologies were used to collect the information, such as RNA sequencing, mass spectrometry, and immunoassays.

(g) The ProteomicsDB [18] is a mainly mass spectrometry–focused resource which is continuously expanded and updated. It contains a wide range of information for the human protein isoforms including peptides identified by mass spectrometry and proteolytic cleavage sites. Importantly, the database visualizes tissue-specific protein expression for every protein and provides a feature viewer which includes PTM, domain, and structure annotations.

(h) The Protein Abundance Database PaxDb [19] collects protein quantification datasets to provide a unified resource of protein abundances across different tissues. Integrated datasets in the database provide very high coverage of the human proteome, and tissue-specific datasets are also available.

(i) SheddomeDB [20] is a recently updated database of known ectodomain shedding proteins based on the curation of scientific literature. Albeit containing less than 500 entries across multiple organisms, SheddomeDB is a valuable resource as ectodomain shedding is an underrepresented annotation despite its importance for biomarker research [20].

(j) Vesiclepedia [21, 22] is a resource that collects studies on extracellular vesicles (EVs) and the identified cargo, including proteins. Each protein entry contains information about the studies which confirmed its association with extracellular vesicles. Details on study design, sample material, and vesicle type are also provided.

(k) Antibodypedia [23] is a resource to search for antibodies and the results of their validation experiments. Data can be provided by antibody suppliers as well as researchers but is reviewed before inclusion. Importantly, validation data can be searched for a specific application which allows researchers to find antibodies most suitable for their intended use.

(l) The CiteAb database [24] ranks included antibodies based on their references in scientific publications. This approach lets researchers discover the reagents that are most trusted by the research community and provides a simple strategy to find all the relevant literature which can also be filtered regarding the desired application type.

(m) The database iPTMnet [25] contains detailed information on known protein modification sites. These PTM sites are visualized across variants and isoforms and are annotated with confidence scores based on the available evidence. As iPTMnet integrates knowledge from several other databases, it contains a comprehensive overview of current information on PTMs. Note that a manifold of other PTM databases exists, often specific for one modification type. These databases have been reviewed and collected previously and offer a starting point to extensively investigate available PTM information [26].

(n) InterPro [27] integrates data from several other resources. The database classifies proteins into families based on annotations regarding domains and motifs and visualizes those along the protein sequence. This information is provided across all isoforms of a protein. InterPro can also be used to discover proteins similar to the protein of interest.

(o) The Basic Local Alignment Search Tool (BLAST) [28] can identify regions of high sequence similarity both at a nucleotide and protein level. The tool can be used to find similar proteins within a proteome set, such as the Swiss-Prot human proteome, for a protein sequence or sequence fragment of interest.

(p) The AlphaFold Database [29] is the largest collection of high-quality protein structures that were predicted by a state-of-the-art machine learning model [30] (*see* **Note 2**). Almost the entire human proteome is covered. For each protein entry, the predicted structure is displayed, which includes information regarding the confidence of the predictions. The tool menu allows us to easily highlight protein regions of interest, and both the structure models and screenshots thereof can be easily downloaded.

(q) MobiDB [31] is a database focused on protein structure and binding annotations. It provides information on disordered and structured regions as well as linear interacting peptides within a protein. These properties are visualized along the sequence, which facilitates the identification of longer stretches of disordered coil regions within a protein. The annotations are based both on experimental observations and machine learning predictions (*see* **Note 2**).

3 Methods

Here, a protocol is presented to facilitate the rationalized selection of biomarker candidates using available online bioinformatics tools, e.g., curated protein databases. Decision trees illustrate and summarize how these tools can be utilized to select proteins that are strong disease biomarker candidates (Fig. 1) and to select proteins that are suitable targets in antibody-based immunoassays (Fig. 2). The decision trees show, for each protocol step, which supportive or discouraging conclusions regarding a protein's suitability as a biomarker can be made. Supportive results suggest favoring a protein, while discouraging results might lean toward excluding a protein from further biomarker development or suggest that special care needs to be applied to address a potential issue. If contradictory results exist either between different online resources or between online resources and one's own experimental results, care must be taken about which information source to trust the most. Especially results of prediction tools might be systematically biased and should thus not be trusted at the same level as available

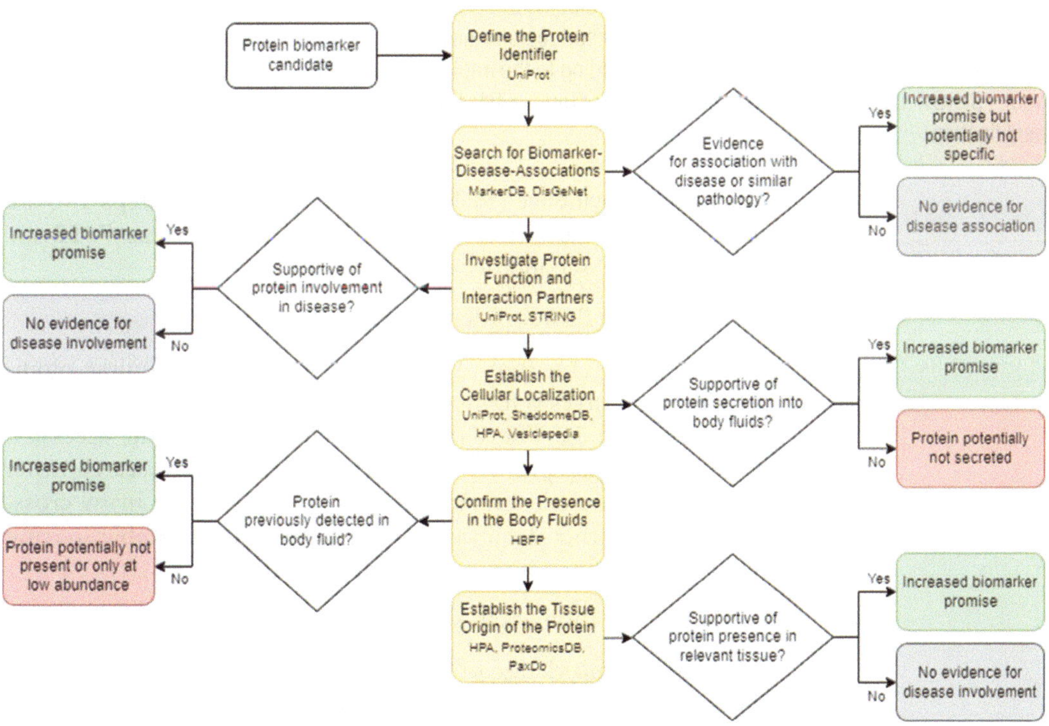

Fig. 1 Decision tree to establish a protein candidate as a marker of disease. Steps to evaluate the suitability of a protein as a specific disease marker are shown as a decision tree flow chart. Yellow boxes indicate the steps to follow as well as the advised tools to use. Each step attempts to answer the indicated question. The outcome boxes are colored according to the implied assessment: supportive (green), opposing (red), or neutral (grey). *HBFP* Human Body Fluid Proteome, *HPA* Human Protein Atlas

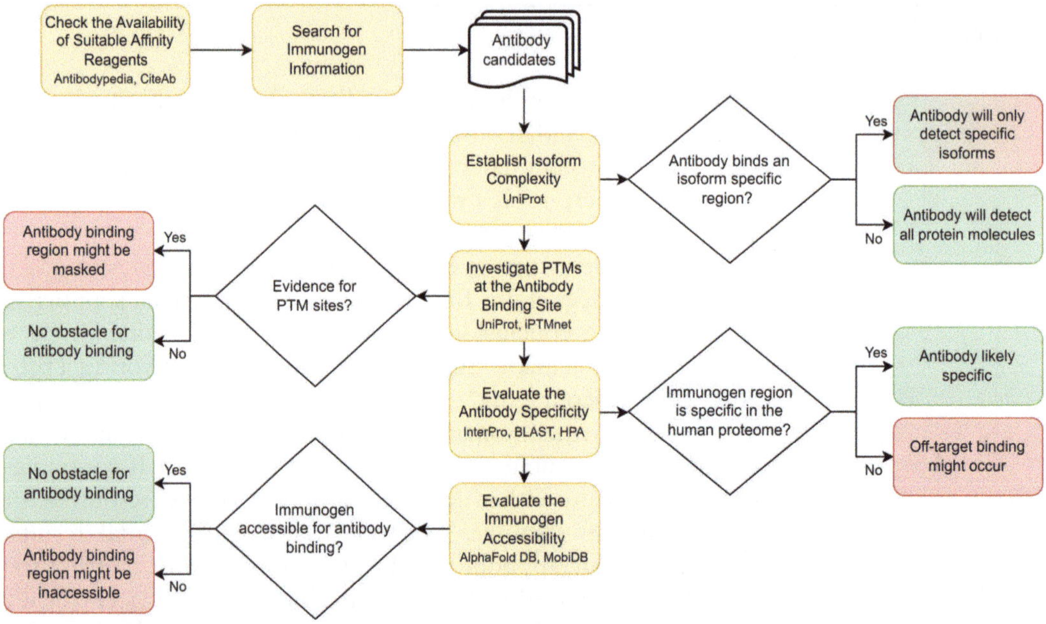

Fig. 2 Decision tree to establish a protein as a suitable assay target. Steps to evaluate the suitability of a protein as an immunoassay target for antibody binding are shown as a decision tree flow chart. Yellow boxes indicate the steps to follow as well as the advised tools to use. Each step attempts to answer the indicated question. The outcome boxes are colored according to the implied assessment: supportive (green) or opposing (red). *BLAST* Basic Local Alignment Search Tool, *HPA* Human Protein Atlas, *PTM* Post-translational modification

experimental evidence (*see* **Notes 1** and **2**). Protocol **steps 1–6** are part of the first decision tree (Fig. 1); protocol **steps 7–12** are included in the second decision tree (Fig. 2).

1. Define the Protein Identifier

The first important step to examine a protein using online bioinformatics resources is to identify its associated identifier in UniProt [11] (*see* **Note 3**). The UniProt ID (also known as the accession number) is highly preserved and is often required to search other protein database entries. Additionally, because of the many cross-references to other databases, UniProt provides a suitable starting point to examine online information for a protein.

2. Search for Biomarker-Disease Associations

Any existing knowledge of a biomarker candidate's association on the protein or gene level with pathologies should be investigated. Next to a conventional search in databases of scientific literature, e.g., PubMed or Google Scholar, specific databases for biomarkers and disease associations, i.e., MarkerDB [12] and DisGeNet [13, 14], can be explored for a more concise overview. A protein's association with a disease of similar pathology or the same involved pathways as the disease of interest might constitute a strong biomarker candidate. On the other hand, associations with other diseases than the one of interest might limit the disease specificity of a biomarker, especially for diseases which are known to co-occur substantially.

3. Investigate Protein Function and Interaction Partners

A protein's functions and interactions increase the understanding of the mechanisms through which a protein is potentially involved in disease pathology. Protein function can be examined on UniProt [11] in the form of gene ontology terms. The involvement of a protein in a pathway that is affected by the investigated pathology bolsters its role as a biomarker. Protein-protein interaction partners listed in the STRING database [15] that are known to be involved in the disease of interest also strengthen a protein's biomarker capacity.

4. Establish the Cellular Localization of the Protein

The subcellular localization of a protein might also be of interest as it informs on the protein's role within the cell. Furthermore, it has been shown that the location within the cell influences a protein's probability of being secreted into the body fluids [32]; a protein found at the membrane or extracellular space might thus be more suitable as a fluid biomarker. UniProt [11] and the Human Protein Atlas [17] can be explored for this information. Ectodomains which can be shed into the body fluids as well as association with extracellular vesicles are relevant properties that might affect a protein's biomarker suitability [32]. Both circumstances can lead to a protein being present in a body fluid. Specialized databases, SheddomeDB [20] and Vesiclepedia [21], include this type of information.

5. Confirm the Presence of the Protein in the Body Fluids

Protein biomarkers are preferably measured in body fluids because of the vast amount of secreted proteins present and the non-invasive sample extraction [5]. Important body fluids for biomarker discovery include plasma, urine, and cerebrospinal fluid [16]. The detection of a biomarker candidate in the body fluid of interest can be confirmed in the Human Body Fluid Proteome database [16].

6. Establish the Tissue Origin of the Protein

It is beneficial to establish the protein's likely tissue origin if it is measured in a body fluid. A protein that is highly and specifically expressed in the tissue affected by the pathology of interest has strong biomarker potential. For example, the brain is of the highest interest regarding the search for novel biomarkers for dementia such as Alzheimer's disease. Ubiquitous expression convolutes the determination of a biomarker's origin. If the biomarker candidate is very lowly expressed in the tissue of interest, it might only be detectable in body fluids with ultrasensitive methods [33]. Information regarding protein expression and abundance in the different human tissues can be checked in the databases of the Human Protein Atlas [17], ProteomicsDB [18], and PaxDb [19].

7. Check the Availability of Suitable Affinity Reagents

After strong candidates for disease protein biomarkers have been identified following **steps 1–6**, their suitability as a target molecule for detection by antibodies in an immunoassay setup, e.g., an enzyme-linked immunosorbent assay (ELISA), is evaluated in the last six protocol steps. Note that these steps should be done individually for each protein that constitutes a strong biomarker candidate according to the previous analysis.

Firstly, the availability of affinity reagents, i.e., commercial antibodies or immunoassay kits, should be reviewed for each protein. The antibody databases Antibodypedia [23] and CiteAb [24] as well as other online resources, e.g., Antibodies Online (https://www.antibodies-online.com), Linscott's Directory (https://www.linscottsdirectory.com/), and Biocompare (https://www.biocompare.com/), can be searched. Attention should be paid to information regarding each antibody's clonality, the host organism, the immunogen used for animal immunization, and for which application the antibody was already validated [34]. This step leads to a list of potential antibodies for immunoassay development.

8. Search for Immunogen Information

Available information regarding the antibody's immunogen, i.e., the peptide used for antibody generation, should be identified. The more well defined the immunogen is, the better it is known what site on the protein the antibody recognizes and binds in the assay. This allows further assessment of the antibody's suitability for the desired immunoassay application. If the immunogen is not disclosed on the supplier's webpage, requesting the information directly from the supplier is possible but might not be successful [34].

9. Establish Isoform Complexity

It is important to consider if an antibody will only detect a subset of isoforms of a biomarker candidate. If the immunogen of a considered antibody lies in a region affected by alternative splicing, the isoforms bound and detected by this antibody are limited. This is an advantage if only a specific isoform of a biomarker is desired to be detected but is detrimental if all protein molecules should be measured in the assay. UniProt [11] provides all known isoforms for a protein.

10. Investigate Post-translational Modifications at the Antibody Binding Site

PTMs within the binding region of an antibody could mask the recognition site and thus hinder the biomarker-antibody interaction required for a successful assay. UniProt [11] provides location-specific information on known PTMs. Specified PTM databases, e.g., iPTMnet [25], often contain additional knowledge, and their combined annotations will thus provide the most complete picture. PTM-specific antibodies exist and are in certain cases specifically required.

11. Evaluate the Antibody Specificity

Off-target binding of commercial antibodies is a consistently reported issue [35]. As validation of antibody specificity in complex fluids is a strenuous process, it is helpful to exclude likely unspecific antibody recognition sites beforehand using bioinformatics tools. If the immunogen is an entire domain of the protein of interest, InterPro [27] annotations can be checked to identify how common this domain is across the human proteome. If a shorter immunogenic peptide was utilized as the immunogen, the specificity of this sequence can be evaluated using BLAST [28]. The Human Protein Atlas [17] provides information for each human protein on its maximum sequence identity with the rest of the human proteome using both a 10- and 50-residue window approach. This calculation allows us to easily identify short and longer regions within a protein's amino acid sequence that are highly similar to other proteins. If an antibody binds in a region of high sequence similarity, cross-reactivity might be an issue.

12. Evaluate the Immunogen Accessibility

If an immunogenic peptide was used for immunization, there is no guarantee that it will be accessible to the antibody within the full-length, folded protein structure. While this is less of an issue in applications that contain a denaturation step, e.g., western blots, this can hinder antibody binding in applications detecting the native protein, e.g., ELISAs [36]. If the immunogenic peptide is disclosed, its location within the folded protein structure can be inspected to evaluate its accessibility for antibody binding. The AlphaFold Database [37] can be accessed to visualize the immunogen region within the predicted structure. An immunogen located at the surface and preferably in a disordered coil region might be preferred over a peptide positioned within the core of the protein and thus not accessible for antibody binding in an immunoassay application. MobiDB [31] can be accessed to identify if an immunogen is located within a region annotated as disordered. Higher disorder implies the protein is in an unfolded coil structure which is preferred for antibodies that were raised against immunogenic peptides.

4 Notes

1. Public bioinformatics databases differ in the extent to which they are updated and curated. The most trustworthy resources are maintained or supported by major research institutes and organizations, including ELIXIR, the European Bioinformatics Institute (EMBL-EBI), or the National Institute of Health (NIH). Involved organizations are often displayed by banners on the homepage which are a strong indicator that a resource contains accurate and up-to-date information. Table 1 informs on the last known update for all included tools.
2. When using the results of prediction tools instead of experimentally validated annotations, caution regarding the reliability of this type of information is important. While extremely helpful, machine learning–based predictors are never fully accurate, and the performance of an included tool should always be contemplated. It is advised to research the general performance of prediction tools within a specific field, e.g., protein structure prediction, to assess how trustworthy they are.
3. If UniProt is searched by protein or gene name, several protein entries will frequently be returned. To identify the correct entry for the protein of interest, it is important to filter for human proteins and preferably for Swiss-Prot entries. These entries are manually and extensively curated in comparison to the TrEMBL section of UniProt. For the human proteome, it should be expected to always find a Swiss-Prot entry as these proteins are well researched and of high importance to the research community.

Acknowledgements

This work is supported by the European Union's Horizon 2020 research and innovation program under the Marie Skłodowska-Curie grant agreement No. 860197, the MIRIADE project. The author would like to acknowledge the fellow researchers from the MIRIADE consortium who provided valuable insights and interesting discussions regarding fluid biomarker development over the years.

References

1. Califf RM (2018) Biomarker definitions and their applications. Exp Biol Med (Maywood) 243(3):213–221
2. Blennow K, Hampel H, Weiner M, Zetterberg H (2010) Cerebrospinal fluid and plasma biomarkers in Alzheimer disease. Nat Rev Neurol 6(3):131–144
3. Hernández J, Thompson IM (2004) Prostate-specific antigen: a review of the validation of the most commonly used cancer biomarker. Cancer 101(5):894–904

4. Mavrina E, Kimble L, Waury K, Gogishvili D, De San G, José N, Das S et al (2022) Multi-omics interdisciplinary research integration to accelerate dementia biomarker development (MIRIADE). Front Neurol 13:890638
5. Dayon L, Cominetti O, Affolter M (2022) Proteomics of human biological fluids for biomarker discoveries: technical advances and recent applications. Expert Rev Proteomics 19(2):131–151
6. Rifai N, Gillette MA, Carr SA (2006) Protein biomarker discovery and validation: the long and uncertain path to clinical utility. Nat Biotechnol 24(8):971–983
7. Waury K, Willemse EAJ, Vanmechelen E, Zetterberg H, Teunissen CE, Abeln S (2022) Bioinformatics tools and data resources for assay development of fluid protein biomarkers. Biomark Res 10(1):83
8. Zhou X, Zhang W, Ouyang Z (2022) Recent advances in on-site mass spectrometry analysis for clinical applications. TrAC Trends Anal Chem 149:116548
9. Del Campo M, Jongbloed W, Twaalfhoven HAM, Veerhuis R, Blankenstein MA, Teunissen CE (2015) Facilitating the validation of novel protein biomarkers for dementia: an optimal workflow for the development of sandwich immunoassays. Front Neurol 6(202)
10. Lin Y, Qian F, Shen L, Chen F, Chen J, Shen B (2019) Computer-aided biomarker discovery for precision medicine: data resources, models and applications. Brief Bioinform 20(3): 952–975
11. The UniProt Consortium, Bateman A, Martin MJ, Orchard S, Magrane M, Ahmad S et al (2023) UniProt: the universal protein knowledgebase in 2023. Nucleic Acids Res 51(D1): D523–D531
12. Wishart DS, Bartok B, Oler E, Liang KYH, Budinski Z, Berjanskii M et al (2021) MarkerDB: an online database of molecular biomarkers. Nucleic Acids Res 49(D1):D1259–D1267
13. Piñero J, Ramírez-Anguita JM, Saüch-Pitarch J, Ronzano F, Centeno E, Sanz F et al (2019) The DisGeNET knowledge platform for disease genomics: 2019 update. Nucleic Acids Res:gkz1021
14. Piñero J, Rodriguez Fraga PS, Valls-Margarit J, Ronzano F, Accuosto P, Lambea Jane R et al (2023) Genomic and proteomic biomarker landscape in clinical trials. Comput Struct Biotechnol J 21:2110–2118
15. Szklarczyk D, Kirsch R, Koutrouli M, Nastou K, Mehryary F, Hachilif R et al (2023) The STRING database in 2023: protein–protein association networks and functional enrichment analyses for any sequenced genome of interest. Nucleic Acids Res 51(D1): D638–D646
16. Shao D, Huang L, Wang Y, Cui X, Li Y, Wang Y et al (2021) HBFP: a new repository for human body fluid proteome. Database
17. Uhlén M, Fagerberg L, Hallström BM, Lindskog C, Oksvold P, Mardinoglu A et al (2015) Tissue-based map of the human proteome. Science 347(6220):1260419
18. Lautenbacher L, Samaras P, Muller J, Grafberger A, Shraideh M, Rank J et al (2022) ProteomicsDB: toward a FAIR open-source resource for life-science research. Nucleic Acids Res 50(D1):D1541–D1552
19. Wang M, Herrmann CJ, Simonovic M, Szklarczyk D, Mering C (2015) Version 4.0 of PaxDb: protein abundance data, integrated across model organisms, tissues, and cell-lines. Proteomics 15(18):3163–3168
20. Huang WY, Wu KP (2023) SheddomeDB 2023: a revision of an ectodomain shedding database based on a comprehensive literature review and online resources. J Proteome Res 8. acs.jproteome.3c00001
21. Pathan M, Fonseka P, Chitti SV, Kang T, Sanwlani R, Van Deun J et al (2019) Vesiclepedia 2019: a compendium of RNA, proteins, lipids and metabolites in extracellular vesicles. Nucleic Acids Res 47(D1):D516–D519
22. Chitti SV, Gummadi S, Kang T, Shahi S, Marzan AL, Nedeva C et al (2024) Vesiclepedia 2024: an extracellular vesicles and extracellular particles repository. Nucleic Acids Res 52(D1): D1694–D1698
23. Björling E, Uhlén M (2008) Antibodypedia, a portal for sharing antibody and antigen validation data. Mol Cell Proteomics 7(10): 2028–2037
24. Helsby MA, Leader PM, Fenn JR, Gulsen T, Bryant C, Doughton G et al (2014) CiteAb: a searchable antibody database that ranks antibodies by the number of times they have been cited. BMC Cell Biol 15(1):6
25. Huang H, Arighi CN, Ross KE, Ren J, Li G, Chen SC et al (2018) iPTMnet: an integrated resource for protein post-translational modification network discovery. Nucleic Acids Res 46 (D1):D542–D550
26. Ramazi S, Zahiri J (2021) Post-translational modifications in proteins: resources, tools and prediction methods. Database 2021:baab012
27. Paysan-Lafosse T, Blum M, Chuguransky S, Grego T, Pinto BL, Salazar GA et al (2023) InterPro in 2022. Nucleic Acids Res 51(D1): D418–D427

28. Altschul S (1997) Gapped BLAST and PSI-BLAST: a new generation of protein database search programs. Nucleic Acids Res 25(17):3389–3402
29. Varadi M, Bertoni D, Magana P, Paramval U, Pidruchna I, Radhakrishnan M et al (2024) AlphaFold protein structure database in 2024: providing structure coverage for over 214 million protein sequences. Nucleic Acids Res 52 (D1):D368–D375
30. Jumper J, Evans R, Pritzel A, Green T, Figurnov M, Ronneberger O et al (2021) Highly accurate protein structure prediction with AlphaFold. Nature 596(7873):583–589
31. Piovesan D, Del Conte A, Clementel D, Monzon AM, Bevilacqua M, Aspromonte MC et al (2023) MobiDB: 10 years of intrinsically disordered proteins. Nucleic Acids Res 51(D1): D438–D444
32. Waury K, De Wit R, Verberk IMW, Teunissen CE, Abeln S (2023) Deciphering protein secretion from the brain to cerebrospinal fluid for biomarker discovery. J Proteome Res 22(9): 3068–3080
33. Teunissen CE, Kimble L, Bayoumy S, Bolsewig K, Burtscher F, Coppens S et al (2023) Methods to discover and validate biofluid-based biomarkers in neurodegenerative dementias. Mol Cell Proteomics 100629
34. Roncador G, Engel P, Maestre L, Anderson AP, Cordell JL, Cragg MS et al (2016) The European antibody network's practical guide to finding and validating suitable antibodies for research. MAbs 8(1):27–36
35. Prassas I, Brinc D, Farkona S, Leung F, Dimitromanolakis A, Chrystoja CC et al (2014) False biomarker discovery due to reactivity of a commercial ELISA for CUZD1 with cancer antigen CA125. Clin Chem 60(2): 381–388
36. Trier NH, Hansen PR, Houen G (2012) Production and characterization of peptide antibodies. Methods 56(2):136–144
37. Varadi M, Anyango S, Deshpande M, Nair S, Natassia C, Yordanova G et al (2022) AlphaFold protein structure database: massively expanding the structural coverage of protein-sequence space with high-accuracy models. Nucleic Acids Res 50(D1):D439–D444

Chapter 22

Computational Tools and Methods for the Study of Systemic Amyloidosis at the Clinical and Molecular Level

Dario Di Silvestre, Francesca Brambilla, Giampaolo Merlini, and Pierluigi Mauri

Abstract

Amyloidosis diseases are characterized by protein misfolding, which forms insoluble beta-sheet fibrils progressively deposited in tissues. Deposition in the form of amyloid aggregates can occur in various organs, damaging their structure and function. The hallmark of amyloidosis is aberrant interactions leading to protein aggregation and proteotoxicity. Accordingly, amyloidosis-related samples represent a valuable source of information to generate new knowledge useful for diagnostic, prognostic, and therapeutic purposes. In this scenario, we outline the path to apply computational methods and strategies based on the combination of proteomics and systems biology approaches. In addition to algorithms useful for subtyping amyloid deposits or assessing proteome recovery after drug treatment, our chapter provides workflows based on protein-protein interaction and protein co-expression network models. In particular, the main steps to reconstruct and analyze them at both functional and topological levels are described. Our chapter aims to provide tools and instructions to identify and monitor prognostic, diagnostic, and therapeutic markers and to shed light on the processes, pathways, and functions affected by amyloidogenic proteins.

Key words Amyloidosis, Proteomics, PPI, Protein co-expression, Hubs, Proteome recovery, Diagnosis

1 Introduction

Systemic amyloidosis is a disease whose hallmark is the misfolding of circulating proteins, which are progressively accumulated at the extracellular level in the form of insoluble amyloid beta-sheet fibrils. They are used for classifying amyloidosis according to the amyloid protein, and *to date*, more than 40 amyloidogenic proteins have

---

All authors are part of European Proteomics Amyloid Network (EPAN).

Authors Dario Di Silvestre, Francesca Brambilla, and Pierluigi Mauri are part of PROMETEO, the CNR proteome network.

Taufika Islam Williams (ed.), *Tissue Proteomics: Methods and Protocols*, Methods in Molecular Biology, vol. 2884, https://doi.org/10.1007/978-1-0716-4298-6_22,

been identified in humans, causing systemic or localized amyloidosis [1]. Amyloidosis disease can involve all organs, including the heart, kidney, liver, lung, and brain, affecting the organ's structure and function [2]. In addition to distortion of tissue structures caused by large amounts of amyloid deposits, aberrant interactions of prefibrillar aggregates with key cellular components contribute, in different proportions in different types of amyloidosis, to cell death and organ dysfunction [3].

Staining with Congo red dye is one of the major methods used to detect the amyloid structure of protein aggregates to diagnose amyloidosis in tissue sections [4]. Amyloid deposits are often laser microdissected from biopsies and then processed by MS analysis [5]. Although microdissection aims to improve diagnosis, the analysis of whole tissues/organs is paramount to illuminating molecular and histological alterations induced by amyloid aggregates in tissue-resident cells. Consequently, integrating comprehensive omics data in proper in silico models represents a promising strategy to dissect the biological mechanisms triggered by the pathological state [6]. In this context, data-derived systems biology approaches based on graph theory allow a holistic evaluation of biological samples [7]. They are more and more adopted to investigate human diseases; they are also starting to be used for investigating the mechanisms underlying the aggregation of amyloidogenic proteins, as well as those induced by amyloid deposits [6, 8–11].

Starting from the proteome characterization, we provide a workflow that will help researchers follow the steps to subtype amyloidosis and evaluate the effectiveness of pharmacological therapies. In addition, to be used for these purposes, the knowledge acquired through protein profiling and quantitation is further processed to support new insights inferred by their modeling as protein-protein interaction (PPI) and co-expression networks (Fig. 1). These models can be individually analyzed or combined for evaluating how amyloid aggregates, prefibrillar and fibrillar, affect the physiological protein relationships. In addition to characterizing functional and topological modules [7], the correlation between the network's structure and function is further exploited to identify topologically relevant proteins, called hubs [7, 12]. They occupy a specific position in the network model; therefore, we assume that they could be key in regulating amyloidosis-related processes [6].

2 Material

2.1 Molecular Data

1. Molecular profiles from amyloidosis patients and controls (*see* **Note 1**).

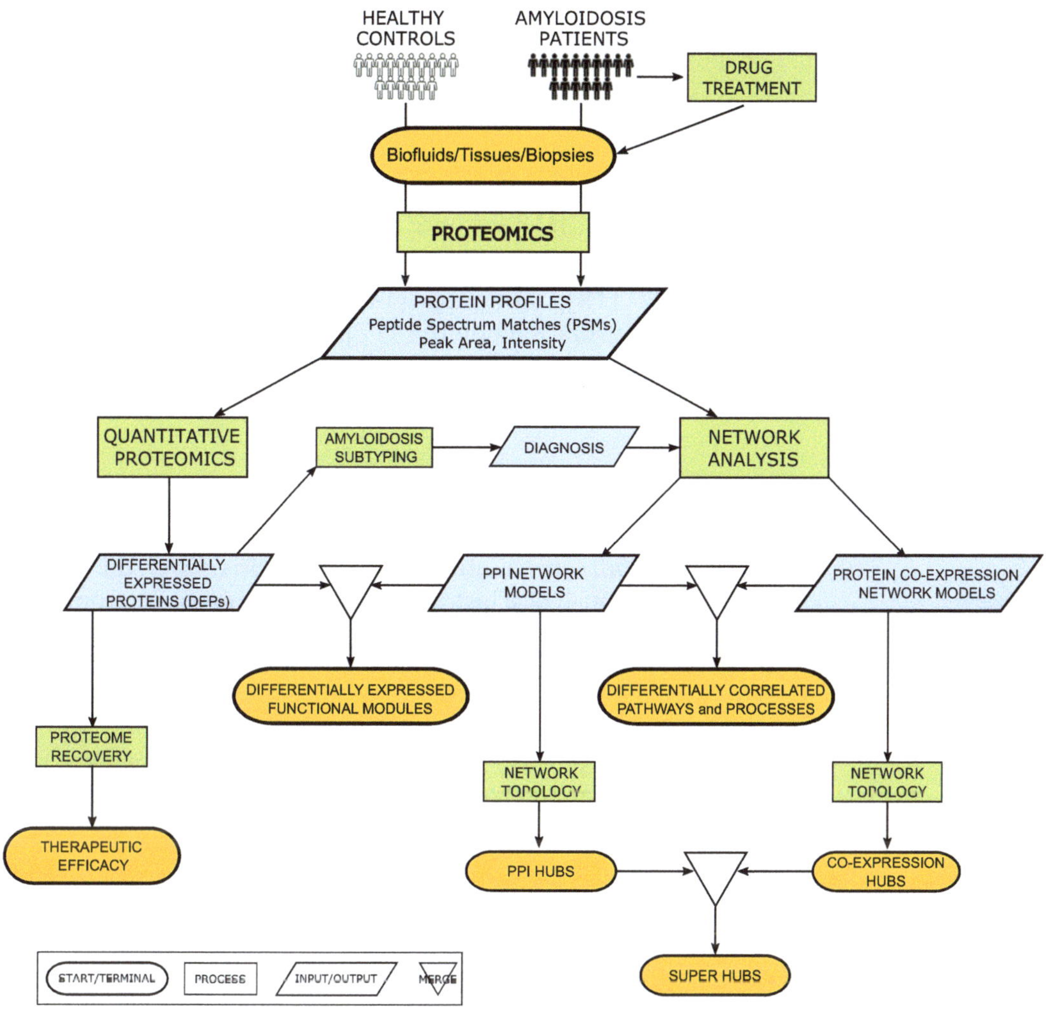

Fig. 1 Workflow showing the main steps for evaluating amyloidosis samples by means of the combination of proteomics and systems biology approaches. In addition to identifying relevant proteins and defined hubs, it allows also for amyloidosis subtyping and the evaluation of drug treatment effectiveness through proteome recovery. The achievement of goals requires specific inputs in combination with the use of software, algorithms, and computational tools which are presented and described in their fundamental aspects

2. Protein lists with quantitative parameters (arbitrary unit (AU), i.e., SpC, PSM, or intensity) (*see* **Note 2**).

2.2 Power PCs

1. Power PCs with at least 32 gb of RAM, 1 TB Hard Disk, and an Intel Core i5 Processor (equipped with Windows, Linux, macOS).

2.3 Computational Software, Tools, and Algorithms

1. Proteome Discoverer 2.5 (www.thermofisher.com), MaxQuant (www.maxquant.org), CompOmics (www.compomics.com), Proline (www.profiproteomics.fr/proline/), PEAKS Studio 11 (www.bioinfor.com/peaks-studio/).

2. Alpha-value (α-value).
3. Proteome recovery index (PR_i).
4. MAProMa software.
5. QGIS platform (www.qgis.org/de/site/).
6. R studio (github.com/rstudio/rstudio), Python (www.python.org), JMP software (www.jmp.com).
7. Cytoscape (www.cytoscape.org/) Cytoscape's APPs: Centiscape2.2 [13], stringApp [14], BiNGO [15], WordCloud [16], Network Randomizer [17], clusterMaker [16].

3 Methods

The methods described in this section will provide a guideline for processing the profiles characterized following proteomic analysis of samples from subjects affected by amyloidosis. Given the heterogeneity of the analytical instrumentation at the mass spectrometers, methods, and management software level, we assume that the quantitative analysis from protein lists is already defined as the starting point of our workflow. Major details on protocols adopted for sample preparation and mass spectrometry analysis may be retrieved in Di Silvestre et al. 2012 [18], Haver et al. 2023 [19], Spodzieja et al. 2019 [20], He et al. 2023 [21], Canetti et al. [22], and Picken et al. 2015 [23].

3.1 Quantitative Proteomics

The evaluation of proteome changes, both qualitative and quantitative, can be achieved by strategies relying on the combination of liquid chromatography and mass spectrometry (LC-MS). They encompass a wide array of experimental designs varying in structure and properties, including sample selection, sample processing (including labeling), MS acquisition methods (DIA, DDA), and bioinformatic pipelines [24]. This diversity also emerges from an overwhelming number of bioinformatic pipelines for data analysis, each using multiple tools, algorithms, and statistics. Combining results from different workflows remains a simple way to improve the performance of LC-MS, maximizing the exploitation of the experimental MS spectra [25]. However, mainly in the context of clinical proteomics, such heterogeneity highlights the challenges that still need to be addressed to enable more efficient, accurate, reproducible, and standardized analysis [26].

In this section, we summarize the label-free quantitative approach based on quantitative evaluation by arbitrary units obtained from search engines, such as spectral count (SpC), peptide spectrum match (PSM), and intensity. In addition to DAve and DCI formulas [27] and *g*-test [28], for comparing a number of limited protein profiles, linear discriminant analysis (LDA) is proposed as a statistic to evaluate protein profiles (>4 biological replicates per group) by groups characterized by specific phenotypes.

1. Starting from raw protein lists, format them by the Autoformat tool of the MAProMa software. For each identified protein, it allows for the selection of Reference, Accession number, Score, Spectral count (SpC) or Peptide Spectrum Matches (PSMs), p*I*, and MW. Of note, it may be possible to use intensity as a quantitative parameter.
2. Using the Multiconsensus Comparison tool of MAProMa, select and compare the protein lists arranged previously to obtain a $p \times s$ data matrix (p represents proteins and s samples) reporting the Score and SpC values for each protein identified (*see* **Note 3**).
3. To evaluate the analytical reproducibility in biological and technical replicates, calculate the correlation among the corresponding protein profiles in $p \times s$ data matrix (where p represents proteins and s samples where they were identified) by multivariate analysis (Fig. 2a).
4. Import the $p \times s$ data matrix in JMP 15.2, and by the menu *Analyze→Multivariate methods→Multivariate*, upload the columns corresponding to biological and technical replicates in *Y*, *Columns* box.
5. If a low analytical reproducibility is observed (correlation≤0.8), normalize the SpC values using the Total Signal normalization procedure (*see* **Note 4**). The normalized SpC values (*norm*SpC) are obtained by the following formula:

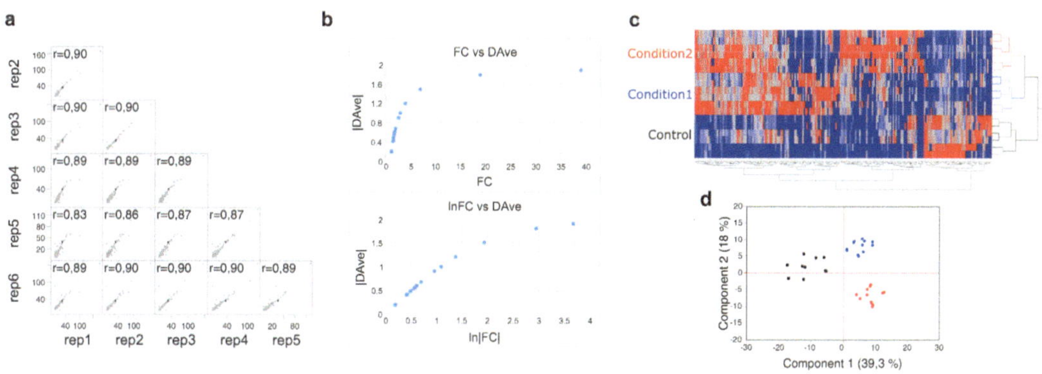

Fig. 2 Quantitative proteomics. (**a**) Correlation (Spearman) among replicate analyses using SpC values (JMP15.2). (**b**) Comparison between |DAve| and fold change (|FC|), including its natural logarithm (ln|FC|). Of note, in some cases, the fold-change (FC) calculation returns non-sense values (such as n/0 or 0/n). The DAve index presents a logarithmic trend, and it linearly correlates with FC up to 5. (**c**) Hierarchical clustering and (**d**) principal component analysis (PCA) performed using differentially expressed proteins

$$normSpC = pSpC/\sum_{p=1}^{n} SpC$$

where p and n represent respectively a specific protein and the total number of proteins identified in a given sample.

In other words, normalization is obtained by dividing the SpC value of each protein (pSpC) by the sum of the SpCs corresponding to all identified proteins (ΣSpC).

6. Pairwise comparisons to extract differentially expressed proteins (DEPs) between samples at different conditions may be performed by applying DAve (differential average) and DCI (differential confidence index) indices in MAProMa software (*see* **Note 5**). Specifically

$$\text{DAve} = \frac{(Xp - Yp)}{(Xp + Yp)}/0.5$$

$$\text{DCI} = \frac{(Xp + Yp) * (Xp - Yp)}{2}$$

where Xp is the normalized SpC of a given protein p in the first sample/condition, while Yp represents the normalized SpC of the same protein p in the second sample/condition.

7. DAve index assumes values ranging from −2 to 2 (Fig. 2b), while the DCI index assumes values from 0 to $\pm\infty$. A set of confident threshold values include ≥ 500 (or ≤ -500) for DCI and ≥ 0.5 (or ≤ -0.5) for DAve (*see* **Note 6**).

8. Pairwise comparison of samples may be further evaluated by G test according to Sokal and Rohlf [29]:

$$\text{Gtest} = 2f1\ln\left(\frac{f1}{f\hat{}1}\right) + 2f2\ln\left(\frac{f2}{f\hat{}2}\right)$$

where f_1 and f_2 are the normalized SpC for a given protein in sample 1 and sample 2, respectively, while $\hat{f}_1$ and $\hat{f}_2$ are expected SpCs. Assuming the protein is equally expressed in sample 1 and sample 2, $\hat{f}_1 = \hat{f}_2 = (f_1 + f_2)/2$.

According to the chi-square distribution table with one degree of freedom, proteins with *G test value* ≥ 3.841 are differentially expressed with $P \leq 0.05$ (*see* **Note 7**).

9. When multiple replicate analyses per condition are available (both biological, $n \geq 4$, and technical, $n \geq 2$), DEPs may be extracted by linear discriminant analysis (LDA). This analysis is performed by R or discriminant methods, including linear, quadratic, regularized, and wide linear, offered by JMP 15.2 software.

10. A $p \times s$ data matrix (p represents proteins and s samples where they were identified) is formatted so that all samples s corresponding to a specific phenotype/condition report the same tag.

11. Using the Transpose function in JMP 15.2, transpose the $p \times s$ data matrix in $s \times p$, with p as columns and s as rows.
12. By the menu *Analyze→Multivariate methods→Discriminant*, upload proteins p in Y, covariate box, and sample s in X, categories box of JMP 15.2.
13. The output result will provide a data matrix reporting the F ratio and P value associated with each protein processed. Proteins with $P < 0.05$ are selected as DEPs.
14. Using hierarchical clustering or PCA analysis, it is possible to verify the phenotype discrimination level obtained by the selected DEPs (Fig. 2c, d). If biological and technical replicates result poorly grouped by phenotype/condition, selection of DEPs with $P < 0.01$ or less is recommended.

3.2 Amyloidosis Subtyping

The α-value algorithm, proposed by Brambilla et al. [30], allows subtyping of patients with amyloidosis by estimating the relative abundance of amyloid fibril proteins [1]. Since its development, it has been increasingly used by the scientific community. Indeed, recently, in addition to subcutaneous adipose tissues, it has been successfully applied to proteomic data from cardiac tissues [31]. Of note, to be run, it requires the evaluation of the amyloidogenic-related protein levels in the specific reference tissues (healthy control) for determining the baseline noise. Although it is not always possible to characterize the baseline noise for selected amyloidogenic proteins, it is possible to deduce it from non-specific subtype affected samples.

1. Starting from the list of proteins identified in a sample collected from an amyloidosis patient, extract the quantitative parameter (arbitrary unit (AU), i.e., SpC, PSM, or intensity) corresponding to the selected amyloid fibril proteins.
2. For each amyloid fibril protein (i), the AU ratio is calculated:

$$Pi = Ij/Ib$$

   where Ij and Ib represent the AU value obtained, respectively, in samples from patient and baseline control (average of healthy subjects).
3. The α-value of the selected amyloid fibril proteins, including IgLCk, IgLCλ, TTR, and SAA, is calculated according to the following formula:

$$\alpha_{i,j} = 100^{*}(P_{i,j}/\Sigma(i = 1 \rightarrow n)P_{i,j})$$

   where:

i = selected amyloid fibril protein.

n = number of amyloid fibril proteins useful for amyloidosis classification,

$P_{i,j}$ = Ratio B_i/B_b, of an amyloid fibril protein i in a subject j.

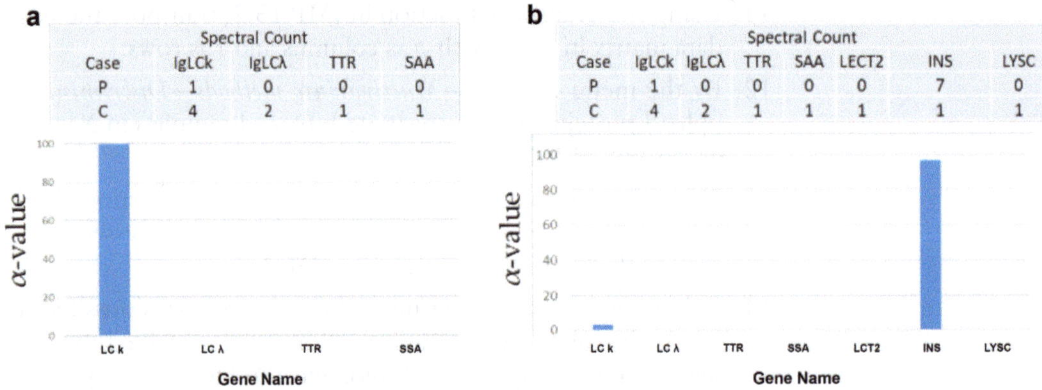

Fig. 3 Amyloidosis subtyping. (**a**) α-value based on the main four amyloidogenic proteins (IgLCk, IgLCλ, TTR, SAA). (**b**) α-value extended to additional amyloidogenic proteins (LECT2, INS, and LYSC)

4. Because the sum of the calculated α-values per subject is equal to 100, a level of α-value >50 may be used to characterize the specific amyloidosis subtype related to the predominant amyloid fibril protein (biomarker).
5. Although the α-value algorithm was proposed to classify the four main amyloidosis subtypes, it can be updated with newly discovered amyloidogenic proteins [1], as previously reported [32] (Fig. 3).

3.3 Proteome Recovery

Following drug treatment, proteome recovery may be used as a measure for evaluating the drug effectiveness [33]. In addition to samples treated with specific drugs, this kind of analysis requires protein profiles from healthy control and untreated samples. Specifically, by proteome recovery index (PR_i), proteomic changes are simultaneously evaluated in pre- and post-treatment of disease subjects, extracting those proteins whose expression returns to the reference condition (healthy control) following drug treatment (Fig. 4).

1. Starting from the average AU (SpC, PSM, or intensity) of proteins identified in control and disease subjects (pre- and after treatment with drugs), the extraction of so-called *recovered proteins* is carried out by the proteome remodeling index (PR_i). It is calculated by the following equation:

$$PR_i = \frac{M(b-p)_j}{M(p-t)_j}$$

where j indicates a given identified protein, $M(b-p)_j$ the perturbation module (b: baseline (healthy control); p: perturbed (disease pre-treatment)), while $M(p-t)_J$ the treatment module (p: perturbed (disease pre-treatment); t: treatment (disease after treatment)) (*see* **Note 8**).

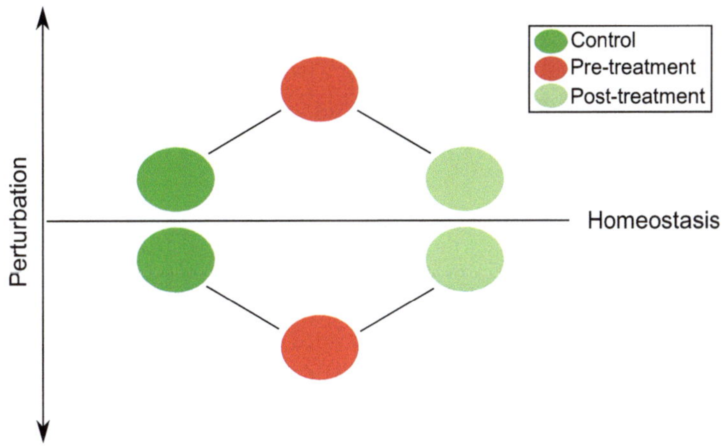

Fig. 4 Proteome recovery index (PR_i) scheme

2. PR_i typically ranges from −0.5 to −2, and more significant recoveries fall from −0.66 to −1.5 (*see* **Note 9**).
3. PR_i can be useful to evaluate the minimal residual disease (MRD), expressed as recovery at a healthy level of amyloidosis proteins biomarkers, as the evaluation of treatment effect.

3.4 Network Analysis

In this section, we provide strategies and tips for transforming and analyzing proteomics data as network models. In particular, we will describe the steps to build and process at functional and topological levels protein-protein interaction (PPI) and protein co-expression networks (Fig. 5a–c). The goal of these procedures is to support new insights concerning the molecular mechanisms underlying the investigated pathophysiological states. It happens through the identification of functional and topological modules characterizing the investigated phenotypes [6, 7]. In the case of big network models (>500 nodes), their representation of functional modules may be also simplified by a series of pie charts renamed "Bubbles" [33, 34] (Fig. 5d, e). In addition, the analysis of the network's structure allows the identification of topologically relevant proteins, called hubs and bottlenecks (Fig. 5f). Their position suggests a key role in regulating disease-related processes, including amyloidosis [6]. Thus, equal to DEPs, they can be used as biomarkers.

3.4.1 Protein-Protein Interaction (PPI) Network Reconstruction

1. Open the stringApp in Cytoscape (*File→Import→Network from Public databases.*) (Fig. 5b).
2. In the Data Source box, select STRING:protein query. Then, select species and paste the gene names corresponding to the proteins of interest (complete protein profiles or DEPs).
3. Set the confidence score cutoff (0.10) and maximum additional interactors to 0.

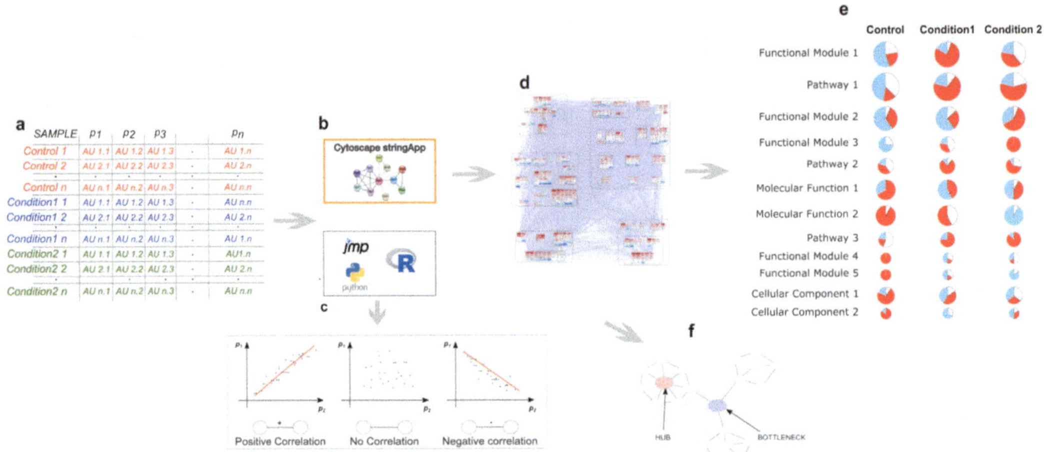

Fig. 5 Network analysis. (**a**) Typical experimental protein data matrix reporting AU (SpC, PSM, intensity). (**b**) Protein-protein interaction (PPI) and (**c**) protein co-expression network reconstruction. (**d**) PPI functional modules. (**e**) Bubble charts from the DEPs grouped in PPI of protein co-expression functional modules. Up- and down-regulated proteins are reported as a percentage of pie (red, normalized AU > 75%; blue, normalized AU < 25%), including the number of proteins unchanged (white, 25% < normalized AU < 75%). (**f**) Hubs and bottlenecks identified following topological analysis (by Betweenness, Centroid, Bridging, Degree, etc. centralities) of the PPI and protein co-expression networks

4. Once the network model has been reconstructed, filter the interaction type by the Filter tool in Cytoscape.
5. Mainly for topological analysis, only interactions annotated as Edge: experiments (score ≥0.3) or Edge: databases (score ≥0.3) are maintained, while all the others are discarded.

3.4.2 Protein Co-expression Network Reconstruction

1. Starting from a $p \times s$ data matrix (p represents proteins and s samples where they were identified), transpose it in a matrix $s \times p$ (with p as columns and s as rows) by Transpose function in JMP 15.2 (Fig. 5a).
2. By the menu *Analyze→Multivariate methods→Multivariate*, upload proteins p in Y, Columns box of JMP 15.2.
3. Select *Multivariate→Nonparametric Correlations→Spearman (or Kendall, or Hoeffding)* (*see* **Note 10**) (Fig. 5c).
4. The output result will return a data matrix reporting Protein1, Protein2, Spearman p, and Probability p.
5. Filter correlations and maintain only those with *Spearman* $p > |0.75|$ and *Probability* $p \leq 0.05$.
6. To reconstruct a protein co-expression network, upload in Cytoscape the data matrix reporting the correlation values filtered.
7. Select the correlation data matrix by the menu *File→Import→Network from File.*

8. Finally, in *Import Network from Table*, select the first column of proteins as the source node and the second column as the target node while uploading the *Spearman* score and the *Probability* as attributes.

3.4.3 PPI vs. Co-expression from Functional to Topological Analysis

Functional Enrichment

1. Both PPI and protein co-expression network models are processed at the functional level to retrieve the most enriched pathways, biological processes, molecular functions, and cellular components.
2. After a PPI or a protein co-expression network is uploaded in Cytoscape, by stringApp, select the menu *STRING Enrichment→Retrieve functional enrichment tool.*
3. At the end of the process, a new window in the table panel will show the enriched biological terms, including GO biological process, GO molecular function, GO cellular component, KEGG pathways, Reactome pathways, and WikiPathways.
4. The enriched biological terms can be used to group our nodes into functional modules whose expression variation can be assessed under the analyzed conditions/phenotypes (Fig. 5d, e).
5. After functional module reconstruction, from the *Node Table* in Cytoscape, add a *New Single Column (String)* and assign the same tag for all nodes belonging to the same functional module.
6. From the menu of Cytoscape *File→Import→Table from File*, for each node, upload the expression value it assumes in each considered condition/phenotype.
7. Prior to the association between nodes and expression values, normalize them in the range of 0–100 starting from the average value (per condition) of the original AU (i.e., SpC/PSM, peak area, intensity).
8. For each protein, set the highest value of AU to 100 and normalize the AU value corresponding to the other conditions proportionally.
9. From the menu *File→Export→Table to File*, select and export the "Network_Name default node" table containing, for each node, the corresponding Tag and the normalized expression values (0–100) in the analyzed conditions/phenotypes.
10. For each functional module and condition, the total proteins and those up- (normalized AU > 75) and down-regulated (normalized AU < 25) are counted.
11. To simplify interpretation and visualization, the table is transformed into Bubble charts, where, for each condition, the functional modules are represented as a circle, and the diameter is proportional to the total number of proteins belonging to them [33, 34] (Fig. 5e).

Network Topology

1. Both PPI and protein co-expression network models are processed at the topological level to retrieve topologically relevant nodes, called hubs (Fig. 5f).
2. After a PPI or a protein co-expression network reconstruction in the *Layout* menu of Cytoscape, select *Apply Preferred Layout.*
3. Remove all disconnected nodes and maintain only the largest connected component.
4. From Cytoscape, open the *Centiscape2.2 App* (*see* **Note 11**).
5. Select the centralities of interest, including Diameter, Average Distance, Degree, and Radiality.
6. Closeness, Stress, Betweenness, Centroid Value, Eccentricity, EigenVector, Bridging and Edge Betweenness.
7. Select *Undirected Network* or *Directed Network.* Select *Weighted Edges (as distance)* (*see* **Note 12**).
8. Calculate the average value of each centrality measured.
9. To select hub nodes, filter and discard those showing centrality values lower than the average calculated on the whole network.
10. To validate the selection of hubs, starting from the reference network, reconstruct the random network models by means of the *Network Randomizer App* in Cytoscape.
11. Network randomization may be done by preserving the Degree. Moreover, the parametric Erdos-Renyi model, Watt-Strogatz model, Barabasi-Albert model, Multiplication model, Lattice model, Community affiliation model, and Multiple network generation may be chosen in parametric randomization (*see* **Note 13**).
12. Alternatively, build an *in-house* R script based on VertexSort (to build random models), igraph (to compute centralities), and ggplot2 (to plot results) libraries (*see* **Note 14**).
13. To confirm the selection of hubs, verify that the average value for a given centrality (i.e., Betweenness) used to select the hubs is different from the average values of the same centrality (Betweenness) in random networks (Fig. 6).
14. Finally, compare the list of PPI hubs with the list of co-expression hubs (*see* **Note 15**). The proteins found to be hubs in both models are defined as super hubs; in order of priority, they are those to be considered most relevant, also for validation or further investigation.

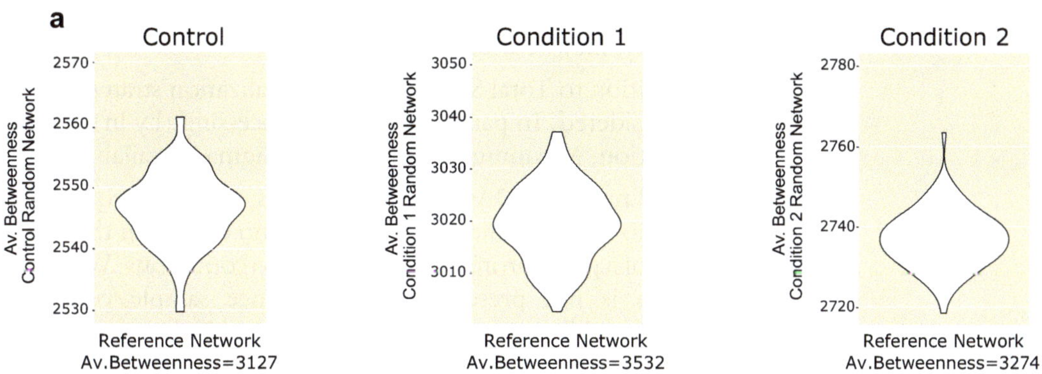

b

Number of nodes	930	Number of nodes	1148	Number of nodes	1040
Number of edges	9608	Number of edges	11242	Number of edges	11259
Avg. number of neighbors	20.7	Avg. number of neighbors	19.8	Avg. number of neighbors	21.4
Network diameter	9	Network diameter	9	Network diameter	8
Network radius	5	Network radius	5	Network radius	4
Characteristic path length	3.29	Characteristic path length	3.20	Characteristic path length	3.28
Clustering coefficient	0.417	Clustering coefficient	0.390	Clustering coefficient	0.412
Network density	0.023	Network density	0.017	Network density	0.021
Network heterogeneity	1.137	Network heterogeneity	1.127	Network heterogeneity	1.149
Network centralization	0.123	Network centralization	0.191	Network centralization	0.124

Fig. 6 Network topology and hub validation. (**a**) Examples of violin plots reconstructed with the average betweenness values (Av. Betweenness) from PPI random networks. To validate the selection of hubs, these distributions have to show values different from the Av. Betweenness from the reference network. (**b**) Example of topological parameters calculated by *Tools→Analyze Network* tool in Cytoscape

4 Notes

1. Molecular profiles, mainly proteomics and transcriptomics, may derive from tissues, organs, or biological fluids. Since Congo red staining of subcutaneous adipose tissue is a low-invasive procedure adopted as the gold standard for diagnosis of systemic amyloidosis, this tissue, both entire or after laser microdissection, is extensively analyzed by LC-MS for reliable amyloidosis typing [30]. In addition, LC-MS profiles have been characterized from formalin-fixed paraffin-embedded (FFPE) [19] and fresh [31] samples, including the heart [31], liver, and kidney [35].
2. The availability of LC-MS platforms with specific characteristics, including nanoflow HPLC, high-resolution and high-speed scanning, methods, and software, has produced a heterogeneity of pipelines leading to the definition of protein profiles. However, they all provide as output a list of proteins with associated quantitative parameters (arbitrary unit (AU), i.e., SpC, PSM, or intensity). As an example of instrumentation that provided reliable results, we refer to the latest generation of MS mainly based on Orbitrap or qTOF mass analyzers.

3. In addition to SpC, other arbitrary units (AU), such as PSM or intensity, may be used.
4. In addition to Total Signal, other normalization strategies may be considered. In particular, log preprocessing (by ln), Z normalization, Maximum Signal, or Row Sigma is available [36].
5. According to the DAve and DCI indices, when X is equal to Y, the ratio is zero, which corresponds to no change in the abundance of a given protein between the two conditions. When the protein is not present in the reference sample/condition (X = 0), while it is exclusively identified in Y, DAve will be −2.0. On the contrary, if a protein was not identified in the second sample/condition (Y = 0), while it was exclusively identified in X, DAve will be +2.0. Intermediate DAve values indicate different amounts of that protein under the two different samples/conditions. Specifically, positive DAve values indicate proteins up-regulated in X (and down-regulated in Y), while negative DAve values indicate proteins up-regulated in Y (and down-regulated in X).
6. To score only significant variations and avoid the possibility of intercepting variations due to experimental errors, a confidence threshold of DAve may be set to |0.5|. It corresponds to a variation of 50%, which is higher than the calculated experimental variation of 10–15%. Of note, this threshold may be decreased up to |0.2| according to discriminant analysis (LDA, PCA, etc.) on replicated analyses. The DAve values overlap with ln(FC), where FC stands for fold change (Fig. 2b).
7. *G test values* ≥6.635 and ≥10.828 indicate proteins differentially expressed with $P \leq 0.01$ and $P \leq 0.001$, respectively.
8. Perturbation and treatment modules correspond to DAve values, healthy control vs. disease pre-treatment and disease pre-treatment vs. disease after treatment, respectively.
9. Following treatment, if the protein level is recovered forward, healthy control, perturbation, and treatment modules have similar values and opposite signs; hence, the PR_i will be negative and close to unit (−1).
10. Since SpCs or PSMs data are usually not normally distributed, *Spearman, Kendall, or Hoeffding* may be adopted because they allow non-parametric correlations. Hence, before applying parametric or non-parametric correlations, it is necessary to check the distribution of the data. In case of normal distribution, other correlation coefficients should be adopted, including *Pearson*'s correlation.
11. Alternatively, in the *Tools* menu of Cytoscape, run the *Analyze Network* tool.

12. PPI and co-expression networks are undirected, while signaling networks are directed. In co-expression networks, the correlation coefficient can be used to weight the edges. Since it is a measure of distance, while the correlation coefficient is a measure of similarity ranging from 0 to 1, transform the weight as (1-correlation coefficient).
13. In the *Network Randomizer App*, a statistical module based on the two-sample Kolmogorov-Smirnov test compares networks to assess if their numerical attributes come from the same or different distribution. However, we do not know which way is better for creating random networks in order to compare them with real data [22]. In addition, generating random networks will always make the randomization non-random at some levels, but it is important to consider that, when creating an in silico dataset for validating real data, the bigger the random dataset, the finer the comparison will be.
14. Example of R code for randomizing an undirected PPI network and calculating the betweenness.

```
library(gplots)
library(igraph)
library(readxl)
library(VertexSort)
library(ggplot2)
library(CINNA)
library(centiserve)
library(svglite)

--------------------------------Network randomiza-
tion----------------------------

Dataset1 <- read_xlsx("Dataset1.xlsx") #Dataset1 should con-
tains 2 columns reporting for each row Node1 and Node2
network_Dataset1 <- graph_from_data_frame(Dataset1[, c(1,2)],
directed=FALSE)
set.seed(123) #By setting set.seed (any number, but always the
same) we will obtain reproducible results
random <- dpr(network_Dataset1, 100, vparallel = T, vcpus =
4) #dpr function create random models preserving the original
node degree per node #100 is the n of random networks that will
be created.
text <- list()
---------------------------------Betweenness from random net-
works----------------------------

for (i in 1:100){
   text[[i]] <-  igraph::betweenness(random[[i]])
```

```
  addNamesToList <- function(myList){
     listNames <- vector()
     for(i in 1:length(myList)){
       if(identical(names(myList[i]),NULL)){
         listNames <- c(listNames,paste("item",i,sep=""))
       }else{
          listNames <- c(listNames,names(myList[i]))
       }
    }
    names(myList) <- listNames
    return (myList)
  }
  text_def <- addNamesToList(text)
}
mean_random_bet <- list()
for(i in 1:100){
mean_random_bet[[i]] <- mean(text_def[[i]])
}
bet_ref<- mean(betweenness(network_Dataset1))   #betweenness
from the reference network

-------------------------------------Betweenness Violin plot
------------------------------------

svglite(width=4, height=7)
dbdb <- as.data.frame(unlist(mean_random_bet))
colnames(dbdb)<- "mean_random_bet"
hw_p <- ggplot(dbdb, aes(x = "betweenness Dataset1", y =mean_-
random_bet))
hw_p +geom_violin()+
annotate("text", label = " mean betweenness reference Data-
set1:X", x =1 , y = 890, size = 3, colour = "black")
dev.off()
```

15. This last step can be carried out only if hubs are calculated by both network models (PPI and protein co-expression). On note, both PPI and protein co-expression networks have to be reconstructed starting from the same experimental data.

Acknowledgments

We thank the kind help of Marta Grazia Bitonti for the development of MAProMa software and the adaptation of the QGIS platform to biological data.

Funding

This work was supported by PON ELIXIR CNRBiOMICS (PIR01_00017), Elixir proteomics Implementation Study 2019–2021, Elixir Implementation Study Proteomics 2021–23, ELIXIRxNextGenIT (PNRR, Prot. IR0000010), and Cariplo-Telethon GJC2021 n. 2022–0578, Italian Ministry of Health (RF2019–12370396).

References

1. Buxbaum JN, Dispenzieri A, Eisenberg DS, Fändrich M, Merlini G, Saraiva MJM, Sekijima Y, Westermark P (2022) Amyloid nomenclature 2022: update, novel proteins, and recommendations by the International Society of Amyloidosis (ISA) Nomenclature Committee. Amyloid 29(4):213–219. https://doi.org/10.1080/13506129.2022.2147636
2. Brehme M, Voisine C (2016) Model systems of protein-misfolding diseases reveal chaperone modifiers of proteotoxicity. Dis Model Mech 9(8):823–838. https://doi.org/10.1242/dmm.024703
3. Merlini G, Bellotti V (2003) Molecular mechanisms of amyloidosis. N Engl J Med 349(6):583–596. https://doi.org/10.1056/NEJMra023144
4. Yakupova EI, Bobyleva LG, Vikhlyantsev IM, Bobylev AG (2019) Congo Red and amyloids: history and relationship. Biosci Rep 39(1): BSR20181415. https://doi.org/10.1042/BSR20181415
5. Kourelis TV, Dasari SS, Dispenzieri A, Maleszewski JJ, Redfield MM, Fayyaz AU, Grogan M, Ramirez-Alvarado M, Abou Ezzeddine OF, McPhail ED (2020) A proteomic atlas of cardiac amyloid plaques. JACC CardioOncol 2(4):632–643. https://doi.org/10.1016/j.jaccao.2020.08.013
6. Di Silvestre D, Brambilla F, Lavatelli F, Chirivì M, Canetti D, Bearzi C, Rizzi R, Bijzet J, Hazenberg BPC, Bellotti V, Gillmore JD, Mauri P (2023) The protein network in subcutaneous fat biopsies from patients with AL amyloidosis: more than diagnosis? Cells 12(5):699. https://doi.org/10.3390/cells12050699
7. Vella D, Zoppis I, Mauri G, Mauri P, Di Silvestre D (2017) From protein-protein interactions to protein co-expression networks: a new perspective to evaluate large-scale proteomic data. EURASIP J Bioinform Syst Biol 2017(1):6. https://doi.org/10.1186/s13637-017-0059-z
8. Brambilla F, Lavatelli F, Di Silvestre D, Valentini V, Palladini G, Merlini G, Mauri P (2013) Shotgun protein profile of human adipose tissue and its changes in relation to systemic amyloidoses. J Proteome Res 12(12): 5642–5655. https://doi.org/10.1021/pr400583h
9. Cai D, Li Y, Zhou C, Jiang Y, Jiao J, Wu L (2017) Comparative proteomics analysis of primary cutaneous amyloidosis. Exp Ther Med 14(4):3004–3012. https://doi.org/10.3892/etm.2017.4852
10. Bai W, Wang H, Bai H (2019) Identification of candidate genes and therapeutic agents for light chain amyloidosis based on bioinformatics approach. Pharmgenomics Pers Med 12:387–396. https://doi.org/10.2147/PGPM.S228574
11. Genova F, Nonnis S, Maffioli E, Tedeschi G, Strillacci MG, Carisetti M, Sironi G, Cupaioli FA, Di Nanni N, Mezzelani A, Mosca E, Helps CR, PAJ L, Dorso L, 99 Lives Consortium, Longeri M (2021) Multi-omic analyses in Abyssinian cats with primary renal amyloid deposits. Sci Rep 11(1):8339. https://doi.org/10.1038/s41598-021-87168-0
12. Jeong H, Mason SP, Barabási AL, Oltvai ZN (2001) Lethality and centrality in protein networks. Nature 411(6833):41–42. https://doi.org/10.1038/35075138
13. Scardoni G, Tosadori G, Faizan M, Spoto F, Fabbri F, Laudanna C (2014) Biological network analysis with CentiScaPe: centralities and experimental dataset integration. F1000Res 1(3):139. https://doi.org/10.12688/f1000research.4477.2
14. Doncheva NT, Morris JH, Gorodkin J, Jensen LJ (2019) Cytoscape StringApp: network analysis and visualization of proteomics data. J Proteome Res 18(2):623–632. https://doi.org/10.1021/acs.jproteome.8b00702

15. Maere S, Heymans K, Kuiper M (2005) BiNGO: a Cytoscape plugin to assess overrepresentation of gene ontology categories in biological networks. Bioinformatics 21(16): 3448–3449. https://doi.org/10.1093/bioinformatics/bti551
16. Reimand J, Isserlin R, Voisin V, Kucera M, Tannus-Lopes C, Rostamianfar A, Wadi L, Meyer M, Wong J, Xu C, Merico D, Bader GD (2019) Pathway enrichment analysis and visualization of omics data using g:Profiler, GSEA. Cytoscape and EnrichmentMap Nat Protoc 14(2):482–517. https://doi.org/10.1038/s41596-018-0103-9
17. Tosadori G, Bestvina I, Spoto F, Laudanna C, Scardoni G (2016) Creating, generating and comparing random network models with NetworkRandomizer. F1000Res 17(5):2524. https://doi.org/10.12688/f1000research.9203.3
18. Di Silvestre D, Brambilla F, Mauri PL (2013) Multidimensional protein identification technology for direct-tissue proteomics of heart. Methods Mol Biol 1005:25–38. https://doi.org/10.1007/978-1-62703-386-2_3
19. Van Haver D, Dendooven A, Impens F (2023) Proteomics-based analysis and diagnosis of formalin-fixed paraffin-embedded amyloidosis samples. Methods Mol Biol 2718:213–233. https://doi.org/10.1007/978-1-0716-3457-8_12
20. Spodzieja M, Rodziewicz-Motowidło S, Szymanska A (2019) Hyphenated mass spectrometry techniques in the diagnosis of amyloidosis. Curr Med Chem 26(1):104–120. https://doi.org/10.2174/0929867324666171003113019
21. He S, He X, Pan R, Pan L, Lv X, Jin Y, Fan Y, Wang Y, Tian Z, Zhang S (2023) Data-independent acquisition mass spectrometry reveals comprehensive plasma protein profiles in the natural history of patients with hereditary transthyretin amyloidosis (ATTRv). Expert Rev Proteomics 20(1–3):57–69. https://doi.org/10.1080/14789450.2023.2195096
22. Canetti D, Rendell NB, Gilbertson JA, Botcher N, Nocerino P, Blanco A, Di Vagno L, Rowczenio D, Verona G, Mangione PP, Bellotti V, Hawkins PN, Gillmore JD, Taylor GW (2020) Diagnostic amyloid proteomics: experience of the UK National Amyloidosis Centre. Clin Chem Lab Med 58(6):948–957. https://doi.org/10.1515/cclm-2019-1007
23. Picken MM (2015) Proteomics and mass spectrometry in the diagnosis of renal amyloidosis. Clin Kidney J 8(6):665–672. https://doi.org/10.1093/ckj/sfv087
24. Fernández-Costa C, Martínez-Bartolomé S, McClatchy DB, Saviola AJ, Yu NK, Yates JR 3rd (2020) Impact of the identification strategy on the reproducibility of the DDA and DIA results. J Proteome Res 19(8): 3153–3161. https://doi.org/10.1021/acs.jproteome.0c00153
25. Zhao P, Zhong J, Liu W, Zhao J, Zhang G (2017) Protein-level integration strategy of multiengine MS spectra search results for higher confidence and sequence coverage. J Proteome Res 16(12):4446–4454. https://doi.org/10.1021/acs.jproteome.7b00463
26. Woo J, Zhang Q (2023) A streamlined high-throughput plasma proteomics platform for clinical proteomics with improved proteome coverage, reproducibility, and robustness. J Am Soc Mass Spectrom 34(4):754–762. https://doi.org/10.1021/jasms.3c00022
27. Mauri P, Scarpa A, Nascimbeni AC, Benazzi L, Parmagnani E, Mafficini A, Della Peruta M, Bassi C, Miyazaki K, Sorio C (2005) Identification of proteins released by pancreatic cancer cells by multidimensional protein identification technology: a strategy for identification of novel cancer markers. FASEB J 19(9): 1125–1127. https://doi.org/10.1096/fj.04-3000fje
28. Zhang B, VerBerkmoes NC, Langston MA, Uberbacher E, Hettich RL, Samatova NF (2006) Detecting differential and correlated protein expression in label-free shotgun proteomics. J Proteome Res 5(11):2909–2918. https://doi.org/10.1021/pr0600273
29. Sokal RR, Rohlf FJ (1981) Biometry: the principles and practice of statistics in biological research, 2nd edn. W.H. Freeman & Company, New York
30. Brambilla F, Lavatelli F, Di Silvestre D, Valentini V, Rossi R, Palladini G, Obici L, Verga L, Mauri P, Merlini G (2012) Reliable typing of systemic amyloidoses through proteomic analysis of subcutaneous adipose tissue. Blood 119(8):1844–1847. https://doi.org/10.1182/blood-2011-07-365510
31. Noborn F, Thomsen C, Vorontsov E, Bobbio E, Sihlbom C, Nilsson J, Polte CL, Bollano E, Vukusic K, Sandstedt J, Dellgren G, Karason K, Oldfors A, Larson G (2023) Subtyping of cardiac amyloidosis by mass spectrometry-based proteomics of endomyocardial biopsies. Amyloid 30(1):96–108. https://doi.org/10.1080/13506129.2022.2127088
32. Canetti D, Brambilla F, Rendell NB, Nocerino P, Gilbertson JA, Di Silvestre D,

Bergamaschi A, Lavatelli F, Merlini G, Gillmore JD, Bellotti V, Mauri P, Taylor GW (2021) Clinical amyloid typing by proteomics: performance evaluation and data sharing between two centres. Molecules 26(7):1913. https://doi.org/10.3390/molecules26071913

33. Sereni L, Castiello MC, Di Silvestre D, Della Valle P, Brombin C, Ferrua F, Cicalese MP, Pozzi L, Migliavacca M, Bernardo ME, Pignata C, Farah R, Notarangelo LD, Marcus N, Cattaneo L, Spinelli M, Giannelli S, Bosticardo M, van Rossem K, D'Angelo A, Aiuti A, Mauri P, Villa A (2019) Lentiviral gene therapy corrects platelet phenotype and function in patients with Wiskott-Aldrich syndrome. J Allergy Clin Immunol 144(3):825–838. https://doi.org/10.1016/j.jaci.2019.03.012

34. Magrassi L, Brambilla F, Viganò R, Di Silvestre D, Benazzi L, Bellantoni G, Danesino GM, Comincini S, Mauri P (2023) Proteomic analysis on sequential samples of cystic fluid obtained from human brain tumors. Cancers (Basel) 15(16):4070. https://doi.org/10.3390/cancers15164070

35. Said SM, Sethi S, Valeri AM, Leung N, Cornell LD, Fidler ME, Herrera Hernandez L, Vrana JA, Theis JD, Quint PS, Dogan A, Nasr SH (2013) Renal amyloidosis: origin and clinicopathologic correlations of 474 recent cases. Clin J Am Soc Nephrol 8(9):1515–1523. https://doi.org/10.2215/CJN.10491012

36. Carvalho PC, Hewel J, Barbosa VC, Yates JR 3rd (2008) Identifying differences in protein expression levels by spectral counting and feature selection. Genet Mol Res 7(2):342–356. https://doi.org/10.4238/vol7-2gmr426

Index

Taufika Islam Williams (ed.), *Tissue Proteomics: Methods and Protocols*, Methods in Molecular Biology, vol. 2884, https://doi.org/10.1007/978-1-0716-4298-6,

MIX
Papier aus verantwortungsvollen Quellen
Paper from responsible sources
FSC® C105338

If you have any concerns about our products,
you can contact us on
ProductSafety@springernature.com

In case Publisher is established outside the EU,
the EU authorized representative is:
Springer Nature Customer Service Center GmbH
Europaplatz 3, 69115 Heidelberg, Germany

Printed by Libri Plureos GmbH
in Hamburg, Germany